BRE Building Elements

Walls, windows and doors

Performance, diagnosis, maintenance, repair and the avoidance of defects

H W Harrison, ISO, Dip Arch, RIBA

R C de Vekey, PhD, MRIC, CChem, DIC

BRE
Garston
Watford
WD2 7JR

Prices for all available BRE
publications can
be obtained from:
Construction Research
Communications Ltd
151 Rosebery Avenue
London, EC1R 4GB
Telephone
0171 505 6622
Facsimile
0171 505 6606
E-mail
crc@construct.emap.co.uk

BR 352
ISBN 1 86081 235 X

Contents

Preface

This book is about all the main vertical elements of buildings, both external, including walls, windows and doors, and internal, including separating walls, partitions and internal doors. It deals in outline with the achieved performances and deficiencies of the fabric over the whole age range of the national building stock. Of course, many performance requirements are common to buildings of whatever age, the main difference being that occupants tend to make allowances for deficiencies when the building is old.

The construction of a wall and its constituent materials must have adequate strength, be fire resistant, offer the necessary acoustic and thermal properties, and resist erosion and corrosion for the life of the building, so that the occupants can use the building safely and conveniently. In addition, the external envelope should have adequate resistance to the elements, and must allow the maintenance of a suitable indoor environment.

Information on faults in buildings of all types show that nearly half concern walls, windows and doors. Despite the advice that has been available to the industry from the 1920s, faults which formerly occurred on a substantial scale, such as rain penetration through windows and doors, still recur frequently. Some of these faults may appear to be elementary in character, but this only reflects what happens in practice. Recognition of faults before they occur on site is obviously much more preferable to correction after the event. In some cases an underlying cause is lack of knowledge, in others a lack of care.

Readership

Walls, windows and doors is addressed primarily to building surveyors and other professionals performing similar functions – such as architects and builders – who maintain, repair, extend and renew the national building stock. It will also find application in the education field.

In spite of the current explosion of information, or perhaps because of it, people do not use the guidance that exists. In order to try to remedy that situation, the advice given in Chapters 2 to 10 of this book concentrates on practical details. However, there also needs to be sufficient discussion of principles to impart understanding of the reason for certain practices, and much of this information is given in Chapter 1.

Scope of the book

All kinds of external walls, encompassing loadbearing and non-loadbearing, curtain walls, and overcladding are dealt with first; then windows and external doors, including thresholds. Later chapters include separating walls, partitions, and internal doors and stair enclosures including protected shafts. In principle, all types of buildings are included, though obvious practical considerations of space decree that information on heritage buildings is limited in scope.

The book is not a manual of construction practice, nor does it provide the reader with the information necessary to design a wall. Both good and bad features of walls, windows and doors and the joints between them are described, and sources of further information and advice are offered. The drawings are not working drawings but merely show either those aspects to which the particular attention of readers needs to be drawn or simply provide typical details to support text.

Excluded from the scope of this book is consideration of foundations for walls, or basements, which will be dealt with in another publication. Wall-to-roof junctions at eaves and verges, and fabric or flexible plastics sheathings to buildings were dealt with in the companion book, *Roofs and roofing*.

As with the other books in this series, the text concentrates on those aspects of construction which, in the experience of BRE, lead to the greatest number of problems or greatest potential expense, if carried out unsatisfactorily. It follows that these problems will be picked up most frequently by maintenance surveyors and others carrying out remedial work on walls, windows and doors. Occasionally there is information relating to a fault which is infrequently encountered, and about which it may in consequence be difficult to locate information. Although most of the information relates to older buildings, much material concerning observations by BRE of new buildings under construction in the period from 1985 to 1995 is also included.

Many of the difficulties which are referred to BRE for advice stem from too hasty an assumption about the causes of a particular defect. Very often the symptom is treated, not the cause, and the defect recurs. It is to be hoped that this book will encourage a systematic approach to the diagnosis of walls and walling defects.

The case studies provided in some of the chapters are selected from the files of the BRE Advisory Service and the former Housing Defects Prevention Unit, and represent the most frequent kinds of problems on which BRE has been consulted.

The standard headings within the chapters are repeated only where there is a need to refer the reader to earlier statements or where there is something relevant to add to what has gone before.

Since it is necessary to consider the enclosures for stairways in conjunction with stair flights, enclosures for stairways (including such items as protected shafts) are therefore considered as elements of walls and are dealt with in this book in Chapter 6.3.

In the United Kingdom, there are three different sets of building regulations: *The Building Regulations 1991* which apply to England and Wales; *The Building Standards (Scotland) Regulations 1990*; and *The Building Regulations (Northern Ireland) 1994*. There are many common provisions between the three sets, but there are also major differences. This book has been written against the background of the building regulations for England and Wales since, although there has been an active Advisory Service for Scotland and Northern Ireland, the highest proportion of site inspections has been carried out in England and Wales. In addition, the technical aspects of the book are affected more by exposure due to

location and height above sea level than by national or administrative boundaries. The fact that the majority of references to building regulations are to those for England and Wales, should not make the book inapplicable to Scotland and Northern Ireland.

There is insufficient space in this book to deal with the highly sophisticated new forms of external walls, such as the so-called 'smart' or intelligent skins, employing variable external fabric, to improve solar control and daylight utilisation. It is intended that these form part of a further book in this series. Information relating to these techniques may be sought from BRE.

Some important definitions

Since the book is mainly about the problems that can arise in walls, windows and doors, two words, 'fault' and 'defect', need precise definition. Fault describes a departure from good practice in design or execution of design: it is used for any departure from requirements specified in building regulations, British Standards and Codes of practice, and the published recommendations of authoritative organisations. A defect – a shortfall in performance – is the product of a fault, but while such a consequence cannot always be predicted with certainty, all faults have the potential for leading to defects. The word failure has occasionally been used to signify the more serious defects (and catastrophes!).

Where the term 'investigator' has been used, it covers a variety of roles including a member of BRE's Advisory Service, a BRE researcher or a consultant working under contract to BRE.

Because the term 'separating wall' has been used in the construction industry from the earliest days, and is still in current use, we prefer to use it in this book as a generic term despite the comparable term 'compartment wall' which is found in the national building regulations.

Acknowledgements
Photographs which do not bear an attribution have been provided from our own collections or from the BRE Photographic Archive, a unique collection dating from the early 1920s.

To the following colleagues, and former colleagues, who have suggested material for this book or commented on drafts, or both, we offer our thanks:

M J Atkins, P Bonfield, R Cox, E J Daniels, Maggie Davidson, C Grimwood, W H (Bill) Harrison, C Holland, M Howarth, Dr P Littlefair, Penny Morgan, F Nowak, Dr R Orsler, M Pound, P W Pye, R E H Read, J Reid, Justine Redshaw, B Reeves, J Seller, A J Stevens, C M Stirling, N Tinsdeall, P M Trotman, P Walton, and Dr T Yates, all of the BRE.

We wish to acknowledge a special debt to P M Trotman for providing the majority of the information in Chapter 2.6.

In addition, acknowledgement is given to the original though anonymous authors of *Principles of modern building*, Volume 1, from which several passages have been adapted and updated.

H W H
R C de V
July 1998

Chapter 0   **Introduction**

The majority of walls, windows and doors, of most kinds of structure or surface finish, perform well (Figure 0.1). However, there is evidence that avoidable defects in walls, windows and doors occur too often. Such evidence is contained in BRE surveys of housing, of both new construction and rehabilitation, and in the United Kingdom house condition surveys (undertaken every five years); also, particularly for building types other than housing, from the past commissions of the BRE Advisory Service.

Records of failures and faults in buildings

BRE Advisory Service records
Over the years, approximately half the investigations carried out by BRE have been on housing and approximately half on other building types. A typical example is for the period 1987–89, where the figures (Figure 0.2) have been extracted covering all site investigations on internal and external walls, windows and doors. The chart shows that during this period, housing and non-housing cases were approximately equal in numbers, and that offices and public buildings accounted for half the latter category.

Between 1970–74, the BRE Advisory Service carried out 510 investigations, and, of this total, about 1 in 5 concerned external walls. Rain penetration was the defect most frequently investigated, with about half the number of cases occurring in cavity filled walls via DPMs and trays; the other half in solid walls,

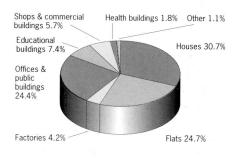

Figure 0.2
Distribution of BRE Advisory Service investigations on internal and external walls, windows and doors by building type, 1987–89

concrete cladding and other kinds of external wall. Rising damp in masonry walls alone accounted for 18 cases. Rain penetration through windows, as opposed to the window-to-wall joints, occurred in about the same numbers as rising damp. There were around twice as many cases of surface condensation as interstitial occurring in external walls. No chart is shown for this period.

In the years 1987–89, 549 investigations were carried out, of which 115 were on external walls. During the period mid-1992 to mid-1995, a total of 506 investigations were carried out, of which 118 were on external walls. Both the total numbers and those for external walls are remarkably similar for all three time periods chosen for comparison.

Comparing the investigations for 1987–89 and 1992–95 (Figure 0.3 on page 2) for external walls according to building type, more houses and flats were investigated in the period beginning in 1992 than in that beginning in 1987, and far fewer shops, commercial buildings and

Figure 0.1
Brick, stone, timber and glass: traditional materials in this Wren masterpiece at Hampton Court Palace

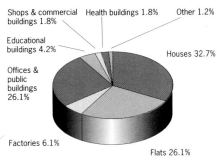

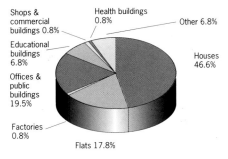

Figure 0.3

Distribution of BRE Advisory Service investigations on external walls, by building type, 1987–89 and 1992–95 compared

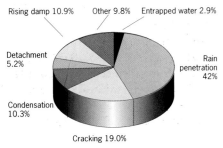

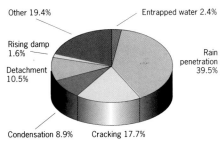

Figure 0.4

Distribution of BRE Advisory Service investigations on external walls, by type of defect, 1987–89 and 1992–95 compared

factories. It is difficult to account for this change, but may be either to do with the workload of the construction industry as a whole, or because the more experienced designers and contractors tend to work on the larger projects.

Of the faults and defects diagnosed (Figure 0.4), rising damp cases were far fewer in the later period, but other cases were very similar in numbers. Probably a greater proportion of dwellings by this time had inserted DPCs than was the case in former years; installers had undoubtedly improved their quality control. However, rain penetration still forms the largest single category of defect.

Responsibility for defects shows that the figures (Figure 0.5) in the 1992 period (for external walls) in relation to design and materials and components had gone down, with those for site going up. Had site skills become too eroded by the recession?

Defect categories for internal walls for 1987–89 have been plotted (31 cases investigated) (Figure 0.6), and it is somewhat surprising that dampness figures so significantly. These cases largely consisted of condensation occurring on internal walls at junctions with external walls. The small number of cases investigated in the later period does not justify a chart. However, by the end of the period, rising damp cases had practically disappeared from the BRE Advisory Service workload, and cases of condensation investigated were fewer than in former years. BRE Advisory Service were also carrying out far fewer sound insulation investigations, largely due to resource limitations.

Very few cases involving windows were examined in 1987–89, 13 in all; but numbers have increased since then, particularly with respect to rain penetration and condensation (49 cases) (Figure 0.7).

In the 25 years, 1970–95, it cannot be said that problems with the vertical external envelope have increased relative to other elements, even though there have been some changes in the nature of the defects encountered. One feature has changed relatively little, however:

BRE assistance with the diagnosis of rain penetration is still being sought by the industry on a substantial scale.

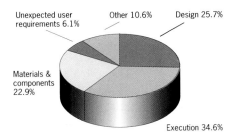

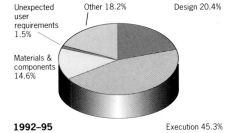

Figure 0.5

Distribution of BRE Advisory Service investigations on external walls, by responsibility for defect, 1987–89 and 1992–95 compared

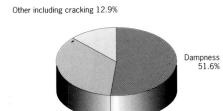

Figure 0.6

Distribution of BRE Advisory Service investigations on internal walls, by type of defect, 1987–89

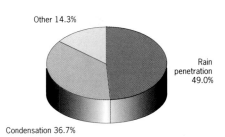

Figure 0.7

Distribution of BRE Advisory Service investigations on windows, by type of defect, 1992–95

BRE Defects Database records

A large amount of information on faults occurring in walls, windows and doors in housing is available in the BRE Defects Prevention Unit's Quality in Housing database[1] which records non-compliance with requirements whatever their origin, whether building regulations, codes of practice, British and industry standards or other authoritative requirements. This database records inspections by BRE or by consultants working under BRE supervision, gathered over the years 1980–90. The data have been analysed according to date of original construction of the dwelling, broadly into two categories: those dwellings being built new at the time of inspection and those being refurbished. The 4,442 records gathered during these site studies provide much of the material for the checklists, but only faults occurring frequently in the database are shown.

An analysis is shown in Figure 0.8

Figure 0.10
A window-to-wall joint in a cavity wall under construction. Thermal insulation will be deficient, in spite of the plastics closer. The solid brick in the inner leaf may make little difference in the presence of unfilled perpends and deficiencies in the cavity fill

of all items relating to new housing construction gathered in 1990. This book itself deals with nearly half of all the items relating to the whole list of elements, and the other two books in this series already published brings the total for the series up to three quarters of all items relating to all elements. The category 'other' includes, for example, fixtures and fittings, porches, canopies etc.

If the type of external wall is considered, whether it be solid or cavity, in rehab or new-build, nearly two thirds of all the faults recorded were in new-build cavity walls – a surprisingly high figure (Figure 0.9). Should not everyone know how to build an ordinary cavity wall to current standards? (Figure 0.10).

Proportion of faults occurring within the element or at the junctions with other elements
During the investigations, records were kept of whether an identified fault or its associated defect occurred within an element, or at the junction of that element with an adjacent element. To keep the external envelope in perspective in relation to all other elements, Figure 0.11 shows that of the one quarter of all the faults in the database which refer to external walls, slightly greater numbers occur in the wall element alone than occur at the junctions with other elements.

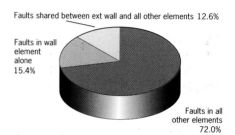

Figure 0.11
Distribution of faults in all external walls identified in BRE DPU site investigations of both rehab and new-build housing, according to whether the fault occurred within the element or at its junction with another element, 1990

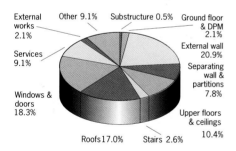

Figure 0.8
Distribution of faults identified in BRE DPU site investigations of new-build housing, according to element, 1990

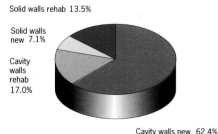

Figure 0.9
Distribution of faults identified in BRE DPU site investigations of rehabilitated and new-build housing, according to type of external wall, 1990

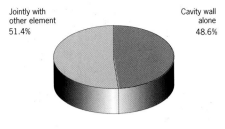

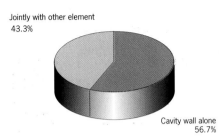

Figure 0.12
Distribution of faults identified in BRE DPU site investigations of rehabilitated and new-build housing: cavity walls according to whether the fault occurred within the element or at its junction with another element, 1990

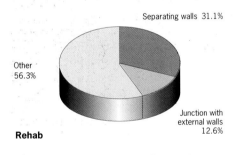

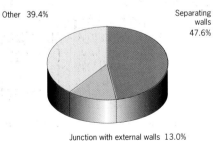

Figure 0.13
Distribution of faults identified in BRE DPU site investigations of rehabilitated and new-build housing: separating walls according to whether the fault occurred within the element or at its junction with another element, 1990

The charts for cavity walls (rehabilitation and new build) are reproduced in Figure 0.12; these indicate that the proportions are still roughly half-and-half. Since solid walls show very little difference from these figures, it is not worth showing charts.

With respect to separating walls, the figures (Figure 0.13) indicate that in the case of both rehab and new build, nearly 1 in 8 faults occurs at the junction between the separating wall and the external wall; 1 in 3 faults occurs within the separating wall itself in the case of rehab, but the figure is 1 in 2 for new build. For the most part this relates to inadequate sound and fire performance; for example with unfilled perpends and dry-lined walls. (The remainders include junctions with services, fixtures and fittings etc.)

For partitions, the charts (Figure 0.14) show strong similarities between rehab and new build, with the head junctions with soffits yielding about 1 in 10 of the totals, about the same or a little less than those with the external walls. Again, the remainders include junctions with services, fixtures and fittings etc.

In the cases of windows and doors, just over half the faults occur within the window or door or doorset itself, and around 1 in 4 occurs at the junction with the external wall (Figure 0.15). This is perhaps somewhat lower than might have been expected from other evidence where the perimeter joint, more often than not, can be a source of problems.

Performance attribute affected
Strength and stability, durability and weathertightness together account for over three quarters of all the items recorded in the case of solid walls in rehab schemes (Figure 0.16), with thermal insulation deficiencies replacing durability in the list of significant items in the case of new build, as one might expect.

For cavity walls (Figure 0.17), most of the categories such as weathertightness are broadly similar for new build and rehab, the main

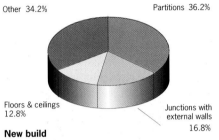

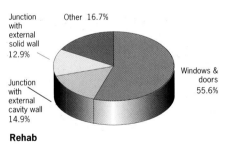

Figure 0.14
Distribution of faults identified in BRE DPU site investigations of rehabilitated and new-build housing: partitions according to whether the fault occurred within the element or at its junction with another element, 1990

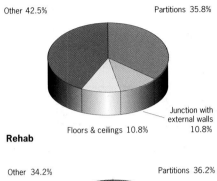

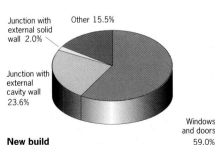

Figure 0.15
Distribution of faults identified in BRE DPU site investigations of rehabilitated and new-build housing: windows and doors according to whether the fault occurred within the element or at its junction with another element, 1990

exceptions being sound and fire, thermal insulation and durability. The higher figure for thermal insulation in new construction is most likely to relate to installation of cavity insulation, but the higher incidence of items relating to sound and fire in rehab is interesting. It could well be related, at least in part, to the use of the wrong density of material for inner leaves where they abut or cross separating walls, or to the lack of necessary cavity barriers.

So far as separating walls are concerned (Figure 0.18), the figures for rehab and new build are broadly comparable – as might be expected, sound and fire items provide nearly half and just over half respectively of all the items.

For partitions, strength and stability is the most numerous category, with hardly any difference between new-build and rehab (Figure 0.19).

There is some variation in the figures for doors and windows between rehab and new-build (Figure 0.20), and it is disappointing to note the higher incidence of durability items in new construction. Is quality really deteriorating?

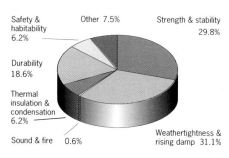

Rehab

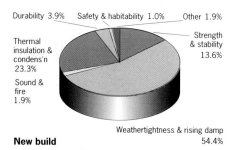

New build

Figure 0.16
Distribution of faults identified in BRE DPU site investigations of rehabilitated and new-build housing: solid walls according to performance, 1990

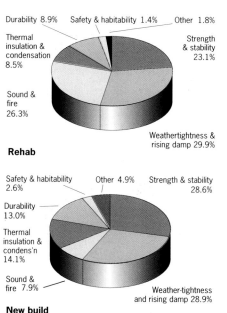

Rehab

New build

Figure 0.17
Distribution of faults identified in BRE DPU site investigations of rehabilitated and new-build housing: cavity walls according to performance, 1990

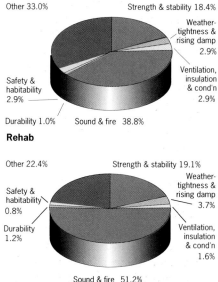

Rehab

New build

Figure 0.18
Distribution of faults identified in BRE DPU site investigations of rehabilitated and new-build housing: separating walls according to performance, 1990

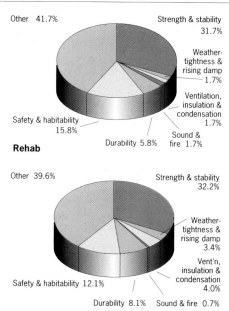

Rehab

New build

Figure 0.19
Distribution of faults identified in BRE DPU site investigations of rehabilitated and new-build housing: partitions according to performance, 1990

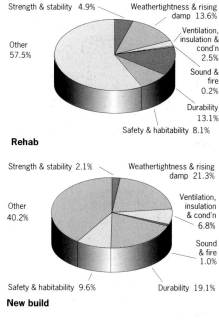

Rehab

New build

Figure 0.20
Distribution of faults identified in BRE DPU site investigations of rehabilitated and new-build housing: windows and doors according to performance, 1990

Figure 0.21
Distribution of faults identified in BRE DPU site investigations of rehabilitated and new-build housing: solid walls according to authority contravened, 1990

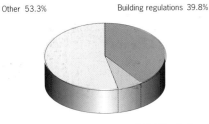

Figure 0.22
Distribution of faults identified in BRE DPU site investigations of rehabilitated and new-build housing: separating walls according to authority contravened, 1990

Authority contravened

Although it is only comparatively infrequently that building regulations are invoked for older properties, the criteria included a record of when the construction was unlikely to have complied with building regulations (in this case sometimes the earlier deemed-to-satisfy construction rather than provisions of later Approved Documents) in force at the time the work was being carried out. Also included was a record of non-compliance with British Standards and Codes of practice. The majority of faults, however, occur in relation to manufacturers' instructions, BRE and industry recommendations, and client requirements, and these are all included under the label 'other'. It must be emphasised that non-compliance was an opinion of site investigators, though care was taken to verify the findings with designers and contractors.

Comparatively few solid external walls were found in new construction, and perhaps care should be taken in comparing these with solid walls in rehab projects. However, the main issues here in both cases are weathertightness, strength and stability, and durability (Figure 0.21).

So far as cavity external walls are concerned there seems to be slightly better compliance with regulations and standards in the case of new-build than with rehab, but the figures are so similar to those for solid walls that it is not worth providing charts.

As might be expected, it is sound and fire matters which show in separating walls (Figure 0.22), whether or not they occur in rehab or in new-build, and transgressions of the building regulations and codes of practice were commonly and frequently observed.

Partitions in rehab projects show a higher incidence of non-compliance with building regulations than do new-build partitions (Figure 0.23), relating in the main to stability matters when altering older construction.

Windows and doors follow a similar pattern (Figure 0.24), clearly relating more to weathertightness and durability, though rehab projects provided far more examples of contraventions of regulations and codes than did new-build.

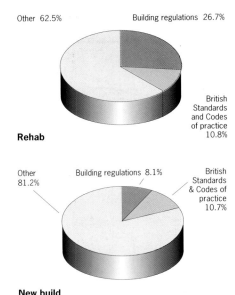

Figure 0.23
Distribution of faults identified in BRE DPU site investigations of rehabilitated and new-build housing: partitions according to authority contravened, 1990

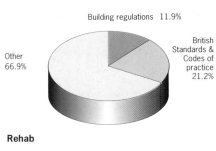

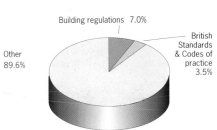

Figure 0.24
Distribution of faults identified in BRE DPU site investigations of rehabilitated and new-build housing: windows and doors according to authority contravened, 1990

Responsibility for faults

With respect to solid walls (Figure 0.25), the site seems to bear a greater share of responsibility in the case of rehab – twice as many as for new-build but still less than half that for design. The explanation for this most probably lies in the difficulties of specifying and carrying out alterations to old brickwork.

For cavity walls (Figure 0.26), the figures for design and site are reversed – in the case of rehab nearly twice as many more due to the designer than to the site; and in the case of new-build nearly twice as many more to the site than to the designer. It is difficult to account with confidence for this reversal, though the most likely explanation for the latter could well lie in a higher incidence of potential weathertightness problems in the installation of cavity insulation.

Separating walls follow a similar pattern (Figure 0.27), though in this case the disparity is even greater – with rehab nearly four times as many due to the designer than to the site; and in the case of new-build nearly three times as many down to the site than to the designer. Again, there is no simple explanation, but it is probably due to designers' failures to specify adequate upgrading in respect of sound and fire in rehab cases, and workmanship items such as unfilled mortar joints and other unfilled voids in new-build cases.

Yet again, partitions follow the pattern (Figure 0.28), though the figures are similar to those for cavity walls.

For doors and windows in new-build, the design and the site bear about equal responsibility, but for rehab the greater share is attributed to the designer (Figure 0.29 on page 8). This could well be due to selection of qualities of performance for replacement items which are inappropriate to the situation of use.

All elements included in this study show very small numbers relating to material and component quality, with few over 13% of the totals. As seems also to be the case with the elements covered in the other books in this series, those on roofs and floors, it is what people do with components and materials that is more important than their intrinsic quality when it comes to defects in performance.

For faults with their origin in execution, in common with roofs

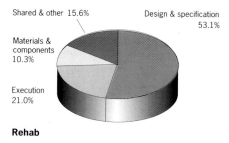

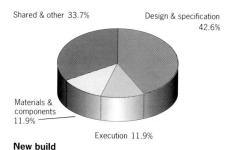

Figure 0.25
Distribution of faults identified in BRE DPU site investigations of rehabilitated and new-build housing: solid walls according to responsibility for the defect, 1990

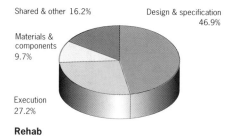

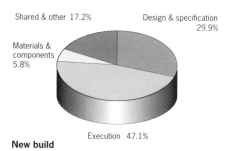

Figure 0.26
Distribution of faults identified in BRE DPU site investigations of rehabilitated and new-build housing: cavity walls according to responsibility for the defect, 1990

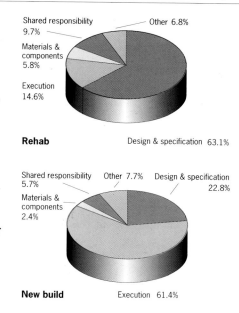

Figure 0.27
Distribution of faults identified in BRE DPU site investigations of rehabilitated and new-build housing: separating walls according to responsibility for the defect, 1990

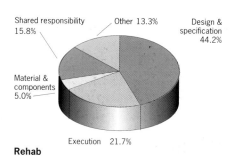

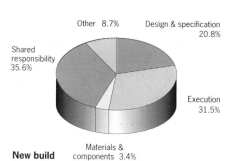

Figure 0.28
Distribution of faults identified in BRE DPU site investigations of rehabilitated and new-build housing: partitions according to responsibility for the defect, 1990

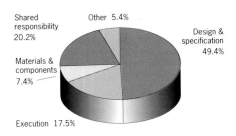

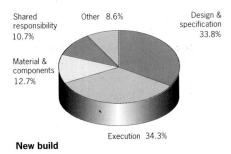

Figure 0.29
Distribution of faults identified in BRE DPU site investigations of rehabilitated and new-build housing: windows and doors according to responsibility for the defect, 1990

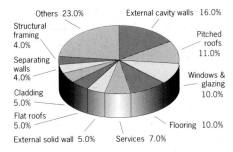

Figure 0.30
Occurrence of defects by element: CQF database, 1997

and floors, it is estimated that a majority of those relating to walls, windows and doors were due to lack of care rather than to lack of knowledge – something like 10 to 1.

Innovation
The data relating to solid and cavity external walls have been examined to determine whether or not the faults related to innovation. It is clear that in both cases around 90% of all items relate to practices which were common before the 1939-45 war. (Actual figures recorded were: new-build 88.6% and rehab 90.6%). It could be concluded from this that it is basic training, education and motivation that are lacking rather than information relating to innovative practices.

Although in the site observations carried out by BRE there has normally been no weighting for the degree of importance of the consequences of failures when they occur, it can be observed that many failures in walls, windows and doors could potentially be of a serious nature.

Thoroughness of surveys
The work of the former Defects Prevention Unit at BRE showed clearly that insufficient time was spent on initial surveys carried out before houses were refurbished. In consequence, many items were unidentified until work was well under way – often too late to make adequate provision for them.

House condition surveys
The 1991 English, 1996 Scottish, 1991 Welsh and 1991 Northern Irish House Condition Surveys[2,3,4,5][†] provide a limited amount of information about the numbers of dwellings having walls of different construction, and the proportions of those which show faults. Special analyses were carried out for this book on the data for England[§], and

the main conclusions follow.

So far as external walls are concerned, accounting for around 13% of all repair costs, approximately a quarter of solid walls and 1 in 6 cavity walls in all ages of dwellings show faults. Approximately one quarter of walls of timber frame houses built prior to 1919 show faults, whereas the incidence drops to 1 in 50 for post-1964 timber frame dwellings.

Windows and doors account for around 14% of all repair costs. Around half of all dwellings having wood casement windows needed some remedial work doing to the windows; the figures were slightly higher for sash windows. Even PVC-U windows needed attention in around 1 in 20 of all dwellings having these windows. Between one third and one half of all dwellings having metal windows needed some work doing to them. Between 1 in 6 and 1 in 3 of all dwellings, depending on age, had windows which were in some degree insecure.

External wood doors needed repairs in around one third of all dwellings, but metal doors fared rather better, with only one in ten needing attention.

Internal walls and partitions in 1 in 16 dwellings showed some signs of rising damp, and it was estimated that around 1 in 50 dwellings built before 1919 required some rebuilding of partitions. Hacking off and replastering was estimated to be needed in around one third of all dwellings of this age.

Internal doors needed repair in around a quarter of all dwellings, and 1 in 6 dwellings needed at least one new internal door.

So far as Scotland is concerned, the 1996 survey results showed that some 3% of external wall structures required repair, 14.3% of finishes, 6.1% of doors and 9.3% of windows[3].

Construction Quality Forum
The main purpose of the Construction Quality Forum (CQF) Database, managed by BRE for the industry, is to identify the most significant defects and their causes. Figure 0.30 gives the breakdown of

† These surveys were the latest available when this book was being written. Since then the other 1996 surveys have become available, and a brief inspection shows that the figures for England have changed relatively little.

§ Although similar information was collected for the remainder of the UK, these have not been analysed for this book.

the contents of the database in January 1997. The database contains nearly twice as many entries dealing with non-residential construction as with residential. It can be seen that this book deals with elements forming around 40% of the totals identified by the CQF.

BRE publications on walls, windows and doors

Principles of modern building
Principles of modern building, Volume 1, Walls, partitions and chimneys, was first published in 1938[6]. It was followed by second and third editions, the latter published in 1959. Even the third edition is now showing its age, not only because the examples date from the 1950s, and all the dimensions in the text are imperial, but also because such performance requirements as thermal insulation are minimal in today's terms. However, since the book deals with principles, in some respects it is less out of date than might be imagined.

Those general principles enunciated in *Principles of modern building* which have stood the test of time form the basis of the introductory sentences of some of the following chapters. Descriptions of agents affecting walls are also taken in the main from BRE texts which have been widely used also for other publications.

BRE Digests and other papers
The various BRE publications, such as Digests, Information Papers, Good Building Guides and Good Repair Guides which are listed in the references and further reading lists, together provide a fairly comprehensive coverage of walls, windows and doors. Since they have been published over a time span of many years, though, they can have received very little cross-referencing between them all. All these publications have been drawn upon to a considerable extent in *Walls, windows and doors* which therefore gives a key to the main BRE publications relevant to the subject.

BRE Reports
A number of BRE reports contain sections relevant to walls, windows and doors which may not be apparent from their titles. Those that are relevant include: *Assessing traditional housing for rehabilitation*[7], *Surveyor's check list for rehabilitation of traditional housing*[8], *Quality in traditional housing*, Volume 2, an aid to design[9], and Volume 3, an aid to site inspection[10]. Information on the construction of non-traditional housing is available in the many BRE reports on particular systems: the current CRC catalogue should be consulted[11].

Reports by other publishers
Probably the most important independent reference for the design of walls for new buildings to be published since the 1970s is *Wall technology*[12], which describes in a systematic way the performance requirements of walls, and the properties of commonly available solutions. This CIRIA publication contains many references to other publications which are not repeated here. The first volume, covering performance requirements, was drafted for CIRIA by BRE, and other volumes in the series were drafted by leading members of the industry and professions.

Changes in construction practice over the years

Walls: a historical note
It was the Romans who exploited the practice of placing lime and pozzolana cemented concrete between shutters to erect durable walls of substantial thicknesses for their buildings. Pozzolanas are cementitious materials occurring naturally, such as pumice or volcanic earth; or which are artificially made, for example, from crushed brick or tile. For the more utilitarian buildings, these concrete walls may have been left as struck from the shutters; but, for the more important buildings, the walls were faced with irregular pieces of ordinary limestone (*opus incertum*) or with regular pyramids of

brick tile pushed into the concrete diamond fashion (*opus testaceum*) or even thin sheets of marble. To retain adequate stability, many walls were provided with through stone or brick bonding to tie the wythes together, a practice which still finds echoes in our contemporary cavity walls. In addition, quoins were formed with bonded stone or brick. External facings were attached by means of iron or bronze cramps, a practice which continued for many centuries, and which still can leave a legacy of spalling façades when rain penetration occurs and corrosion sets in.

Internal finishings, where not the same as those used on the exterior, were of cements and painted stuccoes, or stone, marble or glass tessarae mosaics similar to those used on floors. Again, as with the structure of the walls themselves, these techniques have provided the basis for most of the traditional finishes until the second half of the twentieth century.

Timber framed construction of walls, with various kinds of infilling, is thought to go back even earlier than the Romans, though the material, being organic, has meant that the very oldest examples have decayed and extant remains are scattered. Gradually the practices were refined to improve the protection of the timber, for example by raising it above the wet earth, and in consequence some of our Tudor and Jacobean buildings continue to bear impressive testimony to the longevity of English oak even if altered or restored.

After the Roman occupation ended, many of the building skills were lost; indeed many buildings were deliberately destroyed. When stonemasonry came to be revived in late Saxon times, some of the practices, such as are to be found in long-and-short work, harked back to carpentry practices. Timber strapping, as a means of improving stability, continued to be used in uncoursed or rubble masonry or flint walling until late Victorian times, and still may cause concern when exposed during rehabilitation work.

Figure 0.31
Saxon masonry in the chancel of Escomb
Old Church, County Durham

Figure 0.32
Compton Wynyates, Warwickshire

Ashlar, or 'clene-hewen', rectangular stone in regular courses, was also used in ancient times, and, after some use in Saxon times in the UK (Figure 0.31), was revived on a significant scale in Norman times for major buildings such as cathedrals and castles. As time went by, the materials gradually became economic for use at the domestic scale in areas near where the stone was quarried. Although brick had been available since Roman times, it had largely fallen out of use, with no native manufacture until the fifteenth century. It received added impetus when facing brickwork became fashionable in the first half of the sixteenth century, and several of our great country houses, such as Compton Wynyates in Warwickshire (Figure 0.32), were built in loadbearing brick, albeit with stone dressings.

Timber framing with various kinds of infilling, including wattle and daub and brick, continued in use in those parts of the country where suitable timbers were easily available. Elsewhere locally quarried stone or locally produced brick have contributed to the richness of the architectural scene. It was the Industrial Revolution and the construction of the canal and railway systems which turned the old economic system on its head and enabled materials to be used well away from the areas where they were produced.

Brick walls were almost invariably built in a single leaf until the gradual introduction of cavity walls from early Victorian times. Indeed, single leaf domestic construction, rendered in those geographical areas subject to driving rain, was common until the 1939–45 war; and much of the ribbon housing development of the late interwar period was still being built in 9 inch brickwork.

Cements before 1850 were all similar to Roman, and indeed that was what they were still called – Roman cements. One of the more famous architects of his time, John Nash, first began to use Roman cements in 1796, and he continued to use them until he changed to mastic cements in 1820 (British Patents 3872 and 4033, 1815–16). However, in 1824, Joseph Aspdin took out British Patent 5022 for one of the first of the Portland cements[13].

External rendered finishes to rubble walls were traditionally based on lime mixes, occasionally pigmented (Figure 0.33), until Portland cements became available.

Loadbearing masonry is still a popular method of construction for low-rise buildings. For example, in the last years of the twentieth century in England, the external walls of 9 out of 10 housing association dwellings inspected by Housing Association Property Mutual were still being built in loadbearing clay brick.

Windows and doors: a historical note

Windows were probably thought to be unnecessary in the most primitive of dwellings, as they were right up until the nineteenth century in many a Hebridean black house. When first used they were probably solely for ventilation purposes; indeed, the word 'window' is said to derive from the Anglo-Saxon term 'wind-eye'. To control the ventilation rates through these primitive windows, the sliding shutter, operating as a hit-or-miss, was invented, kept shut on the windward side of the building and open on the lee.

To carry the shutter, a rebated frame was needed. To permit light to enter the building using the narrow sheets of horn which provided the only widely available suitable and durable material then available, mullions were needed to divide up

Figure 0.33
A restored Welsh farmhouse

the space. It would only have taken a hole to be placed in the sliding shutter to accommodate a piece of horn, oiled paper or linen to obtain the best of all possible worlds – the first horizontal sliding window, as we know it, would have come into existence. The glazing of the time was reinforced with a lattice of cleft wood or wicker. The mullions on these early windows were set diamond fashion into mortices in the head and sill, and this automatically gave the best possible angle for the square section to control glare, with no dark shadow line to the building's interior, a desirable architectural detail which is arguably and regrettably ignored today.

Glass had been used by the Romans on a quite extensive scale, and Davey[13] records the use of bronze window frames to hold slabbed glass panes of around 21 inches × 28 inches. After the Romans left, however, the art was lost until medieval times. Lloyd refers to glass being introduced into the royal apartments from around 1250[14]. Crown glass, formed by blowing cylinders of the molten raw material, dominated the manufacturing process until around 1910 when fire finished sheet glass began to be made on a large scale. Although cast plate glass had been introduced in the late seventeenth century, the better quality glasses were not made on a significant industrial scale until the invention of float glass in the late 1950s.

Window openings in trabeated stone walls could easily be formed by just leaving out a few stones underneath a particularly long stone. When wider apertures were needed to light larger buildings such as cathedrals, this coincided with a more extensive use of the arch: semi-circular at first, following Roman practice, but later, from the thirteenth century onwards, extending to the full panoply of multi-centred arches. In the major cathedrals the windows became huge, being subdivided by vertical stone mullions which themselves branched into tracery above arch springing level. Because glass was

available only in comparatively small sizes, it was set into lead cames and wired to wrought iron transomes to provide resistance to wind pressures, a practice which has survived to the present day even though technology has now given us almost limitless possibilities with other forms of glazing. Even so, it was possible to build windows of enormous size using the old techniques: indeed, it has been said that the great east windows of Gloucester and York cathedrals (Figure 0.34), at around 10 m wide by 23 m high, were amongst the largest windows in the world until comparatively late Victorian times.

Even if windows were un-necessary, doors certainly were not. If they were just holes left in the wattle and daub or the cob, they would still need to be covered by a closely woven hurdle to keep out most wind and snow. In time, rough boards were joined together on ledges, and set in pivots in the threshold stone and in the timber lintel, or, in non-domestic buildings, in the stone lintel. Hinges had to await later improvements in metallurgy.

Summary of the main changes in common practices since the 1950s

Cavity walls, at first in twin leaves of brick, facings in the outside leaf and commons in the inside, only became standard practice for domestic work after the 1939-45 war (Figure 0.35 on page 12) though they had been used in certain parts of the country one hundred years earlier. Blocks, at first made with coke breeze or clinker aggregates, then gradually replaced the commons in the internal leaves. Although it was realised that the thermal insulation value of the wall had been improved, such improvements were marginal when considered in today's terms. The main aim of the cavity wall was to improve weathertightness.

Apart from a hiatus in the late 1950s when it was thought that cavity walls might be significantly better for sound insulation, the solid one-brick separating wall has proved to be a popular choice for the

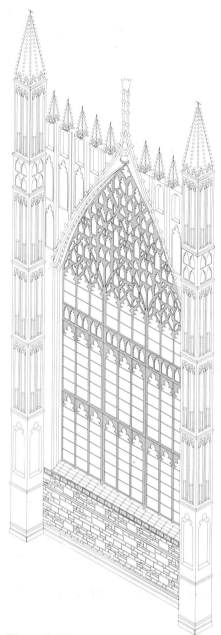

Figure 0.34
The great East Window in York Minster

Figure 0.35
A cavity wall with brick inner and outer leaves. The window has been summarily removed prior to replacement

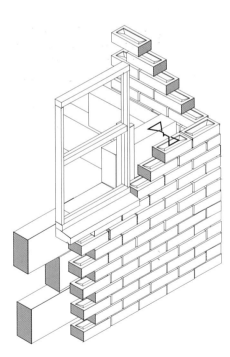

Figure 0.36
'Integral' structure

masonry separating wall, sometimes substituted by dense concrete blocks. Cavity construction has also been widely used. Since the 1970s, the properly designed and built timber framed separating wall has proved to be an effective alternative to masonry.

External walling will normally take one of three characteristic forms of construction so far as its geometry is concerned:
● 'integral structure' Figure 0.36
● 'fitting between' structure, Figure 0.37
● 'running clear' of structure, Figure 0.38
though there are examples of combinations of these (eg some masonry).

Because of inherent problems with thermal insulation and weather-proofing of joints, it is likely that the 'fitting between' category will reduce in importance in the future, though the national building stock contains many examples from the past.

Examples of significant changes include the fashion begun in the 1960s for introducing large glazed areas (as a proportion of the total wall), often for aesthetic reasons, and counterbalancing the excessive amounts of daylighting and solar gain by tinting the glass (Figure 0.39).

There have been relatively few changes in basic walling materials until the later years of the twentieth century, though some innovations have unfortunately proved to be in advance of the industry's ability to cope successfully with them (Figure 0.40).

One of the most far-reaching of changes has been the continuous improvement in the thermal insulation of buildings following increases in fuel costs and rising concern at the amounts of carbon dioxide being released into the environment from heating thermally inefficient buildings. Changes in building regulations have forced designers to use smaller window areas, or to make the glazing thermally more efficient. Greater sophistication of the control of the internal environment, coupled with the greater emphasis on thermal

efficiency (Figure 0.41); the construction of taller buildings more severely exposed to wind and rain; and greater use of specialist cladding systems (the detailed functioning and requirements of the more innovative designs of which may be less well understood by both building designers and builders) – these have led to the increased possibility of inadequate performance of the external skin of many buildings in both the long and the short term.

Nothing lasts for ever, and the problems of deterioration of materials have always been with us, especially those that are organic based and their traditional finishes. However, even some apparently

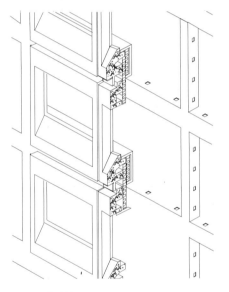

Figure 0.37
'Fitting between' structure

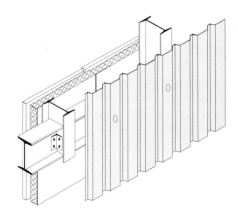

Figure 0.38
'Running clear' of structure

Figure 0.39
The old and the new: Holy Trinity, Hull, one of the largest parish churches in England, reflected in an adjacent brown tinted glazed facade. (Photograph by permission of B T Harrison)

Figure 0.41
Visual imaging of a thermal bridge at the junction of two walls and the ceiling of a room. The blue and the green indicate poor thermal insulation

Figure 0.42
Spraying phenolphthalein on these brick and mortar samples indicates the progress of the carbonation front from each face. The purple colour indicates that the mortar is still largely uncarbonated, and that a path for water penetration exists in the joint in the perpend. (See also Chapter 2.4)

more robust inorganic materials can deteriorate given the wrong combination of factors (Figure 0.42) thereby further stimulating the analysis of 'performance, diagnosis, maintenance, repair and the avoidance of defects' from which this series of books takes its subsidiary title.

Figure 0.40
Failure of masonry paint over both calcium silicate and replacement fletton brick masonry

Chapter 1

The basic functions of the vertical elements

Figure 1.1
Tay House, Glasgow (Photograph by permission of Stoakes Systems Ltd)

This first chapter deals in turn with the basic functions which affect all of the vertical external envelope of a building, whether large (Figure 1.1) or small, and all of its vertical internal subdivisions. In later chapters each of these functions is considered in greater detail where relevant to a particular component. The chapter takes its cue from a passage in *Principles of modern building*: '*It will be convenient to consider the whole range of wall functions together, even though any given wall, external or internal, loadbearing or non-loadbearing, may not have to perform all of them*'.

Chapter 1.1   **Strength and stability**

The predominating factor in the design of a wall is whether it has to carry an imposed load, for example from floors and roofs. Ever since Roman times the loading on a non-framed wall has largely governed its thickness, and over the years conventions became established for their construction.

However, it is only comparatively rarely that the wall can stand by itself without receiving support or restraint from the remainder of the building. Even the smaller domestic scale building of loadbearing brick often relies on the floor to provide lateral restraint to the external wall. All that this chapter can do is to draw attention to some of the more important considerations in relation to the contribution which walls make to the structure as a whole, but not to provide sufficient information to allow the structural design of walls to be carried out.

All walls need to be sufficiently strong to carry the self weight of the structure, together with imposed loads; for example those due to furniture, equipment or the occupants of the building.

Current requirements as far as the structure of walls is concerned are embodied in the various national building regulations. Taking the England and Wales Regulations[15] as an example:

'(1) The building shall be constructed so that the combined dead, imposed and wind loads are sustained and transmitted by it to the ground

(a) safely; and

(b) without causing such deflection or deformation of any part of the building, or such movement of the ground, as will

impair the stability of any part of another building

(2) In assessing whether a building complies with sub paragraph (1) regard shall be had to the imposed and wind loads to which it is likely to be subjected in the ordinary course of its use for the purpose for which it is intended.'

Structural design of walls for buildings is covered by the main British Standard Codes of practice for the various materials:
- steel to BS 5950-1 to 9[16]
- concrete to BS 8110-1 to 3[17]
- timber to BS 5268[18]
- masonry to BS 5628[19], Approved Document 1/2[20] or BS 8103[21]

Loads

There is a wide range of structural considerations that may be required of a walling system, ranging from purely non-loadbearing cladding to the cellular masonry or concrete walling system which ultimately carries all the actions on a structure. The forces are illustrated by Figure 1.2.

Dead loads

The dead weight of any floors, partitions, ceilings, roofs and claddings must be carried to the ground (foundations) via a structural frame or a loadbearing wall. The design dead load at any point is the sum of all the individual dead loads acting at that point from above factored up to allow for the variability of materials. The partial factor of safety is normally around 1.2–1.4 for unfavourable loads and 0.9 for favourable loads. Eccentric loads (loads not acting axially), as illustrated by Figure 1.3, normally

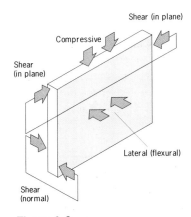

Figure 1.2
Different loadings on a wall

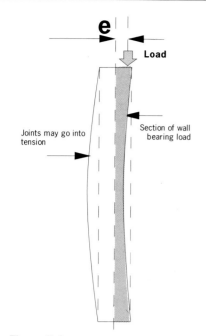

Figure 1.3
Eccentric loads

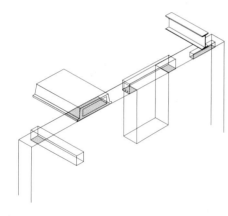

Figure 1.4
Concentrated loads

Figure 1.5
Gale damage to the external leaf of a cavity wall. Few wall ties were in evidence. (Photograph by permission of East Anglian Daily Times)

invoke special design requirements, as do concentrated loads (ie large loads acting over a small area, as in Figure 1.4). Individual loads are the product of the measured or nominal density and the volume of the material. Detailed design advice is given in BS 5628-1, clauses 21, 22 and 23, and in BS 8110-1, section 2.

Applied loads

Vertical applied loads due to furniture, fixtures, fittings, live occupants, snow and wind acting on all surfaces must be carried to the ground (foundations) via a structural frame or a loadbearing wall. The design applied load at any point is the sum of all the applied loads acting at that point from above. γf, the partial factor of safety is normally around 1.2–1.6. Eccentric loads (loads not acting axially) invoke special design requirements. Individual design loads are nominal values based on statistical probabilities drawn from observations of building types. Detailed design advice is the same as given for dead loads in the section above.

Seismic loads

Seismic loads generate similar forces (ie lateral and shear forces on walls) to wind loads but potentially over a much greater dynamic range. They are not explicitly designed for by current UK practice. They do occur, but experience suggests that their intensity in the UK is always less than wind forces.

Wind

Gales in the UK cause around £70 million worth of damage to buildings each year. In the severe gales of October 1987 around 1.3 million houses were damaged; in those of early 1990, around 1.1 million houses were damaged when one third of the costs of the damage related to walls and their components (Figure 1.5). Most damage by wind is of a relatively minor nature; nevertheless there are still substantial sums of money spent on rectification of damage when the failure of glazed areas and wall

cladding results in high internal pressures in buildings[22].

Changes have been made to the rules for design since the 1939–45 war. In particular the 1952 wind loading code in BS CP 3[19] did not take account fully of the effects of building shape and wind directions, and gave, as a result, lower wind loads than would be obtained today using BS 6399-2[23].

Wind speed

Wind speed varies depending on the geographical location of the site, the altitude of the site above sea level, the direction of the wind, and the season of the year. More local effects such as height above ground level, topography and terrain also affect the speed and direction of the wind at a particular site. In the immediate vicinity of a building, or a group of buildings, the wind changes speed and direction rapidly, depending on the form and scale of the building. Wind speeds in coastal regions are generally greater than in inland areas; for instance, the speeds near the coasts of southern England are some 10–25% greater than they are at the same altitudes in places in the centre of southern England. The highest wind speeds occur in the north of the British Isles, the highest basic gust wind speeds used in design being 56 m/s in the far north west of Scotland.

Further information on wind speed, and factors which affect it, is given in Chapter 1.1 of *Roofs and roofing*[24] from which Figure 1.6 (on page 18) is taken.

A building modifies the flow of wind round it. Figure 1.7 (on page 19) shows the case where the wind blows directly face on to the building, but the flows will differ according to the wind direction, and according to whether the building is tall or squat in profile (Figure 1.8).

Tall outline

If the height of the building is more than half the crosswind width B of the outline, the wind tends to flow round the sides of the building, except at the very top of the building. In the case of the tall building, the

Figure 1.6
A map of the UK showing the reference
basic hourly mean wind speed in
metres per second

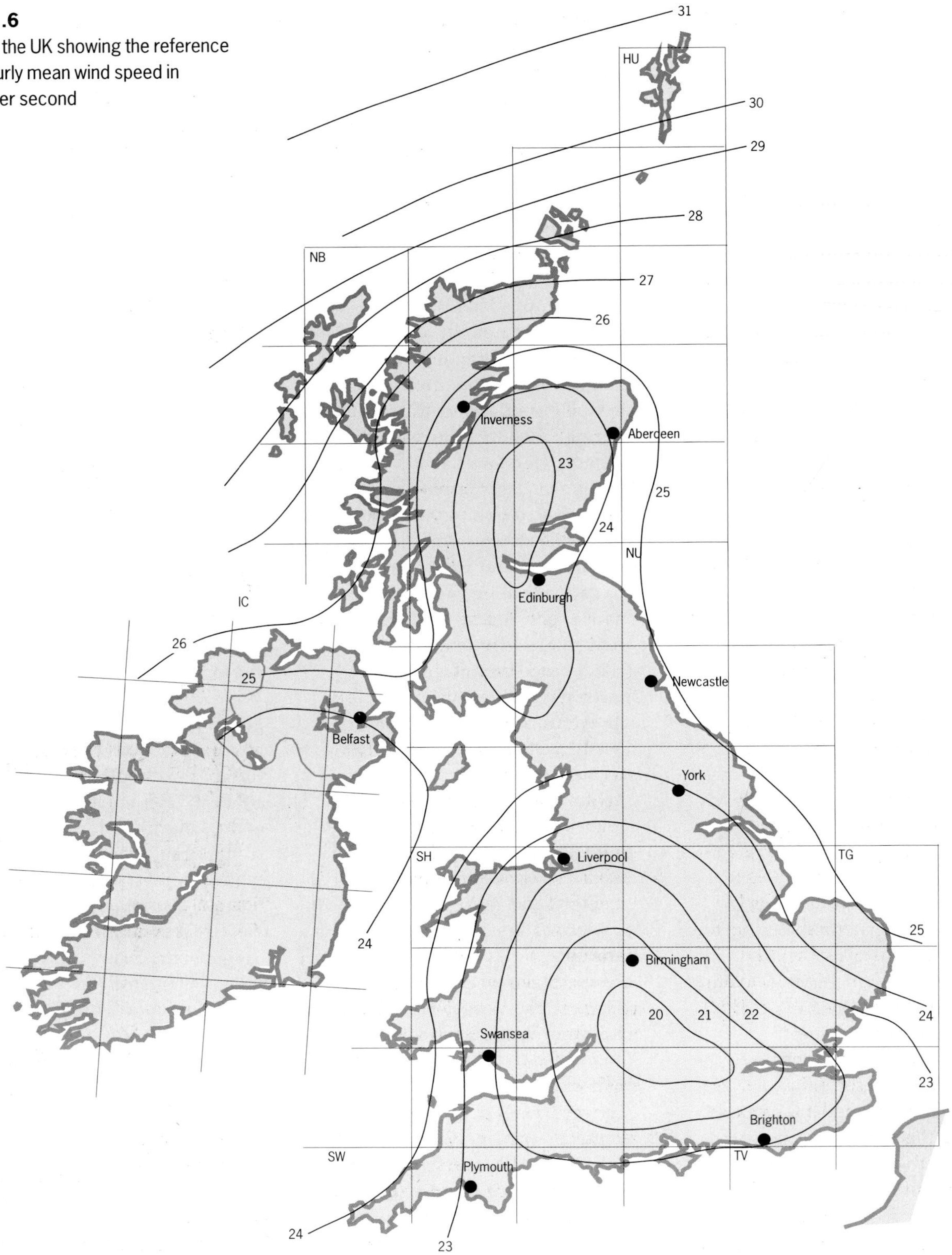

crosswind width B is used as the
critical dimension b for sizing the
pressure zones on the building. (That
is, b = B).

Squat outline
If the height of the building is less
than half the crosswind width B of
the outline, the wind tends to flow

over the top of the building, except at
the ends of the building.

In this case, the critical dimension
b used to size the pressure zones on
the building is twice the height H.
(That is, b = 2H).

The wind pressure increases up
the face of the building, from zero at
ground level to a maximum about

two thirds of the total height. The
pressure also varies horizontally
within each zone. But the empirical
pressure coefficients given in BRE
Digest 346 Part 6[25] are average
values computed to give the correct
total load on each zone. For the
design of small components such as
windows or cladding panels, it is the

peak suctions higher up the building which are critical, and the Digest gives special pressure coefficients for these peak suctions.

In addition to pressures on the external faces of buildings, the design also has to take into account the effects of wind on the inside faces of the external wall (Figure 1.9). The internal pressure inside a building depends on the sizes of the openings in its walls.

A conventional building with all its large doors and windows closed has a low porosity. This has a damping effect. If windows get broken on one face during a severe storm, or if there are other large openings on one face, internal pressures can rapidly approach external pressures, and the net load on the opposite external wall will then greatly increase.

As long as a multi-layer skin acts structurally as a single unit (for example a properly tied cavity wall), the wind distribution through the layers is irrelevant. But there are cases where the pressures between the layers can be important, depending on how permeable the cladding system is, and on whether there are voids and how large these are. In the case of impermeable cladding with voids, another key factor is the pressure at the location to which the voids are vented. For a description of the principal loading cases, see pages 16–18 of the BRE report on overcladding[26].

In many cases, the degree of permeability of the building surface is indeterminate and it is safest to assume that the walling must transmit the full wind loads through any adhesive bond and mechanical fixings. Bond and fixing strengths may be determined by testing small sections or by applying a proof suction load to the prototype panel using a test rig[27].

A further factor to consider will be local deformation of the surface of the walling under wind loads which may alter the geometry of a joint. Certain kinds of walling joints, for example unfilled joints in rainscreen systems, are more tolerant of changes in joint geometry than are

face-sealed systems. BS 8200[28] suggests a limit depending on the material, in the range 1/90 to 1/500 span for deflections in opaque infill panels in secondary framing. It will be possible to calculate the effects on joints of differential movements on adjacent panels, and those responsible for the final detailed design should carry out the necessary calculations.

Seasonal factors

Seasonal factors can be relevant to construction work. The highest extreme winds in the UK are expected in December and January. In the summer months of June and July, extreme winds may be expected to be only about 65% of the winter extremes; for the six month summer period from April to September, normal wind speeds are only expected to be 84% of those in the six month winter period, October to March. Tables of these seasonal factors are published for one month, two month and four month periods starting in each month of the year[25].

BRE can provide realistic test regimes; for example, using computer controlled simulation of actual wind flow patterns (see Chapter 4.2).

Fatigue

There are two types of fatigue that may need to be considered: high-cycle fatigue, and low-cycle fatigue.

High-cycle fatigue

High-cycle fatigue occurs when the structure is subjected to very many thousands of load cycles at a small proportion of the ultimate capacity of the structure caused by oscillations at resonant frequencies. High-cycle fatigue is essentially a serviceability problem, provided the structure is properly inspected and maintained, and that fatigue cracks can be repaired or components replaced before their fatigue life is exhausted. This is common practice, for example, for the holding-down bolts of slender steel chimney stacks.

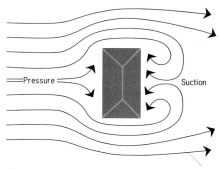

Figure 1.7
Wind blowing directly on the face of a building

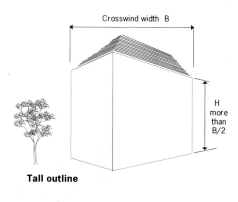

Figure 1.8
Effect of the shape of a building on wind

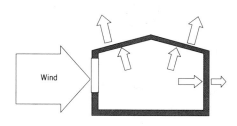

Figure 1.9
Internal pressures due to one dominant opening

Low-cycle fatigue

Low-cycle fatigue occurs when the structure is subjected, by the repeated action of severe gust loading, to relatively few load cycles close to the ultimate strength of the structure; that is, for a few tens to a few thousands cycles. Such high stresses tend to occur in thin metal claddings around the fixing points. There could also be loss of bond between insulation and its substrate in composite sheets. Tests can be specified to check for this.

Owing to the intermittent nature of storms, low-cycle fatigue occurs in a few short periods during the strongest storms, and a satisfactory inspection of a structure after one severe storm may not guarantee survival during the next storm.

Prototypes can be subjected to a simulation test using a suitable suction device. From continuous wind records, BRE has proposed a test regime which simulates the 50-year design life of a component exposed to the wind in the UK. The pressures applied are given as a percentage of the full design pressure at the site in question.

Fluctuations in external surface temperatures can lead to fatigue of material at the interface in metal-skinned sandwich or laminated panels used for cladding external walls. This has been known to cause local delamination which has adversely affected both appearance and durability, and BRE tests have shown that it is possible to predict the risks of delamination.

Compressive strengths of masonry materials

It may be important to be able to obtain some indication of the compressive strengths of masonry materials, particularly in older buildings when change of use is contemplated. A test has been developed by BRE for use on mortars and some masonry units with strengths up to 7 N/mm^2 [29].

Fixing devices and their structural performance

The most common devices and their structural performance requirements are as follows:
- wall ties for connecting one leaf of a cavity wall across the cavity to another leaf or to a frame structure, and cladding restraint ties for connecting cladding elements to frames. They are required to resist tension and compression forces while allowing limited differential lateral movement
- shear ties for connecting two adjacent masonry leaves together, for connecting masonry walls which need to interact to produce composite action and for connecting masonry walls to frame structures. They are required to resist shear, tension and compression forces
- slip ties for connecting two adjacent walls or masonry cladding to frame structures at movement joints and allowing in-plane movements. They are required to resist shear but not tension and compression forces
- straps for connecting masonry walls to other adjacent components such as floors and roofs. They are required to resist tension forces and need to be blocked to resist compression forces
- joist hangers for supporting joists, beams or rafters on masonry walls by direct loading via a flange which is embedded in a mortar joint. They are required to resist vertical loads, but occasionally horizontal loads where lateral restraint is needed (Figure 1.10)
- support (shelf) angles for supporting masonry walls on other structural elements. They are required to resist vertical loads and on some occasions horizontal loads
- brackets comprising an individual support for two adjacent masonry units, which form part of a masonry wall, on other structural elements. These have the same requirements as support angles

Lateral support and restraint

Lateral support is often required to be provided by floors to walls, and such requirements are set out for new buildings, for example, in Approved Document A of the Building Regulations for England and Wales, Table 11. All external, compartment or separating walls greater than 3 m long will require lateral support by every floor forming a junction with the supported wall, and all internal loadbearing walls of whatever length will require support from the floor at the top of each storey.

Straps may not be needed if the floor has an adequate bearing on the wall – for example 90 mm in the cases both of the bearing of timber joists and of the bearing of concrete joists.

BRE site studies have shown that these requirements have not always been complied with in practice, and surveyors should be aware that a proportion of relatively new construction will be deficient in this respect. Information from Housing Association Property Mutual, for example, indicates that lateral restraint strapping was not provided in 1 in 5 cases where such strapping should have been provided in accordance with BS 8103-1. For

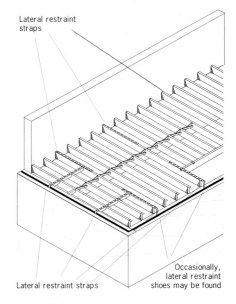

Lateral restraint straps

Lateral restraint straps

Occasionally, lateral restraint shoes may be found

Figure 1.10
Typical provision of lateral restraint for external walls from floors. The ground floor of a two storey building is shown

further information, see Chapter 2.1 in *Floors and flooring*[(30)].

It should be recalled that older buildings, with much thicker loadbearing walls than have become common in later years, may not need to be provided with special provision for lateral restraint. Indeed, where such buildings have adequately withstood the test of time, and no significant alterations are proposed, there is little point in providing strapping.

Loads on fixings

Fixings associated with walls of buildings perform a wide range of tasks and it is helpful to divide them up into broad classes based on the type of loading: ie tension, compression or shear.

Pure tension loading is normally borne by straps, usually thin strips of metal, with a cross-section area typically between 10 and 100 mm². The design loads are likely to be in the range of 1–50 kN and the limiting factor is usually the connections to other components. However, if they are not installed correctly, they cannot work (Figure 1.11).

Ties are used for an enormous range of applications and may be called upon to resist either tension, compression or shear or a combination of any two loads or all three. Typical wall ties are used for tying masonry in tension and compression to columns and floors or across cavities, and for supporting cladding units and facing slabs. Wall ties have a typical design ultimate resistance in tension and

Figure 1.11
A lateral restraint strap installed without supporting nogging, and bent in consequence, and covering only two joists. It cannot possibly provide lateral restraint to the wall

compression between 0.5 and 5 kN depending on the application, and need a low shear resistance to reduce differential movement damage, whereas ties in timber frame systems are also required to transmit shear forces. Ties capable of taking shear are also used to join masonry at butt joints (eg at the edge of panels within frame systems or where a new wall is bonded to an existing wall – for the latter they are termed 'starter ties'). A 60 mm² strip crossing a filled mortar joint is attributed a shear resistance of up to 5 kN. A wide range of very specialised strip ties is available to support wall panels at movement joints at either the head or sides of a panel. These have a requirement for zero compression/tension resistance but a design shear resistance of 5–10 kN.

Vertical support is given by distributed ties in shear, by corbel anchors, by angle or bracket fixings at the base of wall panels, or by brackets suspended by angled ties where the load is in tension only. Ties used in shear in sandwich panels may have very high design loads. Sandwich panels are usually bonded during manufacture with pins through the insulation. Panels may be supported in a number of ways. The forces on angle or corbel type fixings are a complex mixture of bending and shear, with shear and tension on the holding bolts. The force applied may be up to 0.2N/mm² (ie 2 tonnes per metre run of 100 mm masonry – the maximum figure allowed by BS 8625 for three storeys of masonry at a density of 2000 kg/m³) and individual design calculations are necessary.

Change of building use

It is important to carry out an appraisal of the structural design of walls which will carry significant changes in floor loadings where there is a proposed change of use of the building; indeed, compliance with Part A of the Building Regulations is required in England and Wales.

Figure 1.12
This lintel has been supported on totally inadequate packing, which will cause the structure it supports to deflect

Maintenance and other service loads

During service, it can be assumed that the walls of buildings will be subjected to irregular applications of loads of widely varying magnitudes, some of which may be so severe as to cause damage. However, even ordinary loads can cause damage if the construction is defective (Figure 1.12).

Loads may cause damage which ranges from no more than unsightly disfigurement of the wall surface to breaking and dislodgement of finishes with serious risk to the safety of both occupants entering and leaving the building, and also to passers-by.

The main principle to be adopted is to limit damage to as small a section of the wall as possible, to prevent collapse of the roof or floors, and to ensure that sufficient of the remainder of the structure survives to provide an escape route for occupants.

Impacts

Accidental impact loads on the façades of buildings are related to height above ground level and access (ie public or private, pedestrian or vehicular). Although practical experience indicates that accident impact damage to the cladding of high-rise buildings is infrequent, the possibility of vandalism must be foreseen on the lower storeys, and also generally in low-rise buildings: the major source of damage otherwise seems likely to be from access cradles.

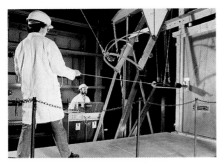

Figure 1.13
A hard body test under way on a doorset.
The impactor is on the end of the weighted
cylinder which is swung against the
specimen

Figure 1.14
A soft body test about to take place on a
doorset. Again, the impactor is swung
against the specimen

At ground level adjacent to a
public footpath or car park, impacts
can arise from road vehicles, human
bodies, cycles, push chairs and
children's toys (eg balls and
skateboards). Walls in secluded areas
with public access are particularly
vulnerable and, where vandalism is a
problem, extensive wilful damage
can be inflicted near ground level. In
contrast, walls within private gardens
may suffer from the same types of
impacts, except for vandalism, since
there is usually greater attention to
care and consequently impacts are
less severe and not so frequent.

At higher levels impact damage

will normally be restricted to that
caused by ladders and mobile
platforms used for maintenance
access. Limited damage by missiles
may also occur where vandalism is
a problem.

Accidental loads are also caused
by explosions of pressurised
containers, dust concentrations,
inflammable vapours and gas leaks
from bottled gas appliances and
piped gas installations to cookers,
heaters and central heating boilers.
Explosions can result in major
damage which ranges from breaking
and displacement of wall
components to total collapse of a
building. See also Chapter 2.3.

Accidental loads are those which
have a very low probability of
occurring for any given structure and
which cover a greater, but uncertain,
range of force than normal design
loads. It may be possible to provide
some kind of protection for certain
kinds of impacts on walls; for
example protection by bollards in
multi-storey car parks. With respect
to explosions and impacts from
objects such as aircraft, because of
the uncertainties it is not economic
to design for no damage in normal
buildings, but special cases, such as
nuclear power stations, may have to
be fully designed for such events. As
an alternative the main principle
adopted is to limit damage to as
small an area as possible such that
the remainder of the structure
survives and provides an escape
route for occupants. For explosions
this might be in the form of cladding
which fails locally and vents the
gases. Other strategies are the
provision of horizontal and vertical
ties or 'protected members'. Walls
clearly will be involved in such
events and loadbearing walls will
have to carry the forces down to the
foundations. Advice on their
magnitude is given in BS 5628-1,
section 5[31] and in BS 8110-1,
clause 2.4.3.2[17].

The performance of walls
subjected to impact loads may be
judged on their ability to remain
safely in place with a satisfactory
appearance and adequate
weathertightness. Vulnerability to

impact damage will depend not only
on the physical properties of the
walling materials used but also on the
type and location of the building.

The stability and integrity of the
whole wall under impact is normally
assessed by two types of impact test:
'hard body' (Figure 1.13) and 'soft
body' (Figure 1.14). The latter
measures the ability to withstand a
heavy blow from a large impactor,
and examines the possibility that
parts of the wall could fall and cause
injury to people. Although damage is
permitted to occur under test, no
part of the cladding should become
dislodged. In tall buildings impacts
are most likely to occur from access
cradles and firemen's ladders
externally, which can be simulated
by the hard body test.

British Standard BS 8200 specifies
requirements for impact resistance in
relation to two criteria. These are
'retention of performance'
concerning, for example,
weathertightness, and 'safety to
persons' which deals with the
probability of breakage and
displacement of materials likely to
cause personal injury. These are
assessed by tests with hard and soft
body impacts at different energy
levels which are related to a range of
wall categories defined in the
standard.

Table 4 of BS 8200 gives impact
energies of 500 Nm up to 1.5 m
above access, and 350 Nm above that
height where access is required for
cleaning. There is no requirement for
soft body impacts above 1.5 m if
access is not required

A 'hard body' (steel ball or
cylinder) impact of 10 Nm is also
required for the whole surface of the
cladding, with the same criterion
that nothing should fall off the
building. In practice it is usual for
cladding specifications to call for two
different solutions: one for the
ground floor (or floor at which
access is available) and one for the
floor above that with respect to the
impact test requirements.

The same two impact tests, though
at different energy levels, are used to
determine whether or not the
cladding will perform its functions

after attack – a smaller soft body and a heavier impact from the steel ball.

The smaller soft body impact of 120 Nm is used up to 1.5 m height only to simulate impact damage (eg from footballs).

A hard body impact of 6 Nm is used up to heights of 6 metres above pedestrian access level, and 3 Nm above that height. The cladding should still be able to perform its function.

The level of impacts on walling materials at which damage becomes visually unacceptable is far less than that at which the material or component ceases in other respects to perform acceptably. Although tests are specified, judgement will need to be used on whether the level of performance achieved is acceptable. In particular, the building owner or his agent must judge whether any damage (eg a dent under test) which does not impair technical performance would be visually acceptable.

It should be noted that some lightweight claddings which have been used in the past will not pass the hard body impact test, and in practice will suffer damage where vandalism is rife. It may, for example, seem unreasonable to demand resistance to the impacts of cross-bow bolts or axes.

Impact resistance is not usually the most critical factor in the performance of walls, and it is probably unrealistic to expect that under accidental impacts the wall should not sustain some damage, though the wall itself should not be in danger of collapse.

For the majority of walls, that is to say category C in BS 8200, excluding extremes of vandalism, and depending on height above access level, the minimum impact resistances for 'retention of performance' are:
● 6 Nm – hard body
● 120 Nm – soft body

For 'safety to persons' the same walls require minimum resistances of:
● 10 Nm – hard body
● 500 Nm – soft body

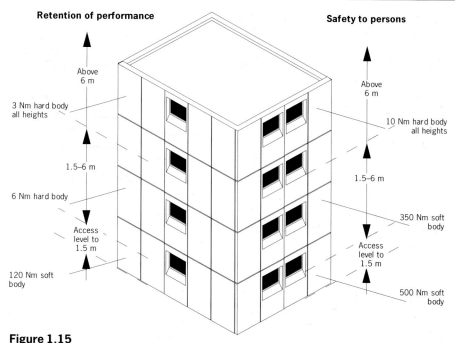

Figure 1.15
Typical impact resistance values for walls

Any walling with which people are likely to come into contact while moving about a building should prevent them from falling from the building. The above requirements will also apply to any wall or partition within a building where it is necessary to prevent people from falling from a height of more than 600 mm. It will be necessary to consult BS 8200 for actual values to be used in particular circumstances.

The wall includes any doors and windows within it, and if appropriate, also includes the guarding of balconies. Additionally, any glazing element (that is glass or plastics sheet materials) with which people are likely to come into contact while in or about the building should be such that it is unlikely to cause cutting or piercing injuries.

The walling should be capable of withstanding accidental impacts produced by body contacts during level walking, falling onto the fabric or during operation of any building element such as window opening or cleaning.

Where the walling is directly contactable, it should provide containment either by
● remaining integral (ie not break and not come away from its fixing) or
● breaking safely (low risk of cutting or piercing injury) but remaining sufficiently intact to prevent a body falling through – this might be suitable laminated glass

For glazing materials, advice about suitability of the material for containment will need to be given by the manufacturer, or advice can be obtained from the Glass and Glazing Federation.

Critical locations for glazing in terms of safety are:
● at medium level in a door and in a side panel, close to either edge of a door
● at low level in walls and partitions and
● in guarding required by virtue of the provisions of Parts K1 and K2 of Schedule 1 of the Building Regulations for England and Wales

See also Chapters 4.1 and 5.1.

Chapter 1.2   **Dimensional stability**

This chapter covers the movements of materials caused by changes in their temperature and moisture levels, and also to some extent with inherent changes in dimension caused by continuous loads – creep, for example – or the absence of support – movements from the ground, for example. Tables of movements for common materials are given at the end of this chapter.

Movements of the fabric

Expansion and contraction of any part of the building fabric subjected to variations of moisture content and temperature will have the potential to cause problems if not accommodated in the design. As a general rule, all common building materials will be subject to thermal

expansion or contraction or both. So far as walls are concerned, it is the larger components which need most consideration, especially where the building is only intermittently heated. However, even small dwellings made from large components can suffer (Figure 1.16).

Linear size changes due to thermal movements are calculated using the expression:

$$R = \alpha\,L\,t$$

where R = change of size,
α = coefficient of linear size change,
L = length of dimension, and
t = temperature difference.

Values of α are to be found in Table 1.1 at the end of the chapter.

Moisture movement is mainly a property of porous materials. Thus aggregate concretes, autoclaved aerated concrete (AAC) and calcium silicate units have reversible movements in the range 0.02–0.2%; softwoods sawn tangentially across the grain in the range 0.6–2.6%, and sawn radially in the range 0.45–2.0%. Concretes, including AAC, also have an irreversible drying or carbonation shrinkage in the range 0.02–0.1%, calcium silicate units in the range 0.02–0.06%, and fired clay has an irreversible volume expansion in the range 0.02–0.1%. This expansion is rapid at first and slows down with time, but can continue for 30 years or more. In most real situations there is some restraint, though rarely complete restraint, afforded to materials undergoing these movements.

Linear size changes due to moisture movements are calculated using the expression:

$$R = \frac{\text{factor} \times \text{dimension}}{100}$$

Where two differing materials are joined, differential movement can occur, which usually exacerbates the problem. The best practice is to try to accommodate movements at the smallest and most elemental level, provided this does not prejudice the structural function. Thus, provided wall components are simply shaped, not too thick, not too large, have movement-tolerant fixings and have joints at their periphery which accommodate movement, the movement strains and the corresponding loads (stresses) are

Figure 1.16
A REEMA large concrete panel dwelling showing signs of movements which are disrupting the joints

If the slab has been cast into frogs in the brickwork, horizontal displacement may show one course below the bearing

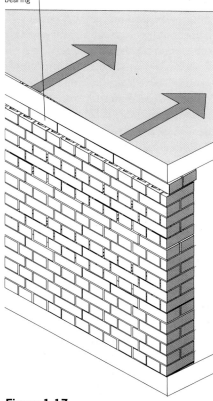

Figure 1.17
Horizontal displacement of the top course of a wall carrying a slab

generally small and can be borne by the element. With large structures, if one or more of these requirements are not met, stresses may accumulate over large areas and damage result. In small structures such as detached domestic houses, it is often possible to omit explicit movement design and depend on restraint from other elements to accommodate movement loads.

Progressive movement may conceivably occur at the edges of floors at the junction with external walls, mainly as a result of ratcheting where movement resulting from expansion (usually thermal) is not fully compensated by contraction. Detritus then builds up, progressive movement occurs, and the material in a wall cracks. However, the crack may also show at the head of the wall just underneath the bearing of the floor, depending on the detail (Figure 1.17). Design data on movements of the structure is given in BRE Digests 227–229[32].

Stress in members
When a potential change of size is fully restrained, the induced stress and force can be calculated using the expressions:

$$\text{Stress} = E \times \text{strain}$$

$$\text{Force} = \text{stress} \times \text{area}$$

For thermal movements the following expressions apply:

$$\text{Strain} = \alpha\, t$$

$$\text{Stress} = \alpha\, t\, E$$

$$\text{Force} = \alpha\, t\, E A$$

Further information will be found in BRE Digest 228.

Vibrations
Ground-born vibration sometimes affects buildings, though there is little evidence that they produce even cosmetic damage, such as small cracks in plaster. Guidance on the measurement of vibrations is given in BRE Digest 403[33] and on human exposure to vibrations in BS 6472[34].

Movement joints
Walls will need to be provided with movement joints which also reflect the positions of movement joints in the remainder of the structure; these joints will need to continue through the wall finishes, and normally will need also to provide continuity in other aspects of the performance of the wall such as its fire resistance (Figure 1.18). If not provided, it is not inevitable that cracking will occur, but it is likely. Problems which arise in this connection are dealt with in later chapters appropriate to various structures and finishes.

Movements from the ground
The characteristics and performance of foundations will be dealt with in another book in this series. However, a brief note is included here on the behaviour of foundations as it affects the walls above. It should be appreciated that in most cases it is uneconomic and unnecessary to design the foundations to reduce the transmitted movements to zero, since most buildings can tolerate a small amount of movement. Although there are no definitive criteria for the majority of buildings,

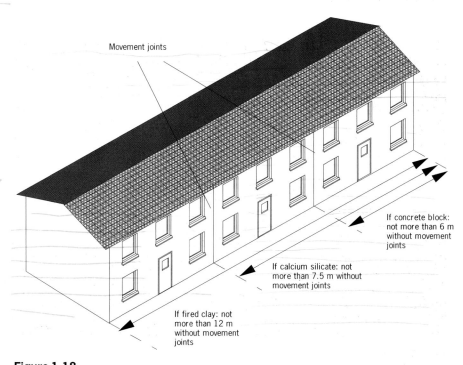

Movement joints

If concrete block: not more than 6 m without movement joints

If calcium silicate: not more than 7.5 m without movement joints

If fired clay: not more than 12 m without movement joints

Figure 1.18
Recommendations for the provision of movement joints in walling

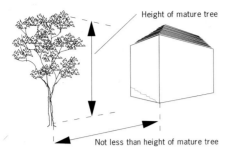

Figure 1.19
For poplar, oak, willow, whitebeam and cherry, the minimum distance should be not less than the mature height of the tree

what movement is allowable will depend on visual appearance and serviceability rather than the stability of the structure. It is to be expected that foundations will move to some degree in service, and this will affect, for example, the formation of and the size of cracks in masonry. The *English house condition survey*[2] records that around 1 in 20 dwellings suffer from some degree of settlement, and around 1 in 40 differential movement.

Serviceability is very subjective and depends on the function of the building, the reaction of the users and the owners; also economic factors such as value, insurance cover and the cost of making good any damage. In many cases, the level of acceptable movement will be dictated by a particular function of the building or one of its services (eg overhead cranes, lifts, precision machinery, drains etc).

Assessment of visual damage is often very subjective, although it is usually desirable wherever possible to avoid cracking of walls and cladding materials, and noticeable deviations from the horizontal or vertical (ie slopes of more than 1/250). However, it is impossible to specify general limits on the movements required to cause cracking in a wall, because these values depend on such factors as:
● the nature of the building material
● the relative size of the wall, in particular its length to height ratio
● the mode of deformation (ie sagging or hogging of the ground).

However, in the absence of definitive limits, the maximum differential movement may be taken to be about 25 mm for buildings on sands and 40 mm for buildings on clays.

The ground movements that occur as a result of the loads applied by the external walls of a building can normally be reduced to an acceptable level if the bearing pressure exerted by the foundations is limited to one third of that required to cause the ground to fail. Maximum permissible bearing pressures for a range of soil types, including rocks, sands and gravels, and clays and silts are given in standards, though no figures are given for very soft clays, peat and organic soils or fill.

Soft clays, peat and organic soils are, generally speaking, too compressible to be suitable for bearing foundation loads, and foundations should be carried down through such layers to a more reliable stratum. Made ground or fill has to be treated with special care because of its potential for large irregular settlement, and the possibility of chemical contamination. There is also a danger that any fill with a high organic content may degrade, generating methane and causing settlement.

Frost heave
In the UK, frost heave can normally be avoided if the foundations are taken to a depth of at least 500 mm.

Moisture movement
Seasonal volume changes in firm clays are likely to influence the ground to a depth of at least 1 m, which is the minimum recommended foundation depth in this type of soil. Where there are adjacent trees, the zone of influence can be considerably deeper.

Effect of trees
Trees can also have a long term effect on clay soils. As trees grow, their demand for moisture increases and the root systems extend both laterally and downwards affecting an increasingly large volume of soil. In many clay soils, rainfall during the

winter is unable to replenish the moisture extracted by the roots during the summer and a zone of drier soil develops under the tree which can be up to 6 m deep. This desiccation causes the surrounding ground to shrink, resulting in possible subsidence damage to any nearby foundations. The depth of desiccation at a particular site can only be established unequivocally by a site investigation, and the foundation designer must decide whether the extra cost of such an investigation is merited. On the other hand, if an existing tree is removed, the moisture will return slowly to the ground causing it to swell. In some clay soils this rehydration can take 20 to 30 years.

The magnitude of the movements associated with a new tree growing or an existing tree being removed depends mainly on the type and size of the tree, the shrinkage potential of the soil, and the prevailing climate. The recommended minimum distances which trees should be allowed to grow near buildings if the effects of root systems are to be minimised, and unless special precautions have been taken with foundation design in shrinkable clays, can be summarised as follows. For poplar, oak, willow, whitebeam and cherry, the minimum distance should be not less than the mature height of the tree (Figure 1.19), which can be up to 24 m in the case of poplars whose roots run mainly horizontally. For species such as plane, lime, ash, beech and birch the minimum distance should be half the mature height of the tree. Apple and pear trees should be at least 5–6 m away.

Creep
There is a possibility that creep may be involved in displacements caused by eccentric loads on tall buildings, although it is more likely that displacements are mainly caused by cyclic temperature variations coupled with ratcheting.

Table 1.1

Coefficients of linear thermal expansion per °C × 10^{-6}

Clay brick	5–8
Calcium silicate brick	8–14
Concrete brick and block	6–12
Granite	8–10
Limestone	3–4
Marble	4–6
Sandstone	7–12
Dense concrete gravel aggregate	12–14
Dense concrete crushed rock aggregate (not limestone)	10–13
Dense concrete limestone aggregate	7–8
Aerated concrete	8
Steel	12
Aluminium	24
Glass	9–11
Polycarbonate	60–70
PVC-U	40–70
Thermoplastics	40–70
GRP	20–35
Hardwoods and softwoods	
with grain	4–6
across grain	30–70
Plasterboards	18–21
Cast gypsum	18–21
Dense sand/cement renders	10–13
Sanded plasters	12–15
Lightweight plasters	16–18
GRG	17–20
Clay tiling	4–6

Table 1.2

Modulus of elasticity E (kN/mm^2)

Steel	210
Aluminium	70
Glass	70
Polycarbonate	2.2–2.5
Thermoplastics	2.1–3.5
PVC-U	2.1–3.5
Hardwoods	7–21
Softwoods	5.5–12.5
Blockboard	
with core	7–11
across core	5–8
Clay brick	4–26
Calcium silicate brick	14–18
Concrete brick and block	10–25
Concrete dense aggregate	10–36
Cast gypsum	16–20
Dense sand/cement renders	20–35
Sanded plasters	8.5–16
Plasterboards	16

Table 1.3

Reversible moisture movement (%)

Clay brick		0.02
Calcium silicate brick		0.01–0.05
Concrete brick and block		0.02–0.06
Limestone		0.01
Sandstone		0.07
Dense concrete gravel aggregate		0.02–0.06
Dense concrete crushed rock aggregate (not limestone)		0.03–0.10
Dense concrete limestone aggregate		0.02–0.03
Aerated concrete		0.02–0.03
Dense sand/cement renders		0.02–0.06
Softwoods	tangential	0.6–2.6
	radial	0.45–2.0
Hardwoods	tangential	0.8–4.0
	radial	0.5–2.5
Plywoods	with grain	0.15–0.2
	across grain	0.2–0.3
Blockboard	with core	0.05–0.07
	across core	0.15–0.35

Table 1.4

Irreversible moisture movement (%)

Clay brick	0.02–0.07	expansion
Calcium silicate brick	0.01–0.04	shrinkage
Concrete brick and block		
dense aggregate	0.02–0.06	shrinkage
medium lightweight	0.03–0.06	shrinkage
ultra lightweight	0.20–0.40	shrinkage
Dense concrete gravel aggregate	0.03–0.08	shrinkage
Dense concrete crushed rock aggregate (not limestone)	0.03–0.08	shrinkage
Dense concrete limestone aggregate	0.03–0.04	shrinkage
Aerated concrete	0.07–0.09	shrinkage
Dense sand/cement renders	0.04–0.10	shrinkage

> **BRE Digest 228**[33] **describes how to use the data in these tables in the calculation of dimensional stability**

Chapter 1.3

Exclusion and disposal of rain and snow

This chapter includes also the prevention of rising damp.

Although keeping out the rain could be considered as one of the prime performance requirements of walls, windows and doors in general, a surprisingly high number of buildings are defective in this way; this has led to considerable research effort over the years.

As *Principles of modern building*[6] pointed out, there are basically three ways of preventing driving rain penetrating a wall to the interior of a building:

- a continuous impermeable skin on the exterior of a wall, or a water resistant membrane or barrier somewhere within the wall thickness
- sufficient thickness of absorbent material on the exterior to allow drying-off between rainstorms
- a continuous cavity somewhere within the wall thickness or combinations of any of the above.

Each method has both advantages and disadvantages. A continuous impermeable skin which will give satisfactory performance over time is very difficult to provide. All buildings move to some extent, and this can show on the face of the wall in the form of cracks or gaps opening. Solid masonry was the prime example of the second method above, but economy in the use of bricks ensured that one brick solid walls became the norm – barely thick enough in the wetter areas of the country to prevent rainwater from reaching the internal face before surface evaporation removed it. The third method was then introduced, but, though fine in principle, the structural needs of the building meant that the cavity had to be compromised with ties and lintels, and some extra protection added back, for example by DPCs and cavity trays.

All three methods have left a legacy of inadequate performance with which the industry has to struggle. Perhaps the method which in principle comes closest to the ideal is the surface composed of overlapping impermeable sheets, without seals, which maintain their integrity in the face of building movements – tile and slate for example. However, even these materials have drawbacks. When they are used on roofs they are normally out of reach of impacts, but on walls they become vulnerable.

Masonry is not the only type of external wall to suffer from rain penetration, however. Many other types also suffer; for example concrete panel and other forms of cladding, if not through the material itself, then through the joints.

From the point of view of the occupants, penetrating damp is regarded as a very serious matter and a sign that something is seriously wrong with the building. In spite of this, it is reported that 1 in 5 of the housing stock in England suffers from damp, which in two thirds of the cases is due to penetrating or rising damp, and the remainder to condensation.

Windows should be designed to control snow and water penetration appropriate to the building location, type and occupancy. Generally, it is thought to be an unrealistic requirement that windows should never leak. Most people will probably accept the need to mop up window leakage that remains on a sill and occurs once a year in a storm. Few, however, will accept leakage occurring more frequently and particularly when it runs off sills and down walls; and no leakage at all can be tolerated in rooms containing electrical or electronic equipment, or clean factory processes. It is the dividing line between these two categories of leakage which is used as the failure criterion in window grading weathertightness tests.

Rain

Within the terms of building regulations, control of moisture penetration is a functional requirement, and buildings need to be designed and built adequately to resist such penetration. For modern walls, rain penetration tends to show as well defined, roughly circular or oval patches on the internal finishes.

Figure 1.20
Uneven and disfiguring weathering caused by erratic run-off

Figure 1.21
A map showing categories of exposure to wind driven rain and corresponding spell indices for worst direction at each location

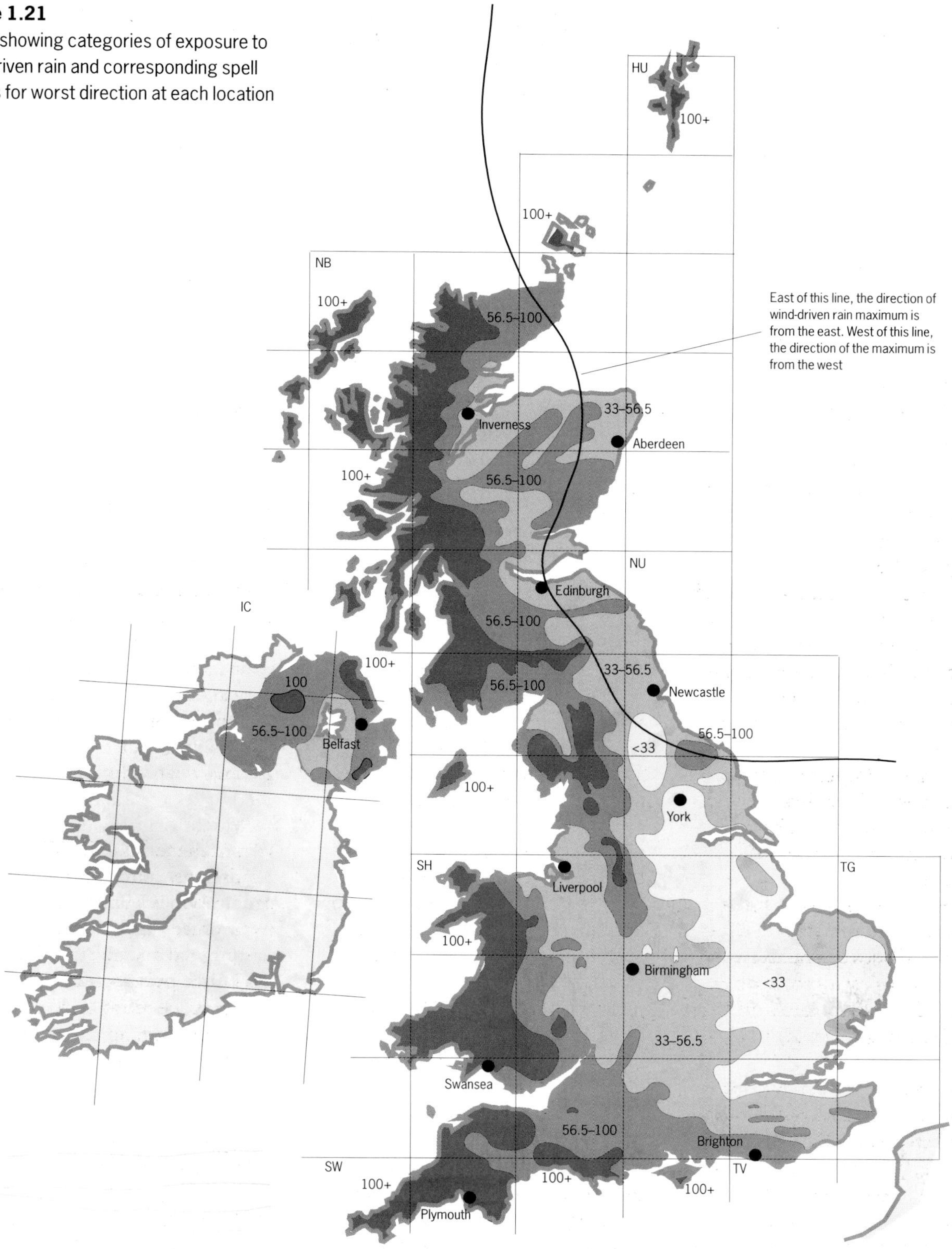

HU
100+

100+

NB
100+

56.5–100

East of this line, the direction of wind-driven rain maximum is from the east. West of this line, the direction of the maximum is from the west

100+

33–56.5
Inverness
Aberdeen

56.5–100

NU

IC

Edinburgh

100+

56.5–100

100

56.5–100

56.5–100
33–56.5
Newcastle

Belfast

56.5–100

<33

100+

York

SH

TG

Liverpool

100+

Birmingham

<33

33–56.5

Swansea

56.5–100

Brighton

SW

TV

100+

100+

100+

Plymouth

Because rain penetration is intermittent it does not normally give rise to mould growth unless it is particularly severe or persistent.

There are two kinds of rainfall intensity which need to be considered in relation to the building as a whole:
● rain falling approximately vertically
● rain driven by wind

Both categories contribute to the total quantities of rainwater needing disposal, but only one directly affects walls, windows and doors – driving rain.

Hourly measurements of rainfall amount, wind speed and wind direction held on a computer archive have been analysed in order to produce national maps of two types of directional driving rain index, namely an average annual index and a once-in-three-years-spell index. Among the more important applications of these indices, the former is relevant to the weathering and staining of building façades (Figure 1.20 on page 28) and the latter is useful for assessing the risk of rain penetration through masonry walls and building features[35].

Design guidance in the UK on the incidence of driving rain has been calculated since the early 1960s from rainfall and all-directional wind speed measurements. This guidance has now been revised to take into account the direction from which the wind is blowing, and the results can be shown in map form as in Figure 1.21 (on page 29). This map has been taken from *Roofs and roofing*[24].

Work by the Meteorological Office on indices of wind-driven rain, or 'driving rain', has overcome some of the deficiencies of earlier approaches. Maps for the annual wind-driven rain index for open sites (described as airfield conditions) have been prepared by combining the hourly

products of wind speed and rainfall during the 33 year period 1959–91. It then becomes possible to determine construction suitable for use in resisting wind-driven rain in all parts of the UK; for example when using thermal insulation within the cavities of external walls. Figure 1.21 shows four zones, the darkest shaded experiencing more than 100 l/m² per hour, and the others in progressively lighter shading showing 56.5–100, 33–56.5, and less than 33 l/m² per hour.

In using this diagram, the exposure zone for the building is first determined from the map, and then modified according to local features such as height above sea level, and proximity to the coast which increases the severity of exposure by the equivalent of one zone, or shelter belts of trees or adjacent buildings which reduces the severity of exposure by one zone. To give one

example, a site situated near to the Pembrokeshire coast is in the zone experiencing the most severe exposure of more than 100 l/m². If the building is in a sheltered location, and more than 8 km from the coast, it can be assumed that it will be subjected to driving rain in the 56.5–100 l/m² category, that is to say one zone less than that indicated by the map. Similar adjustments can be made for other sites according to the above criteria.

The final step is to check that the construction will be satisfactory. This will depend on a number of factors such as the basic form of the construction, the width of the cavity and whether it is clear or filled, whether the outer face is impervious, or is rendered, or if of masonry, whether the joints are tooled or recessed, and whether the sills and copings are flush or projecting. Guidance on the selection of suitable construction is available, for example for the protection of cavity fill it is to be found in *Thermal insulation: avoiding risks*[36].

By way of illustration, impervious cladding for the full height of all walls up to 12 m height would normally be satisfactory for the severest exposures, but such cladding would offer suitable protection above ground floor facing masonry only if cavities in the latter exceed 100 mm in width. For narrower cavities, the wall must be completely clad. At the other end of the spectrum, facing masonry – however the joints and sills are treated, and irrespective of whether the cavity is filled or not – it is only likely to offer satisfactory protection in the less than 33 l/m² zone.

Analysis of available data given in *Climate and building in Britain*[37] suggests that every ten years on average, most locations in the UK experience one hour in which the total driving rainfall is 30 mm. More tentatively, there is one hour in every 100 years when 80 mm falls. Intensities for shorter periods are illustrated in Figure 1.22, though these figures are probably overestimates.

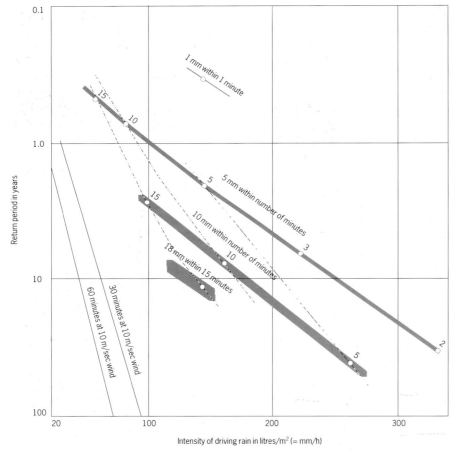

Figure 1.22
Estimated frequencies of driving rain intensities from intense showers (from Holland's rain intensity frequency graphs for Britain, with additional data for storms lasting 30 minutes and 60 minutes, and wind speed of 10 m/s)

Revisions to the wind component of driving rain include the following factors:

- **terrain roughness, R**, to allow for the effect on wind speed of the general terrain upwind of the building. Data are based on similar data for wind loading and vary from about 1.15 for open coastal land to 0.75 for built-up areas

- **topography, T**, similarly based on wind loading data, and varying from 1.2 for features likely to produce funnelling of the wind to 0.8 for sheltered valleys and similar locations. A calculation procedure is available in BS 8104[38] for nearby slopes which are steeper than 1 in 20

- **obstruction, O**, to allow for the localised effects of nearby obstructions giving close shelter to the building being assessed. (Figure 1.23). This relies on the concept of line of sight from the building. For example a building on the edge of an estate facing open land is assumed to receive no shelter.
  Walls may receive shelter from other nearby buildings:
  O = 0.2 for obstructions up to 8 m away
  O = 0.5 for obstructions up to 40 m away
  O = 0.8 for obstructions up to 100 m away
  Walls which are clear of surrounding obstructions have higher values for O

- **wall factor, W**, to allow for the characteristics of the wall in question, such as shape, height, overhangs etc

Factor W allows for behaviour of the wall as an obstacle to wind flow. Rain droplets tend to follow air paths as they flow round the wall and it is only their inertia which carries the droplets onto the surface.

Contrary to expectations, large buildings tend to receive less rainfall per unit area than smaller buildings. This is because, for large buildings, air movements occur on a large scale and water droplets are better able to follow a more gradual curvature of air paths round the walls.

Rainfall on the wall can be calculated as follows:
rainfall on wall = free space driving rainfall $\times$ R $\times$ T $\times$ O $\times$ W

Permeability of surfaces to rainwater

Many materials used for external cladding, such as glass, profiled metal sheets or plastics materials, are impermeable. Once rain has wetted the whole surface it begins to run off, frequently in streams, perhaps driven sideways by wind flow. It is these streams which seek out imperfections in the seals, leading to penetration, and the paths of these streams are frequently very difficult to predict with any degree of certainty.

Other materials used for walls may be absorbent, for example some types of masonry such as brick, stone or concrete. Bricks are commonly available which, when used for the outer leaf of a cavity wall, can give an overall outer leaf absorbence of over 20 l/m^2. This is equivalent to 60% of the rainfall likely to occur in the 'worst spell every three years' in areas of the country with sheltered exposure rating, and 25% of the worst spell in areas with severe exposure.

Run-off

Depending on the porosity of any wall, and the existing degree of its saturation, rainwater will be absorbed until the wall becomes saturated and run-off begins. But it does not follow that all the water runs down in contact with the façade. BRE observations show that even on smooth unbroken surfaces, water flow is not necessarily cumulative. Much bounces or splashes off, to be carried away in the air stream and, depending on the geometry of the surface, usually falls as a curtain of large drops some 300–600 mm away from the façade.

The fact that the shapes of buildings vary so much leads to an uneven distribution of rainfall intensity across the surface of their walls. Even simple shaped elevations show uneven distribution (Figure 1.24). As already indicated, larger walls tend to receive less rainfall per unit area than smaller walls in the same environment.

Calculation for run-off

In cases where the wall is non-absorbent or where it may have become saturated, a quantity of run-off at a lower level must be anticipated. This can be calculated by taking the average wall factor above that level, using this average value to determine the wall index, and then multiplying this index by the height of the wall above the level concerned. As an example, it

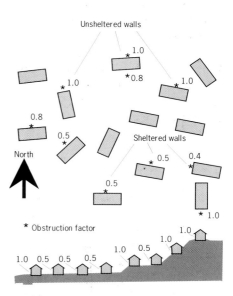

Figure 1.23
Obstruction factor expressing local shelter to a wall

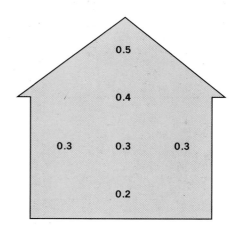

Figure 1.24
Distribution of rainfall over typical two storey gable wall. Shown as a proportion of rainfall in equivalent free space

Figure 1.25
Here, run-off from the impervious window has not cleared the wall, leading to the different appearance of the painted surface below

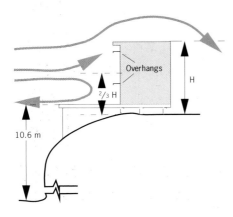

Figure 1.26
Overhangs: cross-section of a BRE test rig showing airflow patterns

can be shown that for an exposed two storey gable wall in, say, the Birmingham area, the run-off at the bottom of the wall during a once-in-three-year storm is approximately 165 l/m width. For a 10-storey block the run-off would be about 412 l/m width.

Surface effects

Surface texture and quite minor surface features can have a significant influence on the behaviour of rainwater run-off. Water on smooth surfaces such as finely textured concrete or smooth renders tends to find preferred run-off or streaming paths. This can lead to unsightly surface staining. Heavily textured finishes such as roughcast rendering can break up streams of water and give a more even wetting pattern. Textured or ribbed surfaces are also better able to prevent surface water blowing sideways, an effect which can produce heavy local loadings on cladding joints.

Some design features tend to concentrate water on certain areas of the wall, in effect increasing the exposure in those locations. An important example of this is the use of flush window sills. Rain water collected from the whole area of the window is shed onto the wall below. Hence a spandrel of absorbent brickwork in sheltered parts of the country can have an exposure equivalent to an area of the country with over ten times the driving rain index. It is quite common to see frost damage in bricks located under flush sills, whereas bricks in other parts of the wall are in perfect condition. A sill projection of at least 25 mm, and having a drip to allow run-off to fall clear of the wall, should be provided in such circumstances (Figure 1.25).

Protection given by overhangs, drip moulds etc

It is sometimes thought that in high winds, water dripping from projections is quickly blown back onto the wall a short distance further down. In fact, air close to the wall forms an almost still boundary layer and to the extent it moves at all, it

flows parallel to the wall surface. Therefore droplets shed from projections tend to fall vertically downwards to the ground. Again, a generous sill projection of at least 25 mm should be provided.

A recommended remedial measure for rain penetration in two storey housing is to provide a render to the first floor level only (including the gable), with a bell finish at the lower edge. This is often a cosmetically more acceptable, and cheaper, solution than complete cladding or render.

The protection given by projections at the top of a freestanding wall, or with flat roof overhangs at verges or eaves, is much less certain. This is because air movement is more turbulent at the top of a wall. In strong winds, rain droplets may even be moving upwards so that projections may become counter productive. There is very little quantitative information on this issue. However BRE have carried out tests on a rig 2.25 m high with a flat roof and positioned near the top of an escarpment 10 m high. (Figure 1.26). Overhangs of different depths were fitted at the top and half way down the exposed test face[39].

Some of the results are shown in Figure 1.27. It can be seen that all the lower overhangs provided significant protection to the wall beneath. But the pattern of protection afforded by the upper overhangs was very different. Below these overhangs, rainfall catches could be several times greater than for a wall with no overhang. The simplified explanation for this can be seen in the wind directions indicated in Figure 1.26, where the upper overhang was situated in the predominantly upwards air flow. An analogous, though less pronounced, effect could arise for verge overhangs on pitched roof gables.

Eaves overhangs to pitched roofs can provide significant protection to the wall below, provided the roof pitch is greater than about 25°. This is because, in terms of wind flow patterns, the roof acts as equivalent to an increase in height of the wall. Therefore the eaves projection is, in effect, positioned some distance from the top of the equivalent wall and is less likely to be in a region of wind uplift.

Rain-screens

Rain-screens, as the name implies, are sheaths designed to keep the majority of driving rain away from buildings. In their simplest form, they can frequently be found in agricultural buildings where timber slats spaced apart provide a degree of ventilation yet exclude the majority of driving rain.

In other building types, rain-screens can protect materials used in the inner layers of external walls which may be vulnerable to wetting (eg thermal insulation). Screens can take many forms: from fine mesh only slightly larger than the size of the average raindrop to continuous sheet with open joints. The joints between jointed sheets can be profiled to give protection against rain being driven sideways (Figure 1.28, plan), as well as to provide for the discharge of vertical flows to the sheets below (Figure 1.28, section).

The design of a satisfactory rain-screen enclosure is not easy since it is normally crucial to obtain very accurate assembly, particularly over joint widths (see feature panel on page 34). Once correctly designed and installed, however, there is less worry over the accommodation of movements and durability of jointing products than would be the case with single stage or face-sealed joints.

Splashing of rainfall

Rain, whether wind-driven or vertical, is not expected conventionally to splash up more than 150 mm from horizontal hard surfaces. This is the height normally used for siting DPCs above paving as well as for flashings above roofing level. However, detritus carried by splashes has been found on vertical surfaces 300 mm and more above horizontal surfaces or paving, indicating that the 150 mm guide height is insufficient to prevent this phenomenon in many circumstances (Figure 1.30 on page 34).

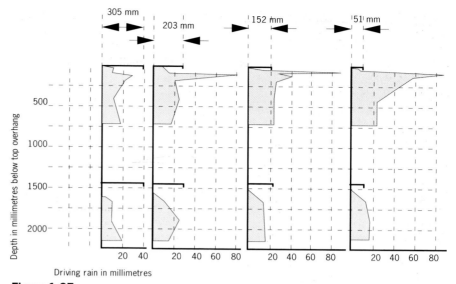

Figure 1.27

Overhangs – comparison of maximum daily catch of driving rain for wind speeds between 16 and 20 m/s

Disposal of rainwater and snow meltwater was covered in Chapter 1.3 of *Roofs and roofing*.

Snow and hail

Wind-driven wet snow will collect in small quantities, for example on ledges and hood moulds, but is not known to cause problems. On the other hand, fine powdered snow will penetrate gaps which may not leak water. Keeping out this snow will depend on the integrity of the air seal.

On rare occasions hailstones over 75 mm across can fall in severe local storms. These can weigh over 100 g and could smash windows when driven by high winds though they are more likely to damage rooflights. On average, one of these severe storms may occur at some place in southern Britain about once in 5 years, but they are very localised and the likelihood of one occurring at any given place is very small indeed.

Rising damp

Another form of water penetration is rising damp and there is a virtually absolute requirement to prevent this reaching the living space. Water which potentially can move from the ground through the base of the wall, or which can be conducted into the remaining parts of the structure from

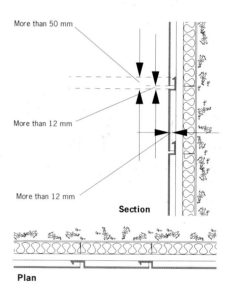

Figure 1.28

One design of edge profiling which gives reasonable protection against driving rain entering the joint. The minimum dimensions shown are to reduce the possibility of rainwater filling the channels

The rain-screen principle

A rain-screen essentially is a relatively thin open-jointed screen spaced away from an inner wall. In its simplest form, the rain-screen can be sheet materials spaced apart, which allow rainwater to drain down the back face of the sheets, with run-off dripping from one edge to another over the horizontal joint. The cavity is fully ventilated and not limited in size to enable rapid drying of any water crossing the cavity. This is known as the drained-and-back-ventilated method.

A more sophisticated version of the rain-screen is that where attempts are made to equalise air pressures both within and outside the cavity by carefully controlling both the sizes of the cavity and the sizes of the open joints. There must also be a complete air seal at the back of this cavity. The rain-screen skin catches most of the droplets, and for those few drops which get past the screen, because the cavity is open to the external air though limited in size, the pressure inside and outside is practically equal and there is therefore neither energy nor air stream available to drive the droplets across it to the inner face.

The width of joints must be accurately controlled, especially where catchment trays at the rear of both vertical and horizontal joints are dispensed with. The widths of these trays are directly related to the width of the joints, and BRE measurements give a basis for determining their dimensions. It may be possible to combine the tray with the vertical members of the support system. Unlapped horizontal joints (Figure 1.29a) are unlikely to fill with water, and air pressure inside and outside the cavity is therefore more or less the same. Lapped horizontal joints will need to be provided with sufficient upstand such that they are not likely to fill with water (Figure 1.29b). It is important that vertical joints do not fill with water, especially at the foot of tall buildings which have continuous vertical joints, since there is a risk that water will overflow inwards instead of outwards, as a result of blocking the ventilation slots.

It is also important to appreciate that water can and will run down the back of some designs of rain-screen panels and any

stiffening or damping applied to the back of the screen will get wet. This will also affect fixings on the back of the screen, and, depending on the design, the underlying insulation which could be protected by a breather membrane.

Since wind action on a rain-screen clad building will produce both positive and negative pressures, it is necessary to close cavities at the corners of the building so that pressures and suctions operating on different faces of the building do not interfere with each other. It is also important to restrict the size of such cavities near to external corners. No absolute limit can be given from experimental evidence, but it has been customary to specify a maximum dimension of around 1.5 m. In any case, from the point of view of minimising wind loads on the cladding it is better to close the cavity at the corner. It is a good idea too to limit the extent of cavities within plane facades, and BS 8200 suggests a maximum dimension of 5 m. (Within reason, the smaller the better)

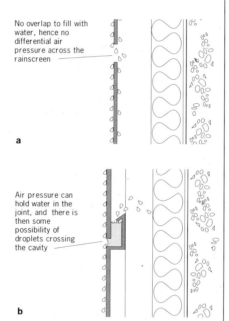

No overlap to fill with water, hence no differential air pressure across the rainscreen

a

Air pressure can hold water in the joint, and there is then some possibility of droplets crossing the cavity

b

Figure 1.29
The rain screen principle

Figure 1.30
Detritus carried to a height of 350 mm on the spandrel of a wooden window above concrete paving

The accurate measurement of rising damp, or strictly speaking the moisture content of walls, is fraught with difficulty. Often there is a need to know when a material is 'dry', although in absolute terms a porous material in a building will nearly always retain some moisture, either from natural hygroscopicity or from the effects of hygroscopic salts contained in the material. A brief description of the most commonly available methods of determining moisture content follows. Other methods are available, but are used mainly in laboratory and specialised situations.

Commercial electrical resistance meters

BRE site investigations have revealed many instances in which systems intended to combat rising damp have been installed in buildings where rising damp is not occurring. A frequent reason for this has been a wrong interpretation of high readings obtained when using an electrical moisture meter.

Electrical resistance meters measure the resistance between two metal pins which are pressed firmly into the surface of the material. The values obtained can be influenced by surface pressure applied to the pins, material which is not

wet external leaves, must be prevented from doing so by effective dampproofing. For new buildings there are provisions in building regulations and British Standards for continuous damp proof courses to be introduced between masonry in contact with the ground and the structure above in order to meet the functional requirement. For traditional solid walls, dampproof

courses have been compulsory in dwelling houses since the Public Health Act 1875.

Despite much widespread concern, true rising damp is not very common. The treatment of rising damp tends to be expensive, so correct diagnosis is very important. Alternative causes of the dampness should always be investigated (Figure 1.31).

homogeneous, and salts. The influence of the surface can be avoided by drilling holes and using shielded probes to measure at some depth in the wall, but even at depth it may be possible for salt concentrations to be sufficiently high to produce a high meter reading in a wall that has a low moisture content.

Commercial resistance moisture meters are useful on timber, and will indicate when a wall which consists of other materials is dry, but a reading on masonry will not distinguish between soluble salts present or actual moisture, and care needs to be taken with understanding the readings from the instrument.

Drilled samples by the gravimetric method

The basis of this method is to drill out damp masonry or mortar and measure both moisture content and hygroscopic moisture content. On-site moisture content can be established by using a carbide meter, a commercial piece of equipment where the damp drillings and carbide are mixed in a pressure vessel. A gauge measures the pressure generated and the calibration is directly related to moisture content. Alternatively, laboratory weighing will achieve the same result.

To extract the samples from the wall, a low speed drill with a masonry bit of about 9 mm diameter is used. Mortar is preferred as in some cases the bricks will have a lower moisture content than the mortar. The sample is collected in a stoppered bottle for subsequent laboratory tests. About two grams is sufficient but if the carbide meter is to be used then six grams will be needed. If successive samples are taken up the height of the wall then a graph can be plotted of moisture content against height.

The heat produced by a sharp drill bit does not cause evaporation of water from the sample, providing, of course, the material is not too hard.

The advantages of the method are:
● it is independent of salts
● it measures moisture content within the material rather than only in the surface layer

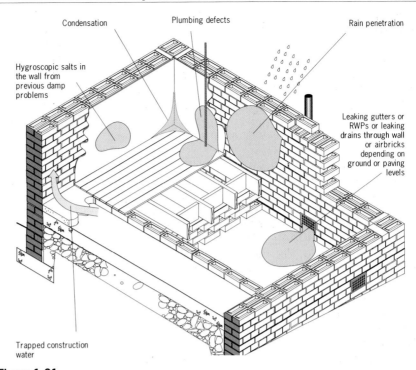

Figure 1.31
Alternative causes of dampness

Labels: Condensation · Plumbing defects · Rain penetration · Hygroscopic salts in the wall from previous damp problems · Leaking gutters or RWPs or leaking drains through wall or airbricks depending on ground or paving levels · Trapped construction water

● a moisture profile can be established by drilling in stages
● measurements are made from only one side of the wall
● it can be used on a wall where previous preparations, such as building in of probes is not possible
● the equipment is inexpensive and may be readily available (eg a chemical balance and dessicator); alternatively a carbide meter can be purchased

The main disadvantage is that the method is semi-destructive, at the minimum requiring a series of 9 mm holes, or, if the wall is plastered, then a chase cut out to locate the mortar joints.

If rising damp has not been found, the trouble may be due to condensation or rain penetration. If hygroscopic salts are the cause, then replacing the plaster in the affected area will be sufficient to overcome this problem if no other source of moisture is present.

If rising damp has been established it does not follow that any existing dampproofing course must have failed. It may have been by-passed. The commonest ways in which by-passing occurs are:

● the ground or paving being at a higher level than the damp proof course. The level should be lowered and any bridging material removed
● the use of an unsuitable mortar as a pointing to hide the edge of the dampproof course from sight. Such pointing should be removed. It is preferable not to point over the edge of a dampproof course, but if it must be done, a mastic should be used
● renderings that have been taken down below DPC level
● internal plaster, occurring for example when a suspended timber floor has been replaced by a concrete floor, and measures taken to link the underfloor dampproof membrane with the dampproof course in the wall have been inadequate

Even when the cause of the by-passing has been removed, the damp conditions will not disappear if hygroscopic salts have been brought up the wall by the rising moisture. In these circumstances it will be necessary to replace the affected plaster.

A DPC inserted too high to protect the structure
A non-traditional dampproofing system was inserted above the level of a timber floor. Assurance was given that the system was effective some way below its physical position. When the floor was found to have a moisture content high enough to induce dry rot it became clear that the position of the installation was too high. Yet it emerged that no clear case of a breach of contract between building owner and contractor was involved.

Hence it is important for specifiers to ascertain precisely what any non-traditional dampproofing system is intended to do in the particular circumstances in question, since systems can sometimes subsequently be found not to be adequate.

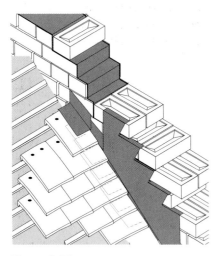

Figure 1.32
A DPC, cavity tray and flashing is needed to prevent rainwater percolating down a wall. Often not easy when the pitch of the roof differs from the rake of the bond

When a building suffering from rising damp has no dampproof course or it has been established beyond doubt that the existing dampproof course has failed, it is necessary to provide some kind of barrier to the rising damp. The most surely effective method is to insert a physical membrane; if possible this should be done despite the slightly higher cost and possible greater inconvenience compared with the use of non-traditional methods.

Some non-traditional DPC systems appear to be more effective or achieve full effectiveness more quickly if the installation is made during the summer when the walls tend to be much drier.

If, subsequent to installation, there are doubts about the effectiveness of a dampproofing system, monitoring of the system may be necessary. It is important to obtain, if possible, readings below the dampproofing systems as well as above. A number of sets of readings should be taken over a period of time. In interpreting the results obtained it should be remembered that, in the absence of a dampproof course, walls tend to become drier in the summer and wetter again in the winter as the water table in the ground moves up and down. A fall in moisture content readings at any one level above the dampproofing system that occurs between summer and winter is good evidence that the system is having an effect. Conversely, a fall in readings between winter and summer is to be expected and does not provide evidence of success. Indeed if the summer readings are the same or higher than those obtained in the previous winter, it can be taken that the dampproofing system is not working.

Materials for DPMs and DPCs
Since most forms of masonry will allow the passage of moisture upwards from wet ground, it will therefore be necessary to provide some form of protection against rising damp. The moisture barrier in a wall usually consists of a membrane laid at a suitable height above the

splash zone, and linked with the horizontal DPM. The ideal material for a DPC would be completely impervious to water in both liquid and vapour forms. In practice, not all materials used in the past have been effective.

Two courses of slates laid in cement mortars to break joint, and two or more courses of engineering or 'blue' bricks laid similarly, have been widely used in the past, especially in one brick thick solid walls, and in many cases have proved effective, especially where the vertical joints were left unfilled. However, the materials in the units themselves are more resistant to rising damp than the mortar used to joint them, and it has become customary, though arguably sometimes unnecessary, to make sure that the dampproofing is still effective by adding a replacement.

Protection against 'falling' damp is another matter; that is to say, protection against rainwater migrating down walls onto window and door heads, and other interruptions such as an abutment against a roof where an external wall becomes internal below the roof (Figure 1.32). In these circumstances masonry solutions are inappropriate, and a completely impervious sheet material must be provided. Lead was the traditional answer, especially effective where it was protected by bituminous paint (BRE Digest 380[40]), but many other materials have proved to be satisfactory, such as copper and bituminous felt, or sandwiches of thin sheets of these materials. Asphalt is occasionally found used as a DPC. Other materials are described in BRE Digest 380.

All these materials should be able to accommodate slight movements in the walling materials. They need to be jointed with care, usually by lapping at least 100 mm, and, in the case of masonry at least, the weight of the wall is usually sufficient to seal the lap without the use of additional jointing materials. The organic felts used in some DPCs may be found to have perished, but this does not necessarily mean that they become ineffective as DPCs provided they are not disturbed by subsequent movements.

Bituminous felt DPCs are vulnerable to being squeezed out slightly under pressure, especially in hot weather, but the amounts exuded are usually insufficient to compromise their performance and durability. Bituminous felt DPCs are also relatively ineffective against horizontal displacement, in effect providing a slip plane on which the walling can move with relative ease. Polythene is common in modern buildings but plain sheet has very low shear bond and the most recent types have a moulded pattern to improve shear and flexural bonds. Higher performance DPCs are often formulated from a blend of pitch and polymer and these have good resistance to squeezing, though doubts have recently been raised on possible risks to health from volatile organic compounds (VOCs).

See Chapter 1.10 of this book, and Chapter 1.5 of *Floors and flooring*[30] for additional comments on the use of materials for DPMs containing VOCs.

Electro-osmotic dampproofing

Of the complaints about electro-osmotic dampproofing that BRE has investigated, some have involved condensation problems that the installation would not be expected to cure, but in others there appeared to be at least a partial failure of the system, suggesting that electro-osmotic systems are not effective in preventing rising damp in walls in all conditions. By far the greater number of systems are of the passive kind, where there is no external source of electricity. Active electro-osmotic systems, which are more rare, employ an external source of electricity. BRE has no evidence which suggests that the two types of electro-osmotic system, active and passive, behave very differently in practice, though some of the active systems may be rather more susceptible to the effects of mechanical damage and electrochemical corrosion.

Soluble salts

Rising damp normally carries soluble salts into the wall and these tend to accumulate in the plaster. Often these salts are hygroscopic and, following the installation of any type of dampproofing system, it is usually necessary to renew the plaster in the previously wet area to remove them. Some salts, however, may remain in the body of the wall and may reappear through the new plaster causing isolated damp patches during periods of high relative humidity. Troubles of this type have often been reported when lightweight plasters have been used for the new work. Any recommendations for replastering made by the installers of the system should be rigorously followed, or their guarantee may be invalidated. There are obvious advantages in arranging for the installers to do the plastering.

Chapter 1.4

Thermal properties and condensation

Figure 1.33
Condensation has been, and still is to some extent, a problem in solid walled dwellings

In the days when fuel was relatively inexpensive, when the effects of the burning of fossil fuels on the environment were largely ignored, and when the majority of our existing stock of buildings was constructed, the conservation of heat was relatively unimportant. It need hardly be said that these conditions no longer obtain. But the legacy of poor insulation lingers on, often manifested in condensation on solid walls (Figure 1.33).

Limitation of heat loss now has significant benefits in the reduction of running costs for heating, and the additional costs of providing improved insulation levels in walls can often be offset by savings in capital costs of heating plant when it is renewed.

Heat is lost from buildings by radiation across cavities, by air leakage through materials and gaps at junctions, and via deliberate ventilation. Fabric transmission losses depend on the total internal air to external air transmission coefficients (U values) of the various parts of the building envelope. In 1966, requirements for the thermal insulation of dwellings were introduced into building regulations. The maximum thermal transmittance (U value) for external walls then was 1.7 W/m^2 °C, a requirement which remained unchanged until 1975, when the level was reduced to 1.0. Since then, levels have continued to fall; for example to 0.6 in 1982 and 0.45 in 1990.

Building regulations now set minimum standards for thermal insulation based on overall 'system' performance of buildings. These standards allow alternative methods of satisfying the requirements, for example in domestic buildings by element, by target U value or by energy rating, and in non-domestic buildings by element, by calculation, or by energy use.

'Trading-off' between the different elements can be made so that, within limits, the specified values can be exceeded at some points provided that this is compensated in other parts of the structure. In practice, therefore, it may happen that particular parts of the external envelope of existing buildings apparently fall below (or above) particular U values with which they are normally associated. However, there is much to be said for aiming to upgrade existing construction to at least these U values, given that it is practicable to do so.

The thermal insulation value of materials reduces with increased moisture content, so that materials which do not absorb water are needed where prolonged wetting is inevitable. In most building, insulants are kept at acceptable moisture contents by protecting them from rain and designing to avoid the build-up of condensation. Limitations on the permeability of layers of material in a wall may be needed to prevent the build-up of condensation within the wall thickness. The limitations range from materials to reduce water vapour flow (such as vapour control layers used towards the inside surfaces of buildings) to materials which are waterproof but vapour permeable (such as breather membranes used towards the

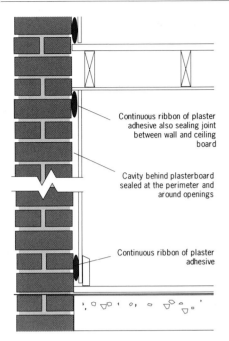

Figure 1.34
Sealing at the perimeter of dry lining

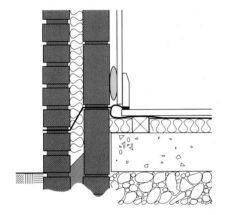

Figure 1.35
Thermal insulation placed within the cavity to cover the edge of the floor

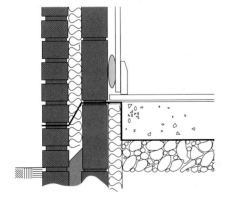

Figure 1.36
Thermal insulation placed inside the inner leaf of the external wall to cover the edge of the floor

outside). Water build-up can also be prevented by ventilating cavities adjacent to the insulation, especially where the wall contains layers with high vapour resistance on the outside of the insulation.

The insulation value of a wall can be degraded by thermal bridges where high thermal transmission materials penetrate layers of low thermal transmission material. Losses due to thermal bridges are often ignored in calculations, especially where thin sections such as mortar joints in lightweight blockwork, metal lintels, fixings and panel edges or timber sections are involved; but these and other materials become more important as thermal insulation standards increase. Pattern staining involving selective deposition of dust, once very common on lath and plaster ceilings with no thermal insulation, will also tend to show on walls which have interrupted insulation.

Air movement within walls, especially across layers of low-density insulation material, can reduce thermal efficiency considerably. Sealing at joints and around areas where services penetrate the insulation is important. Undesigned air movement within walls can also carry water vapour to areas where condensation can cause problems.

The vertical external envelope is a very significant item in the consideration of energy conservation of any building; it is far from being the easiest element to upgrade thermally. In consequence, there may be side effects. BRE has published guidance, *Thermal insulation: avoiding risks* [36] which, amongst other things, draws attention to the potential risks involved in upgrading the thermal insulation of walls and windows. Figure 1.34 reproduces one of the illustrations from this publication.

All the materials of a wall, including any window openings, and even the spaces between layers of the materials (cavities), contribute to its thermal performance. The contribution of the different elements is assessed by testing, but for many purposes values are rounded to standardised published values which

are used in calculations; for example for sizing heating plant and for building regulation requirements. For preformed materials, the usual requirement is for a specified density supported by test data on thermal transmission characteristics. For materials formed in situ, such as blown or foamed cavity fill, British Standards or British Board of Agrément Certificates, as appropriate, specify both the material properties and the workmanship levels to be attained.

U values
This book is not the place to discuss the specific requirements for thermal insulation in terms of thermal transmittance (U values). Indeed, the economic values which may be chosen will vary according to building type, fuel used, its relative cost and a host of other factors. What is important is where that insulation goes within the wall, that it is laid consistently, filling every space and avoiding thermal bridges, and that it is located on the correct side of any item of construction which functions as a vapour control layer.

In cavity walls, air movement into and within the void, and especially through layers of low-density insulation material, can reduce thermal efficiency of the wall considerably.

The position of any thermal insulation will determine the thermal properties of the wall. Where the insulation is placed on the inside of a wall, a quick warm up is achieved. But once the heating is switched off, the temperature will fall rapidly as the wall has little effective heat capacity. If the insulation is placed on the outside of the wall, because of its large heat capacity the wall will be slow to respond to the heating being switched on but relatively slow to cool when it is switched off.

Dwellings built since the 1970s may very well have thermal insulation to the floor set in the plane of the wall, either within the cavity, (Figure 1.35) or inside the inner leaf (Figure 1.36).

In the figure labels:
Continuous ribbon of plaster adhesive also sealing joint between wall and ceiling board

Cavity behind plasterboard sealed at the perimeter and around openings

Continuous ribbon of plaster adhesive

Improvements to thermal insulation

Thermal improvement to an external wall can be undertaken by either internal or external insulation. The cost of internal insulation is independent of the cost of any refurbishment of the external façade: external insulation on the other hand is economically attractive if replacement cladding or overcladding is being undertaken and consequently should always be considered as part of any refurbishment process. In most cases of overcladding, the thermal insulation will be fixed to the outside of the original façade. This is one of the easiest ways of reducing thermal bridges at cross-walls and floors.

External insulation implies that the structure itself will be warmer and that the risk of interstitial condensation within the walling structure is reduced. However, in replacing a cladding or selecting an overcladding system, consideration must be given to assessing the risk and effect of interstitial condensation. In this respect it is useful to divide external insulation systems into four categories:

● Permeable insulation with permeable finish
● Permeable insulation with impermeable finish
● Impermeable insulation with permeable finish
● Impermeable insulation with impermeable finish

With permeable insulation and permeable finish, the principle to be adopted to control interstitial condensation is that the construction should allow any water vapour to migrate to the outside of the building. It is important to ensure that where permeable finishes are installed, subsequent maintenance does not involve any operation which is likely to make the outer skin impermeable.

Where the insulation is permeable and the finish is impermeable there is usually a need to provide a ventilated cavity immediately behind the outer cladding. This is because of the probability of rain penetration through cracks or joints. BS 8200[28]

recommends a minimum cavity width of 10 mm. However, experience indicates that this cavity width is unlikely to be sufficient in systems where there is a risk of the cavity being blocked. Ventilation openings should be provided preferably positioned at not less than storey-height intervals. Care should be taken to ventilate behind large metal sheets. Attention should also be paid to ensuring that cavities between cavity barriers are ventilated. On no account should an impermeable finish with an unventilated cavity be specified.

Other problems may arise when impermeable insulants are used, irrespective of whether the cladding is permeable or not. In order to assess the risk of harmful condensation it is recommended that a calculation as outlined in BS 5250[41] is undertaken. Even with an impermeable insulant the cavity behind any impermeable finish should be ventilated. With some internal insulation systems the incorporation of a vapour control layer is sometimes recommended.

With external types of system there is no need to incorporate a vapour control layer. It is important, however, that no vapour barrier or similar sealing is applied to the outer face of any construction. Any air sealing should be confined to the existing joints.

Cold radiation

When window draughts are complained of by building occupants, it is natural for them to assume that leakage is occurring around opening lights. However, cold radiation through the glass of even well-sealed single glazed windows can cause cold down-draughts of air chilled by the glass. This effect provokes complaints about windows being draughty when they are in fact adequately sealed.

Diffusion of water vapour and vapour pressure

The diffusion of water vapour through external walls can help to reduce internal vapour pressures slightly – the construction is said to

'breathe'. This contribution is minimal, because the amounts diffusing through even the most permeable wall can be very small compared to the normal generation rates and losses by ventilation. However, if the balance of temperature and humidity through the wall is such that condensation occurs on a vulnerable material, problems can result which are all the more severe as they may accumulate out of sight until major damage results. The risk of interstitial condensation should be calculated[41].

Under normal occupancy, in winter conditions, both temperature and vapour pressure will fall from inside to outside. In multi-layered constructions the gradients through the wall will depend on the relative thermal and vapour resistances of each layer. Interstitial condensation can result when the main thermal resistance of the wall is on the warm side of the the main vapour resistance. The extreme situation can arise when an impermeable rain-screen cladding has been put on the outside of a wall, preventing the escape of any vapour.

This may not matter in many circumstances; for example condensation on the the outer leaf of a brick cavity wall will be negligible compared to normal wetting from rainfall. However, materials such as timber and timber based products, which can be susceptible to moisture, or metals which can corrode, are much more vulnerable and can give rise to problems especially if they are structural members. The studs in a timber framed wall or metal reinforcement of a masonry wall may need special protection.

Many porous building materials, timber and brick for example, are hygroscopic. This means that the pores start to fill with water at relatively low humidities and moisture contents may rise high enough to cause problems even though no condensation is taking place.

Condensation

Condensation will occur when the surface temperature of the wall (or, for that matter, window or door) is below the dewpoint temperature for a sustained period of time. The dewpoint will vary according to the air temperature and the relative humidity. Condensation occurs when the relative humidity of the air in direct contact with the cold surface rises to 100%.

The two most common situations in which condensation occurs on the vertical external envelope are:
● single glazed windows
● walls with high thermal capacity where the temperature is unable to follow rapid changes to the air temperature, and can often fall below the dewpoint.

There are five main ways in which the risk of condensation on or in the vertical external envelope can be reduced:
● installing thermal insulation in cavity wall construction to reduce heat losses
● installing a vapour control layer (see next section below) on the warm side of thermal insulation to restrict moisture which diffuses through the insulation from condensing on any colder outer surface
● ensuring cross-ventilation of any cavity or void to remove excessive moisture (Figure 1.37)
● installing extra sheets of glazing in the form of either double (or triple) glazing or double windows
● providing thermally insulated external doors

In BRE experience, run-off of heavy condensation on single glazing can be mistaken by building occupants for, and confused with, rainwater leakage of windows.

Vapour control layers

It will be necessary in many cases to provide a vapour control layer within the construction of the wall to control water vapour movement.

Of the various terms that are used, 'vapour control layer' is preferred to either 'vapour check' or 'vapour barrier', to emphasise that the function of the layer is to control the amount of water vapour entering the construction. As the achieved vapour resistance will depend at least as much on workmanship as on the design and integrity of the materials used, it is not realistic to specify a minimum vapour resistance to be achieved for the layer as a whole, though for the material to qualify as a vapour control layer it should have a vapour resistance greater than 200 MNs/gm.

Plastics films are the most usual materials for forming a vapour control layer in a wall construction. Joints in a flexible sheet vapour control layer should be kept to a minimum. Where they occur, they should either be overlapped by a minimum of 100 mm and taped, or sealed with an appropriate sealant, and should be made over a solid backing. Tears and splits should always be repaired with jointing as above.

Penetrations by services should be kept to a minimum and carefully sealed where they are inevitable. Draughts of moisture laden air through gaps in vapour control layers are more significant than normal still air diffusion through materials, even if there are splits in the vapour control layer, and it is therefore much more important to provide an air seal than to take elaborate precautions for making a total seal of the vapour control layer.

Thermal bridges, which may be so localised as to make little contribution to the total heat loss from the building, can lead though to surface temperatures low enough to promote the growth of troublesome moulds. Walls and their associated windows and doors should be designed with as continuous a layer of thermal insulation as possible, with special consideration being given to junctions between elements such as occur at window to wall joints or at junctions between external walls where more dense materials may overlap or penetrate thermal insulation.

Where a group of buildings is showing problems due to deficiencies in the thermal

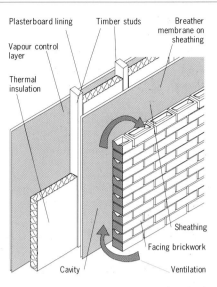

Figure 1.37
Typical timber frame construction, with the cavity, for example, ventilated by open perpends

insulation, it may be worthwhile carrying out a survey using infrared thermography to indicate parts of the structure needing attention (Figure 0.41).

Summer condensation

During cold weather, interstitial condensation occurring within internally insulated solid walls is normally prevented by a vapour control layer on the warm (internal) side of the insulation. However, in spring or early summer, strong sun on unprotected walls which have become wetted by rainfall can drive moisture towards the inside of the building and through permeable insulation or through gaps between less permeable insulation, to condense on the outside face of the vapour control layer[42].

Walls at risk face within 60° of due south. Moisture contents of the external wall need not be high for summer condensation to occur. Once it has occurred it stays for many days, as transfer of moisture out is much slower than transfer in. The condensation can be prevented by internal heating, but this is not an option for the summer. Since the condensation occurs behind a waterproof vapour control layer it is often not noticed. The risk of summer condensation occurring

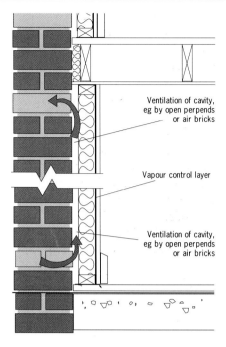

Figure 1.38
Summer condensation occurring in a south facing wall, insulated on the inside

should be assessed before internal insulation is undertaken. Where condensation does occur and there is a cavity, the solution is to ventilate the cavity to the outside – as in Figure 1.38.

Warmth to touch

Although this property of materials may well be of less importance in relation to walls than to floors, in the sense that one might avoid touching walls whereas one cannot avoid contact with floors, albeit indirectly, there may be circumstances when it needs to be given some consideration with respect to walls. A suggested figure for rate of heat transfer with respect to floors which has been discussed at international level, and which should not be exceeded where warmth to the touch is very important, is 45 kJ/m^2 per minute; this figure might well be suitable for use with walls too. The actual rates can be calculated from the thermal conductivity of the various layers of known thicknesses.

Mould

Most of the problems which result from surface condensation are caused by moulds. As well as being unsightly, mould growth is thought to cause respiratory problems in susceptible individuals. It is the reason for many complaints by building occupants, and it should not occur in any part of a building (see feature panel below).

While mould growth can be killed by toxic treatment it is important to realise that this only provides a temporary solution, and that the mould will return unless the source of moisture is removed. An essential first step is to identify the cause of the dampness, and then to remove it.

Mould growths should not be removed by dry brushing or rubbing, as heavy growths release large amounts of spores into the air which may induce allergic reactions. A vacuum cleaner should be used, afterwards dampening the infected area with a 1:4 solution of domestic bleach in water containing a small amount of washing-up liquid. The surfaces should then be wiped down with a damp cloth, rinsed out regularly, and then allowed to dry. Wooden window frames may need several applications. Windows should be kept open to promote dispersion of spores and moisture.

It is not easy to decide if mould is growing upon the surface only or has penetrated further. Where decorations (distemper, wallpaper, polystyrene tiles, flaking paint) can

---

Mould growths and how to prevent them

The minute spores of moulds are always present in the air and are freely deposited on all surfaces. There they can germinate and produce unsightly growths if conditions are suitable. They have two essential requirements: moisture and some form of food for further growth. Food for mould fungi is available in most building and decorating materials and in furnishings. It is present even in house dust.

Mould growth is therefore likely in any situation where there is persistent moisture, ie places where surfaces or air are continuously or frequently damp. Dampness can come from a number of sources: defects in the structure permitting rain penetration; plumbing leaks; rising damp; condensation of water vapour from the internal atmosphere. Of these condensation is by far the most common source.

Moulds can grow on some materials at relative humidities as low as 70%, so the requirement is to keep relative humidity below this level. This can be achieved by a combination of measures, involving some or all of the following:

● generating less water within the building (eg in kitchens by keeping pans closed when cooking, and by avoiding drying clothes indoors)
● ventilating water vapour at its source (eg using extract fans in kitchens and bathrooms, and trickle ventilators in bedrooms)
● insulating walls to increase the temperature of the internal space and of internal surfaces
● in buildings of massive construction, which are occupied only intermittently, fixing internal insulating linings so that the internal surfaces heat up rapidly when the rest of the walls are still cool
● improving heating to increase temperatures
● avoiding or correcting thermal bridging on walls where mould may grow

be stripped off it is often best to do so, but where the growth is slight it may be sufficient to clean down without stripping.

Sterilisation

Before redecoration the stripped or cleaned surface should be sterilised with a toxic wash and kept under observation. At least a week is necessary but longer is advisable. If mould reappears it should be washed down again with the toxic wash to ensure sterilisation is thorough.

Fabrics and soft furnishings should be cleaned by sponging affected areas with a solution of a toxic wash but not bleach. The solution should be applied sparingly and at first tested on a small, insignificant area for any adverse effects, and afterwards thoroughly dried.

There is a considerable number of products for use as toxic washes though not all are widely available. Suitable toxic washes and chemicals which are safe to use include quaternary ammonium compounds and sodium hypochloride. Yellowing of paints may occur with some treatments.

When using proprietary products, only those with labels which state that they have been cleared as safe for this use should be employed; there is some risk in the use of all these materials. The supplier's instructions and recommended precautions should be carefully observed.

If the toxic wash treatment appears to have been successful, redecoration can be undertaken. Fungicides incorporated into the decorative finish protect only the finish itself and do not obviate the need for the preliminary toxic wash treatment. Some manufacturers supply paints and wallpaper adhesives incorporating fungicides, and it is better to use these than to add fungicides to standard paint products (see also Chapter 10).

Chapter 1.5

Ventilation and air leakage

The external envelope of a building allows both intentional and unintentional penetration of air. The unintentional component is often very difficult to predict and can vary substantially between one building and another even though they may be of nominally similar design. In some circumstances this adventitious leakage through the building fabric is a cause of energy waste and even discomfort. There is currently a move towards producing more controlled ventilation of living spaces in buildings. This can be achieved by more airtight construction and by better draughtproofing, or by introducing background and mechanical ventilation, and goes some way towards satisfying the opposing requirements of reducing heat loss but providing sufficient ventilation to prevent condensation and the build-up of indoor pollutants.

Ventilation

Ventilation of new buildings is covered by the building regulations. The Approved Documents for the Building Regulations for England and Wales now include recommendations for background ventilation for habitable rooms and mechanical extract ventilation for bathrooms and kitchens. This book neither deals with mechanical ventilation, nor with natural ventilation methods such as passive stack, which are in essence properties of the whole building. However, some points are made in relation to ventilation simply because most of this is normally provided through windows.

Approved Document F lists opening areas for rapid ventilation according to whether an opening window is present in rooms such as bathrooms, and according to the floor areas of habitable rooms. Background ventilation, however, is listed according to specific open areas, which may be provided as controllable trickle ventilators in windows, or as air bricks in walls (see Chapter 4.1). The *CIBSE Guide*[43] gives design values for various types of building.

Air leakage

An air movement speed of less than 0.1 m/s is regarded as still air, while movements greater than 0.3 m/s are probably unacceptable except in summer. For windows, airtightness requirements are very dependent on building use and degree of environmental control indoors. It is generally agreed that the very highest levels of airtightness from BS 6375 are only necessary with windows where there is full control of temperature and humidity with mechanical ventilation of buildings.

At present there are no recommendations or standards for the airtightness of the outer envelopes of buildings in the UK, although there are some standards for individual components, notably windows. Some countries do have standards for the entire envelope.

The construction features identified as increasing adventitious ventilation rates include gaps between walls and window or door frames, poorly fitting windows or doors, lack of draught proofing, gaps round service pipes through walls,

suspended ground floors, and penetration of upper floor ceilings (eg hatches, pipes and cables). Less often mentioned are gaps in walls (eg at first floor joist level in housing), particularly where joists are built into the inner leaf of a cavity wall. The contribution of air permeation through wall materials themselves is generally small compared with the role of cracks or gaps.

It is sometimes helpful to be able to compare the envelope leakage characteristics of a building with others of different shapes and sizes in order to obtain an idea of what rates can be achieved in practice. Fan pressurisation techniques are now available in which the leakage rates of whole buildings can be measured at a specified pressure differential (Figure 1.39). It has been shown that this leakage rate, divided by the permeable external surface area of the building, gives a measure or 'index' of the construction quality

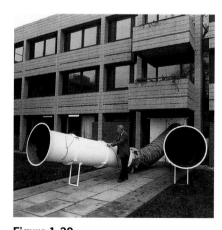

Figure 1.39
Apparatus for measuring air leakage rates for entire buildings

of the building fabric with respect to air leakage.

Figure 1.40 shows results of fan pressurisation measurements on five large commercial buildings in the UK and compares them with buildings tested in Canada, the USA and Sweden. It can be seen that the Low Energy Office (LEO) is twice as air tight as the other UK office built in a more conventional manner, and as tight as the two offices tested in North America.

Broadly speaking, a building with an envelope leakage index of 5 can be classified as having a good, low leakage standard, while index 10 is average. Buildings with index 20 are poor and are likely to give rise to complaints about uncomfortable draughts and difficulties in keeping warm.

Results from two old industrial 'hangar' type buildings showed them to be over twice as leaky as one built within the last decade under current UK building regulations. But comparison with tight Swedish industrial buildings shows that a further five fold reduction is possible.

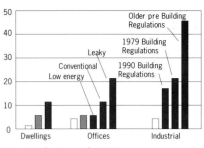

Leakage in m³ per h per m² at 25 Pa

□ Swedish Building Regulations

▨ North American examples

■ United Kingdom

Figure 1.40
Envelope leakage indices of buildings measured by BRE

Chapter 1.6   # Daylighting, reflectivity and glare

Figure 1.41
Hardwick Hall shows, for its time, a remarkably high proportion of glazing to masonry in its external elevations. It was said of it: 'Hardwick Hall – more glass than wall'

Glazing plays a significant, one might almost say predominant, role in forming the external appearance of our buildings (Figure 1.41). This chapter, however, deals with the daylighting requirements rather than with the aesthetic, and comment on other aspects of performance of glazing will be found in subsequent appropriate chapters.

Traditionally, the daylight performance of a building has been described by the average daylight factor (*DF*) (Figure 1.42). Daylight factors can be measured in a scale model using an artificial sky, or can be readily calculated using a variety of methods.

Glazing area
The CIBSE *Window Design Manual*[44] contains a simple method for designing windows to provide daylight and also gives recommended values of *DF* for a number of interiors: 5% for well daylit spaces, and 2% for spaces with combined daylight and electric light. For dwellings, values of 2% for kitchens, 1.5% for living rooms and 1% for bedrooms are recommended.

DF can be predicted using the formula:

$$DF = \frac{MW\emptyset T}{A(1-R^2)}\%$$

In this equation *W* is the total glazed area of windows, *A* is the total area of all the room surfaces (ceiling, floor, walls and windows) and *R*

their area-weighted average reflectance. *M* is a correction factor for dirt and glazing bars[45]. *T* is the glass transmission and Ø the angle of visible sky. Where a room contains both side windows and rooflights, the values of *DF* due to each should be calculated separately, then added together.

The formula can be readily used to determine window sizes at the initial stages of design, as BRE Information Paper IP 15/88[46] explains. From the target value of *DF*, glazing areas can be calculated directly using the formula. Rearranging it we obtain:

$$\text{Glazing area } W = \frac{A(1-R^2)\,DF}{M\emptyset T}$$

Typical values are $R = 0.4$, glass transmission $T = 0.7$, maintenance correction $M = 0.9$ and average $DF = 5$ for a well daylit room.

Angle Ø can be readily measured from drawings (Figure 1.43) as can room surface area *A*. This equation gives a glazing area which can then be used as a starting point in checking the original design and in heat loss and solar gain calculations.

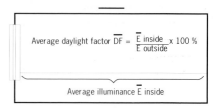

Figure 1.42
The definition of average daylight factor (under standard overcast conditions)

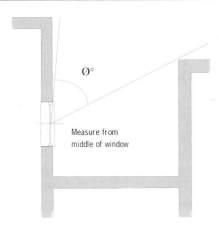

Figure 1.43
The angle of visible sky

Glare from the sky

Glare can occur from excessive contrast between the bright sky and a dark interior. For this reason, the internal surfaces of external window walls and all window reveals should be light coloured. Glare from the sky can also be reduced by:

● providing additional light on the window wall, from other windows or screened rooflights for example
● using adjustable blinds or curtains
● splaying window reveals
● using tinted glazing; since this brings with it a reduction in indoor daylight levels (average *DF* being proportional to glass transmission) this solution should be used only with care

The most important source of glare is direct sunlight, the effects of which should be controlled with shading devices.

Obstructions

Where buildings are sited close together, they may cause mutual obstructions to daylight. As a rule of thumb, to check existing cases or for the provision of additions, the obstruction angle taken at a point 2 m above floor level should be less than 25° to the horizontal (Figure 1.45). If obstructions are closer than this, they should not be continuous.

Where small extensions are being planned, for example in housing, an approximate check on the potential overshadowing of an extension to one side of an existing window can be made by taking angles of 45° to

the horizontal, and 45° to the vertical from the nearest top corner of the extension, and checking whether the resulting shadow obstructs the centre of the adjacent window (Figure 1.46).

Innovative daylighting systems

Innovative daylighting systems are intended to improve the uniformity of natural lighting in an interior and to control and distribute direct sunlight so that it can be used as an effective working illuminant. Sidelighting systems include light shelves, mirrored louvres and prismatic glazing.

The following considerations arise:

● sunlight-only systems are in general inappropriate for the UK
● enough diffused light must enter the window; as daylighting systems redistribute light rather than create it, minimum glazing areas are still required
● the system should cope with sunlight for all possible sun positions
● occupants should not be expected to make daily adjustments to the system
● capital and maintenance costs should be kept low

Appearance

The appearance of a surface finish is often seen to be of paramount importance for the acceptability of a component, both in its initial selection and in its acceptability for continuing use. This is probably true irrespective of situation and building type, with the possible exception of those buildings used for certain industrial and storage purposes. It is no part of the task of this book to set out the range of acceptability of the appearance for the various kinds of finishes, since in the final analysis this is a matter for individual taste and judgement. However, what can be done is to point out some of the factors which affect appearance and the quality and distribution of lighting in interiors, and the effects they might have on the choice of various materials. For comments on particular finishes, see later chapters.

Figure 1.44
A mid-twentieth century school building designed to the 2% minimum *DF* then recommended by the Ministry of Education, equating to an average *DF* of 5% or more. The absence of daylighting at the back of a room could lead to local overheating in summer conditions (Photograph by permission of The Architects Journal)

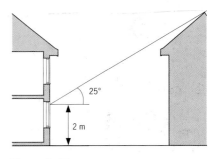

Figure 1.45
Mutual obstruction can be reduced by spacing out buildings

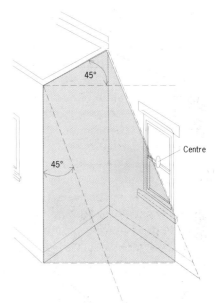

Figure 1.46
Checking for overshadowing

Figure 1.47
Low level sun reflected off vertical glazing

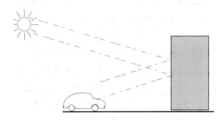

Figure 1.48
Bright sunlight can reflect from polished metal surfaces; vertical glazing can dazzle people in adjacent buildings or motorists

External colour

BS 4904[47] specifies thirty eight preferred colours for external cladding. These are designed to:
● bring the colours of building materials and finishes into a systematic relationship
● combine economy in the number of colours used with sufficient flexibility to meet design and technical requirements. The standard colours are applicable to aluminium, fibre-cement, glass, plastics and steel cladding; and the colour co-ordination framework may also be used for other materials such as concrete, brick, stone and clay tiles

Generally speaking the darker the colour chosen for a wall, the higher the temperature will be reached when exposed to sunlight; this affects durability. Other things being equal, the lighter the colour the longer the life. This should not be taken to extremes, however, since the glare from a light coloured surface, especially on a tall building, on a sunny day can be disabling and a nuisance; lighter colours may also suffer more from soiling. A suitable compromise can often be found in reflectances in the range of 40–65%.

Colour fastness may be identified as a problem. The degree to which this will occur will depend on the material used, and its susceptibility to, for example, ultra-violet radiation. This phenomenon may prove impossible to control in advance by performance specification, and needs to be assessed in conjunction with suppliers. Colour changes have been especially marked in the case of some glass-reinforced polyester (GRP) examples. Specialist advice should be sought.

Internal reflectivity

The reflectance of the wall surfaces will affect the amount and distribution of light in a space. A light coloured floor will increase the illuminances on walls and especially the ceiling, making the room look less gloomy and reducing glare from overhead light fittings. In a daylit space the light levels away from the windows, and hence the overall balance of light in the room, will improve if wall reflectances are increased. Reflectance values will therefore always need to be considered when choosing finishes.

The part played by reflectivity of the surfaces can be demonstrated by calculation of daylight factor. Average daylight factor DF is used as the main criterion of good daylighting – the ratio of indoor to outdoor daylight illuminance under the standard overcast sky. Good housekeeping also demands that windows and especially laylights in ceilings be kept clean. The transmittance of single glazing will normally be around 0.85 but can be as low as 0.6 if the glazing is dirty.

External reflectivity

Light coloured external surfaces can reflect some extra light to windows opposite. This may be important in a courtyard or atrium.

Glare or dazzle can occur when sunlight is reflected from vertical glazing or shiny cladding (Figure 1.47). For most walls and windows this problem usually occurs only when the sun is low in the sky; but some types of modern design incorporate sloping glazed or metallic wall surfaces which can, under certain circumstances, reflect unwanted high altitude sunlight into the eyes of motorists, pedestrians and people in nearby buildings[48] (Figure 1.48).

It is important that the possible effects of unwanted external solar reflection are considered at the design stage; failing that, remedial measures may be needed. From simple input data, the times of the year and of each day at which reflected sunlight might occur, and their duration, can be calculated. It is necessary to obtain the relationship between the angles of incidence and reflection to derive the sun positions at which solar dazzle may be a problem in each particular case.

For a wall design where solar dazzle has been identified as a potential problem, the geometry of the wall may have to be altered. Initial experience suggests that, in Europe and the USA at least, the greatest problems occur with surfaces facing within 90° of due south, sloping back at angles between 5° and 30° to the vertical. Where the surface slopes at more than 40° to the vertical, solar reflections are likely to be less of a problem, unless nearby buildings are very high. It is very unlikely that surfaces which slope forward, so that the top of the building forms an effective overhang, will cause problems in this respect.

Chapter 1.7

Solar heat, control of sunlight and frost

Figure 1.49
A south facing office with a large window area: making optimum use of solar energy, with control by external blinds

Sunlight is normally considered to be a desirable feature in buildings but it can lead to thermal or visual discomfort to users, and to deterioration of materials. However, glare from the sun can be unacceptable and if the sunshine or its reflected image is likely to lie within 45° of the direction of view, shading devices will probably be required. Good control of sunlight is particularly important in working interiors and other rooms where the movements of occupants are restricted, and they cannot choose alternative positions out of the sun (Figure 1.49).

In summer, a building will gain heat through solar radiation falling on its external surfaces and through its windows. Thermal insulation in the external wall will assist in reducing the transfer of this heat to the interior of the building, but there will be a consequent rise in the temperature of the outer surface. Durability of this external covering can be enhanced by providing a highly reflective top surface to reduce heat absorption.

Solar heat gains occur throughout the year. During the summer they may cause overheating and occupant discomfort; but in the winter heating season they can be beneficial, enabling a reduction in space heating load. In most buildings solar heat gains occur principally through windows.

Solar gains depend on:
● glazing area, to which they are roughly proportional
● location and orientation
● external physical obstructions
● type of glazing and shading devices (if any)
● time of day and year (position of the sun)
● weather

The *CIBSE Guide*[43] contains methods for calculating solar heat gains. Section A2 gives solar radiation availability throughout the year; Section A5 contains tables of solar gain factors for a range of glass and blind combinations. Section A8 deals with the prediction of summertime temperatures in buildings; these can also be found using the graphs in the *BRE Environmental Design Manual*[49]. This type of approach can indicate whether measures such as increased solar shading or reductions in glazing area, would be beneficial in reducing unacceptably high interior temperatures.

Solar gains also occur through opaque walls. Their effects can be reduced by using reflective cladding materials, substituting heavyweight for lightweight constructions and by adding extra layers of thermal insulation on or near the outside of the wall.

Shading devices

Shading may be required to reduce solar heat gain, to prevent glare from the sun and localised thermal discomfort due to the sun shining on occupants, and to reduce or eliminate sky glare. Where such problems are likely to arise, all glazing should be fitted with appropriate shading devices (Figure 1.50 on page 50).

Some shading devices may not combine all of these roles effectively. For example, low transmission glazing will reduce solar heat gain and sky glare, but has little effect on direct glare from the sun because of the high intensity of the sun's rays. Conversely, internal shading devices such as venetian blinds can eliminate glare from sky and sun but are less effective at reducing solar heat gain.

For some interiors it may be acceptable to restrict summer sunlight using fixed parts of the building such as balconies or overhanging roofs (for south facing façades) or by fixed louvres or screens. So far as east or west facing elevations are concerned, adjustable shading is often better suited to the low solar altitudes of the UK. Such shading systems should be easily maintained and easy to operate. Adjustable external systems should be robust, or retract when necessary if vulnerable to wind damage.

Shading, especially fixed shading, should not be designed so that it reduces interior daylighting, natural ventilation or a view out to an unacceptable degree.

The CIBSE *Window Design Manual*[44] describes the various types of shading available, and gives shading co-efficients and visible light transmission factors for a range of glass types. The amount of sunlight blocked by fixed shading devices such as external canopies and fins can be estimated using a sunpath diagram such as the BRE Sunlight Availability Protractor[50]; computer programs are also available.

Solar energy
The intensity of solar energy at the earth's surface is about 1 kW/m^2. *Roofs and roofing*[24] gives further information.

Solar radiation affects surface materials in the outer envelope in two distinct ways:
● ultra violet radiation disrupting specific chemical bonds between materials
● solar energy on surfaces causing their temperature to rise

Ultraviolet (UV) radiation, with its high wave energy, can cause chemical change in certain materials by disrupting specific chemical bonds. This is particularly important for organic materials such as plastics and coatings for sheet metals.

UV degradation may cause changes in the appearance or light transmission characteristics of organic materials, and ultimately lead to surface erosion and loss of strength. Components exposed to sunlight behind glass may undergo fading of their original colours. For any given material, the wavelengths responsible are often fairly broad. Measures are usually taken in the design of such materials to lessen the effect of UV radiation; for example by introducing chemical stabilisers.

Solar energy falling on surfaces causes their temperatures to rise significantly above air temperatures. This can occur on indoor surfaces if the rays have been allowed to pass through windows into rooms. Glass is transparent to short wave solar energy. Some of this absorbed heat is re-radiated from the surface, but, because surface temperatures are quite low, re-radiation takes place in the long waveband. There is a buildup of surface temperature as solar energy is absorbed in excess of long wave emission from that surface.

Since visible energy accounts for about half of the total solar energy, it is not surprising that light coloured materials have high reflectance and therefore absorb less solar energy. But light surfaces also have good emittances for long waves. This gives the best combination for reducing solar gain and hence maintaining a cool surface.

Figure 1.50
This example combines solar shading with a walkway for access to the curtain wall (Photograph by permission of Stoakes Systems Ltd)

White materials are cool. They do not absorb much heat from the sun, and they are good radiators

Black materials get much hotter than white materials, and a black surface may experience a very large temperature swing between night and day (more than 50°C for a lightweight black roof)

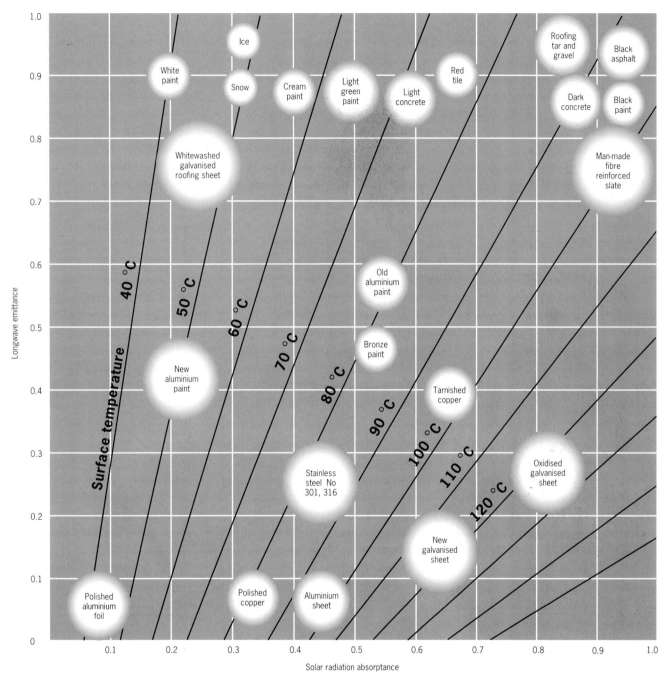

Polished metal surfaces get warmer than white surfaces because, though they absorb a similar amount of heat, they re-radiate less of it

Dull metal surfaces may get hotter than black surfaces. They absorb less heat from the sun but re-radiate little of it

Figure 1.51

Calculated extreme surface temperatures for air temperature of 30 °C

Prediction of temperatures

Bright metal sheet materials (that is to say, not coated) also tend to have high reflectance but in their case long wave emittance is low. This leads to the perhaps surprising result that they can experience significant solar heating.

Figure 1.51 illustrates the temperatures for different surface finishes on a hot summer's day in the UK (eg an air temperature of 30 °C). It shows the large difference which could be expected between, say, two highly insulated walling panels; one with a white and one with a black finish. Because maximum values rather than long term averages are used, Figure 1.51 is likely to be equally valid for roofs and walls. It is also likely to apply to most locations in temperate or equatorial zones.

There is a certain amount of experimental data on measured surface temperatures. For example, in measurements on wall sandwich

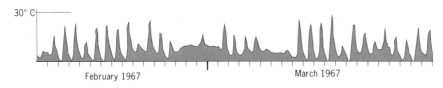

February 1967   March 1967

Figure 1.52
Temperature trace for a SW facing wall in Dorset (painted brick with cavity insulation). From BRE measurements

panels taken by BRE over a two year period, maximum surface temperatures of about 78 °C were recorded for vertical surfaces.

The obverse of solar overheating is cooling by night sky radiation. This occurs when there is a clear night sky so that significant long wave radiation can continue to take place from surfaces, but with no incoming solar contribution. In these circumstances, surface temperatures as much as 8 °C below the outdoor air temperatures can commonly occur with thin sheet cladding materials. Condensation will result.

Measurements of air temperature are made very widely in the UK. There is obviously an enormous number of ways of analysing and representing these data, and meteorological records are available over many years to anyone who wishes to undertake a particular statistical treatment appropriate to their needs.

Air temperature at any one place is continually varying. Variations over short time intervals, often for seconds or even minutes, may be ignored for most purposes, but variations from one hour to another can be significant in their effect on buildings.

Long term average temperatures are important for some purposes – for instance, in calculating seasonal heat losses through walls. For other purposes shorter term averages or average peak volumes may be important – for instance, for sizing of heating, ventilating and air conditioning plant. Finally, threshold or extreme values may sometimes be important – for example where material or mechanical failure may occur. Sufficient meteorological data are

now available to allow analysis for any of these purposes and for almost anywhere in the UK.

Seasonal accumulated temperature difference, ATD or commonly called degree-day totals, varies by a factor of about 1.6 over the United Kingdom between south west and extreme north when values are reduced to sea level. Standard ATD totals give a general indication of the geographic variation in energy needed for space heating but make only a small fixed allowance for solar gains. Thus they do not reflect the full potential of the sun to warm buildings and the spaces around them, or of wind induced losses. The use of ATD totals to building-specific base temperatures, as used in the BRE Domestic Energy Model [51], can take account of some of these factors.

Maps showing minimum and maximum daily temperatures and ATD totals were included in *Roofs and roofing* as Figures 1.22 to 1.24.

The temperature of the exposed surface of a building is rarely the same as the air temperature. One reason for this is that surface temperature is continuously modified by heat exchange between the outdoor air and the occupied internal space of the building. In winter, for example, heat loss from inside the building leads to a slightly elevated temperature for the outer surface of the wall, depending on insulation values.

Figure 1.52 shows a temperature trace for a white painted house wall in Dorset, in sunny but otherwise unexceptional winter conditions. The high peak temperatures reached on many days and the wide diurnal temperature range should be noted.

As well as withstanding the extreme temperatures described above, walls may also be subjected to rapid swings in temperatures caused, for example, by sudden showers of driving rain on an already hot surface, though this is less of a problem than with roofs. The effects of this phenomenon can be tested for in the laboratory.

Solar panels
It is of course possible to make positive use of solar energy, instead of treating it negatively and largely as being detrimental to the building and its occupants. However, both of the two main kinds which have been used in the UK, hot water circuits and photovoltaic installations, come within the definition of building services, and will be dealt with in another publication, but photovoltaics can sometimes be appropriate as an alternative form of cladding.

Frost on saturated construction
Deterioration of building components frequently happens when several meteorological parameters act together. One example of this is frost damage to porous materials. This occurs when freezing conditions coincide with high moisture content, near saturation. The air temperature at which frost damages walling materials usually needs to be substantially below freezing since the wall often receives heat energy from inside the building of which it forms part. It will also have a certain thermal capacity. Furthermore, although water in the largest pores of the materials may freeze at a little below 0 °C, the freezing point in the smallest pores may differ by several degrees.

A number of factors affect the susceptibility of a given wall to frost damage including the precise porous characteristics of the material, the speed and duration of freezing, the number of faces frozen simultaneously, the moisture content and its distribution, and the internal strength of the porous brick or other unit and its parts.

Frost testing

It is scarcely surprising that it has proved difficult to devise a reliable and a representative test for assessment of frost resistance. Any test method must select particular conditions which may then not apply to a specific application of the material. For example freeze and thaw tests on individual bricks are easy to carry out and give very quick results. However, they represent unique conditions which rarely occur in practice in exactly those same circumstances.

It is worth noting that there is no current British Standard test for durability to frost damage, but RILEM recommendations and draft European (prEN) tests for frost resistance of units and mortars are available, both based on UK research. Even so, there are a number of tests which are well established and quite widely used for similar purposes; for example salt crystallisation tests are used in the UK for the assessment of limestones. Reference samples are included in test batches. These tests allow prediction of suitability for different locations on a building; that is, they determine different susceptibilities to frost attack. Crystallisation tests may also be used for other types of stone, although interpretation is less certain.

Chapter 1.8   # Fire

Passive fire protection measures are those features of the fabric, such as compartment walls, that are incorporated into building design to ensure an acceptable level of safety. These features are dealt with in outline in this book. Measures which are brought into action on the occurrence of a fire, such as fire detectors, sprinklers, automatic shutters and smoke extraction systems are referred to as active fire protection, and are not dealt with here.

This chapter makes specific reference to guidance for England and Wales contained in Approved Document B[52] as it affects walls, windows and doors, but not to the mandatory requirements for Scotland which are contained in Part D for structural fire precautions and Part E for means of escape, made under the *Building Standards (Scotland) Regulations 1990*[53], nor to those for Northern Ireland in Technical Booklet E made under the *Building Regulations (Northern Ireland) 1994* [54]. These requirements are different, though no less stringent, and reference should be made to the documents themselves. Until 1986, inner London had its own requirements which could be traced back to the Great Fire of London in 1666. One example is that certain buildings covered by Section 20 of the London Building Acts, every partition had to be fire resisting.

Although building regulations are not retrospective, it may sometimes be necessary in rehabilitation work to review former provisions for the fabric to determine its current relevance.

The development of structural fire protection

In medieval times, ecclesiastical and military buildings had walls of stone, which offered good resistance to fire. The external walls of dwellings were frequently of wattle and daub, which offered reasonable resistance to fire. However, the practice of jettying the external walls meant that the upper storeys of houses facing each other across the street were often so close together that fire could cross the gap with relative ease, and legislation was introduced to reduce the overhangs and so to reduce the risk of fire spreading.

Concern was not limited to external walls; for example separating walls in the late twelfth century in London were specified to be 3 feet thick. The most stringent requirements were, of course, introduced after the Great Fire[55].

Even so, fires still occur with distressing frequency, even in the most highly monitored of buildings (Figure 1.53).

The objectives of fire precautions

The designer of a building or of alterations to a building, whether new or refurbished, will need to know a great deal about the conditions that will be imposed on walls within that building in service. Factors such as loading, environment and durability all have to be understood and assimilated into the design process, and considered in relation to behaviour in fire.

The objectives of fire precautions include:
● to provide adequate facilities for the escape of occupants

● to minimise the spread of fire, both within the building and to nearby buildings
● to reduce the number of outbreaks of fire

Relevant practical aims arising from the second of these objectives are to limit the size of the fire by:
● controlling fire growth and spread
● dividing large buildings, where practicable, into smaller spaces

Fire resistance

The performance of a wall as an element of structure can be judged on its ability to satisfy specified criteria for a given test duration. Examples relevant to particular applications are contained, for example, in Approved Document B of the Building Regulations for England and Wales. It is not proposed to repeat here all the relevant criteria from the Approved Document – suffice it to say that different criteria apply according to, for example, a building type, its

Figure 1.53
St George's Hall, Windsor Castle, after the fire of 1992

Figure 1.54
Fire test on a mockup of the Dublin disco fire

height, and the relative position of the element concerned, both within the building and to the boundary of the site. To quote one example, there are specific requirements for fire resistance of wall areas adjacent to external stairs.

For the purpose of the Approved Document, buildings are divided into purpose groups that reflect their use. Within a purpose group further subdivisions are made depending on the size of the building. In each case, provisions are made with respect to the nature of the exposed surfaces and the fire resistance requirements of elements of structure, and to the spread of flame on the finished surfaces.

It must be remembered that the standard time temperature curve for a fire test does not represent an attempt to simulate the growth of an average fire. Nor can the period achieved in a fire test be related directly to the expected survival time in a real fire. The results obtained from fire tests are relative assessments which enable a quality judgement to be made on a component of a structure. Tests may also be carried out on mockups after fires have occurred, for forensic purposes (Figure 1.54).

The intent of B2 is stated as: *'In order to inhibit the spread of fire within the building, surfaces of materials used on walls and ceilings:*
● *shall offer adequate resistance to the spread of flame*
● *shall have if ignited, a rate of heat release which is reasonable in the circumstances'.*

The intent of B3 is to ensure the stability of the structure in fire. Furthermore, there is a requirement that the building or buildings *'shall be subdivided into compartments where this is necessary to inhibit the spread of fire within a building'.* Further requirements under B3 are concerned with the provisions for the subdivision of concealed spaces and for walls between two or more buildings.

B4 requires that *'the external walls of the building shall offer adequate resistance to the spread of fire over the walls and from one building to another, having regard to the height, use and position of the building'.*

These simple functional requirements are extremely broad. The structural requirement under B3, for example, states that *'the building shall be so constructed that, in the event of fire, its stability shall be maintained for a reasonable period'.*

Guidance on how to achieve the requirements of the Building Regulations is given, for example, in Approved Document B. There is no obligation to adopt any particular solution: if preferred a requirement can be met in some other way. Even to follow exactly the Approved Document will only *'be evidence tending to show compliance with the Regulations'.* Nevertheless, it is generally accepted that to follow the guidance in the Approved Document is to comply with the requirements of the regulations.

All external walls must be fire-resisting to some degree. So far as loadbearing external walls are concerned, as *Principles of modern building*[6] pointed out, they should have a period of fire resistance sufficient to ensure that they will continue to act as loadbearing elements during a fire, but if pierced by numerous window openings through which fire could spread out of or into the building, it would be unreasonable to require the non-window parts of the wall to be of a high standard.

So the concept of 'unprotected areas' is applied to external walls in relation to relevant boundaries of properties, whether notional or not.

Such areas include doors, windows, other kinds of openings, parts of external walls having less than the designated period of fire resistance, and some areas of combustible material. Scope for forming new openings in walls may therefore be limited by this consideration, and, additionally, limitation of certain existing openings may also be appropriate, though small residential buildings are allowed different limits from other purpose groups.

A change in Approved Document B was introduced in 1992, with the deletion of many of the previous provisions for non-combustibility.

One curiosity with regard to legislation may be noted. Although all external walls need to be fire resisting because they are elements of construction, if, because of the rules defining unprotected areas, the wall is allowed to be 100% unprotected, there is no area left which needs to be fire-resisting. This is why completely glazed buildings are allowed.

Surface spread of flame
In addition to needing to meet prescribed standards of fire resistance, surfaces of walls may also need to possess resistance to the spread of flame. The actual criteria to be met depend on the location of the wall if internal, and the size of the room which the wall encloses. Although the requirements vary according to purpose group of the building, typical examples, for most rooms, are Class 1, with Class 0 for circulation areas and protected shafts. Parts of walls may be of Class 2 or Class 3, depending on area of room and total size of wall[52].

If the wall is external there may be other requirements. For all building types there is a general requirement for external walls, whether loadbearing or non-loadbearing, to have suitable periods of fire resistance if they are close to a boundary. Depending on the height of a building, if the walls are within 1 m of the relevant site boundary, they may need to possess

Class 0 surface spread of flame. If they are 1 m or more from the boundary, the requirements differ (Figure 1.55). For other examples see the Approved Document.

Non-combustible material
Under the Approved Document, this is defined as any material which:
● when tested to BS 476-2 does not flame and there is no rise in temperature on either the centre (specimen) or furnace thermocouples
● is classified as non-combustible under BS 476-4
● is inorganic and contains not more than 1% by weight or volume of any organic matter

Material of limited combustibility
Under the Approved Document, this is defined as any material which:
● is non-combustible, or is at least 300 kg/m^3 and does not flame when tested to BS 476-2, nor register a rise of more than 20 °C on the furnace thermocouple
● has a non-combustible core of not less than 8 mm thickness, having combustible facings of not more than 0.5 mm thickness on one or both sides

Insurers' rules
Insurers' rules for the fire protection of industrial and commercial buildings are contained in a book[56] † which is designed to accompany Approved Document B (Fire Safety) of the England and Wales Building Regulations.

The book lists insurers' requirements, provisions of which are sometimes more stringent than those of the Building Regulations, and contains appendices on walls, columns, approved wall claddings, approved lining materials, approved fire stopping materials, and fire break doors and shutters.

Refurbishment
Where refurbishment is being undertaken on buildings controlled under the Fire Precautions Act 1971[57], the fire certificate issued by the fire authority will list conditions which will need to be maintained in the building, provided the use has not changed. If changes are to be made to the building, however, the old certificate may no longer apply, and application should be made to the Fire Officer for the issue of an amended certificate. It therefore behoves all those concerned with refurbishment of such buildings to make themselves aware of the contents of these certificates, and indeed with all other legislation affecting occupied premises.

Fire damage
Inspection of fire damaged structures should be carried out by competent and experienced persons. Tests carried out to BS 476 are not intended to give guidance on the serviceability of such structures after exposure to a fire. Where the opportunity arises, tests may also be carried out on actual buildings. Absence of spalling or absence of change of colour after a fire should not be taken as evidence that any reinforced concrete members are satisfactory, since different aggregates vary in their behaviour in fire.

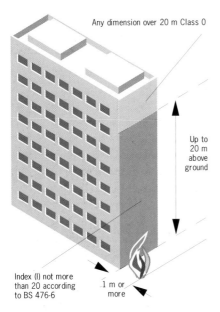

Figure 1.55
Provision for external surfaces of walls of any building more than 20 m high and 1 m or more from the boundary

Any dimension over 20 m Class 0

Up to 20 m above ground

Index (I) not more than 20 according to BS 476-6

1 m or more

Figure 1.56
Butted cavity barriers fixed to the sheathing of a timber framed dwelling before construction of the outer leaf. Gaps are evident

† At the time of writing, this publication was being reprinted in looseleaf form.

Cavity barriers and fire stops

Cavity barriers and fire stops can be the Achilles' heel of fire precautions. Where their importance is not fully appreciated by the installers, work may be only perfunctorily carried out, with potentially disastrous results. Typically, for example, cavity barriers may be loosely butted with consequential gaps being left within the barrier (Figure 1.56). Care is needed, both in briefing the installers and in supervision.

Until 1987, when BS 746-8 was withdrawn, cavity barriers needed to possess stability, integrity and insulation for at least 30 minutes. At that time too, insulation was not specifically required for barriers of lesser dimensions than 1×1 m. What was acceptable at that time remains acceptable. However, since 1987, under BS 746-22 to 24, tests are specified in relation to loadbearing capacity, integrity and insulation. Under the Approved Document, cavity barriers are patently not loadbearing, so are only required to have 30 minutes integrity and 15 minutes insulation.

All joints between elements of structure serving as barriers to fire, together with all openings for services, should be fire-stopped. Where that fire-stopping is required to span an opening greater than 100 mm, it will need to be reinforced, unless the material on its own is shown by test to be satisfactory. Materials suitable for fire-stopping include:
- cement mortar
- gypsum plaster
- vermiculite or perlite mixes
- intumescent mastics
- other proprietary materials which have been shown by test to maintain the fire resistance of the wall

Cavity barriers within voids have an important influence on the growth and development of a fire within an enclosed space, though that is not their prime purpose under the Approved Document. The majority of cavity barriers have probably been designed to BS 476-20 criteria but a new test method is proposed in a forthcoming International Standard.

A common form of seal is intumescent in character, contained within a box, so that expansion under the action of heat effectively seals the void rather than spills out into the surrounding space. Flexible fire-resisting seals are available which can provide continuity in the fire performance of movement joints in compartment walls.

Smoke vents

In certain types of building, there may be a need to provide smoke vents through the external wall. Requirements will be set by building regulations, or there may be requirements under local building legislation. Such vents should not prejudice any other aspect of performance.

Lightning protection

Although the roof carries the main provision for terminals for any lightning protection system, it is the wall and its supporting frame, if any, which has to provide the route for carrying the discharge safely to earth. It will also be necessary to connect all metal parts of walls to any lightning protection system.

BS 6651 recommends that the risk of a building or part of a building being struck by lightning should be calculated. Weighting factors, including the use of the structure, the type of construction, the contents of the building, the degree of isolation, and the type of country in which the building is situated, are used to obtain the overall risk. A decision then needs to be taken on whether the risk is considered to be acceptable without protection, or whether some measure of protection is thought to be necessary. The standard recommends a risk of one in one hundred thousand per year as a suitable threshold for the provision of protection.

There is further description of lightning protection, together with a map giving the incidence of lightning strikes to ground, in *Roofs and roofing*[24].

Where metals are to be used on a significant scale ... advantage may b ... existence of such ... maximise the prov ... protection. Thus, p ... walling systems, str ... concrete reinforcing ... , or cleaning rails may be used as part of the earthing system. Any metal in or on the structure which is close to the lightning protection system (eg cladding or a steel downpipe) may need to be bonded to it to prevent side-flashing. The most important areas of wall and conductor to connect are at the top and bottom of the building. Details should be provided of earthing to, for example, render stop beads wired to metal lathing, and metal flashings fixed with metal fastenings.

There should be provision for connecting the down lead through the wall to the main earthing terminal (see page 50 of BS 6651). Conductors fixed over walling should be of aluminium or copper, the former being preferred as the latter tends to stain surfaces and accelerate corrosion of other metal components. It is possible to provide protection to conductors, for example by coating with PVC.

Care must be taken when overcladding is being specified for an existing building with lightning protection. Existing conductors can be covered with overcladding systems which are composed of incombustible elements. However, it should be noted that continuity of conductors is not easily checked visually after overcladding has been applied.

The checking and repair of lightning protection systems should be delegated to specialist consultants.

Chapter 1.9

Noise and sound insulation

Figure 1.57
'Eight ways to annoy your neighbour'

Sound insulation

Noise descriptors

Noises differ in their level (loudness), frequency content (pitch), and they may vary with time. Consequently different units have been developed to describe different types of noise.

Noise level

Noise level is described on a logarithmic scale in terms of decibels (dB). If the power of a noise source is doubled (eg two compressors instead of one) the level will increase by 3 dB. Subjectively, this increase is noticeable but not large.

Frequency content

The ear can respond to sounds over a wide frequency range (roughly 20 Hz to 20 kHz), but most environmental noises lie between 20 Hz and 5 kHz. The ear is more sensitive to sounds of some frequencies than others, and is particularly sensitive between about 500 Hz and 5 kHz. The 'A weighting' is an electronic circuit built into a sound level meter to make its sensitivity approximate to that of the ear. Measurements made using this weighting are expressed as dB(A). Noises with tonal components (such as fan noise) are particularly annoying, and sometimes this is recognised by adding 5 dB(A) onto a measured level. An increase of 10 dB(A) doubles the perceived loudness of a sound.

Buildings may be required to protect their occupants from noise and vibration from a wide variety of fixed and mobile sources, and sometimes buildings will be required to protect both the rest of the inside and also the outside environment from noisy activities taking place within the building (Figure 1.57).

When a building has to contain noise it is common practice to set a maximum noise level at the boundary of the building plot. Where the building has to protect the occupants from external noise, often a requirement is put on the insulation of the building envelope. The purpose of this section is to explain common noise descriptors, and to set the scene against which the performance of individual components can be assessed.

Airborne sound insulation

Airborne sound insulation of walls depends largely on the type of material used. Expected performances are discussed later in the book, but approximate values are as follows:

- external walls with opening windows   20–28 dB
- external walls without opening windows   45–50 dB
- partitions   30–35 dB
- separating walls   50–55 dB

Existing construction may need to be tested, for example to determine whether performance specification values have been met. Although retrospective testing is rare in the UK, it does happen in other countries, even to the extent of rebuilding when specified conditions are not met.

A minimum level of performance is required under the building regulations, for example under E3 in England and Wales and H3 in Scotland, with respect to separating walls in adjoining dwellings.

Flanking sound transmission

The subject of flanking transmission is a complex one. This type of transmission does not depend solely on the properties of the flanking wall; it depends also on the properties of the separating wall, on whether the flanking wall is continuous, and on how firmly it is bonded to the separating wall (Figure 1.58). A common case arises when a duct passes through sound resisting construction, but the walls of the duct offer only lower resistance.

Single glazed windows, 400 mm apart, when closed, have no effect on the sound insulation between rooms when the sound is limited to 55 dB by the separating wall, and in general the open area of the windows will be more important than the separation between them[58].

Sound insulation of separating walls

In a study of sound insulation in converted and refurbished dwellings, 85% of which had timber floors and 15% concrete, the

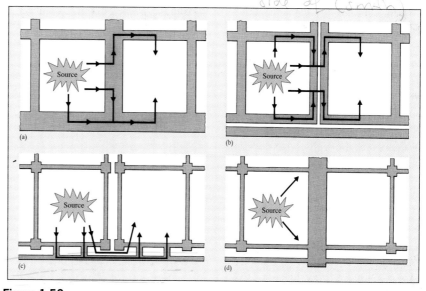

Figure 1.58

Principal flanking paths:
(a) heavy single leaf separating and external walls, (b) heavy double leaf separating and external walls, (c) lightweight double leaf separating and external walls, and (d) heavy single leaf wall, lightweight double leaf external wall (flanking negligible)

measured mean airborne sound insulation $D_{nT,w}$ was 53 dB and the mean impact insulation $L_{nT,w}$ was 56 dB. Of these residents, 92% were satisfied with their flats, though those that did not notice noise from their neighbours (27%) were significantly more satisfied. The most commonly noticed noises (unprompted) included loud music, footsteps and banging doors.

Sound insulation of external walls

A building may be required to offer protection from sounds made by road, rail, and air traffic. Figure 1.34 in Chapter 1.8 of *Roofs and roofing*[24] gives examples of noise spectra from these sources. This noise may enter a building through several routes such as the walls, windows, doors, roof, and ventilation ducts. The contributions of all relevant paths must be considered[59].

Examples of indoor ambient noise levels, that is to say the background noise levels which ought not to be exceeded in various kinds of rooms, and which therefore may impose a requirement on external walls, are given in Table 1.5.

Figure 1.59

Measurement of the sound insulation of a window built into the horizontal transmission suite at BRE

Table 1.5	
Indoor ambient noise level $L_{Aeq,T}$ for various room functions	
Room function	$L_{Aeq,T}$
Lecture theatre	30–35
Wash room, toilet	45–55
Bedroom	30–35
Restaurant	50–40
Executive office	35–40
Museum or Library	40–50
Classroom	35–40
Workshop	55–70
Open plan office	50–45

A masonry external wall can be assumed to have an insulation of at least 45 dB, so the insulation is likely to be governed by windows and doors. If windows forming more than around 10% of the wall area are open, the insulation will be around 10-15 dB whatever the construction of the rest of the wall and dependent on the sound source.

Noise level at the site boundary

Noise levels in and around existing buildings can be measured, but predicting the noise level at the site boundary of, say, a new factory is very complex. The main factors are:
- noise level in the building
- noise from extract fans
- insulation of the building envelope
- distance to the boundary
- barriers and bunds

The noise level in the building depends mainly on the sound power of the equipment in use and the acoustic absorption within the building. The insulation of the envelope can be estimated as described earlier for simple cases. After leaving the building, the sound energy will spread out over a large area and so only a fraction of it will reach the measuring point. The resulting sound level will depend on the distance, the ground cover and the wind speed and direction.

Unwanted noise from the wall itself

Noise from the wall itself, caused by wind, sun, rain or hail, is a possibility which needs to be considered, since it can form a nuisance to occupiers. Such noise is unlikely to emanate from masonry walls, but is far more likely to emanate from metal sheets used, for example, in overcladding of walls.

The factors include hail or rain-induced drumming on thin sheets, wind-induced whistling of ventilation slots, sun-induced 'oil-canning' of edge-stiffened thin sheet metals, and 'stick-slip' intermittent movements on metal-to-metal joints in fixings. Remedial measures are discussed in later chapters appropriate to the construction concerned.

Reverberation

If there is too much reverberant sound in a room, speech may be difficult to distinguish. The sound absorbent qualities of the wall surfaces will certainly help to reduce reverberant sound in rooms. Further information is available in BRE Digest 192[60].

Chapter 1.10

Safety and security

Safety and accidents

Safety

Responsibility for safety of people engaged in building operations is not only confined to the contractor – specifiers also share the responsibility, in the sense that they must consider safety aspects of installing the items they are specifying, consider the risks, and decide whether to select alternatives. The specifiers are also obliged under current legislation to provide information about health and safety issues for all items specified.

Compliance with all relevant items of safety legislation is outside the terms of reference of this book, but reference may be made to the various regulations[15,53,54], to the *Construction safety handbook*[61], and *Designing for health and safety in construction*[62].

Accidents

Accidents often occur as a result of people tripping as they step over door thresholds, sometimes because the threshold is high in relation to the floor level or alternatively because the change in level from the ground to the floor is too great. Serious accidents result when people trip and fall into non-safety glass in doors and windows.

Steps out of buildings may be too high, lacking an intermediate level. Some door threshold designs, particularly PVC-U types, may be bulky and in combination with a deep step create a possible hazard. See also Chapter 5.2.

Ordinary quality thin annealed glass in large areas is a hazard in any situation where it is close to floor level or any horizontal surfaces onto which children might climb (eg large bay windows). Glazed doors and side panels are an obvious hazard but windows positioned low in relation to stairs and landings may also be dangerous.

Open windows should not severely obstruct footpaths (Figure 1.60). Where this is likely either using a different design of window or obstructions such as flower beds or permanent barriers to keep pedestrians clear should be considered.

Window handles and locks should not require excessive force to operate and they should be easily reached without unreasonable stretching over baths, sinks and work surfaces, or the need to stand on steps (see Chapter 1.12). Where there is a risk of people falling through open windows above ground floor level, external barriers should be provided, or windows with safety devices, called restrictors, should be fitted to prevent openings greater than 100 mm. Such devices should not be easy to disengage except by adults during cleaning. Unless there is a balcony, cleaning should be possible from inside without the use of steps or the need to lean out through the open window. Glazing close to floor level (less than 800 mm) should be of toughened[63] or laminated glass to minimise the risk of personal injury from falls and impacts. This covers glass in windows, as well as in doors and door side panels.

The risk of accident will also be present when doors are opened and closed, particularly where they are power driven.

Figure 1.60
Hazard caused by an open casement window

Where hinged or pivotting doors are assessed as prone to risk of slamming, for example in windy conditions, consideration should be given to the fitting of closing devices which should operate at a closing speed of not more than 0.5 m/s.

Power driven sliding or rotating doors should operate at opening and closing speeds not exceeding 0.4 m/s, and should be fitted with safety devices.

Protection against radon and methane

Protection against radon and methane from the ground is a relatively new phenomenon in buildings. Although the problem mainly affects floors rather than walls, it may be necessary to consider the possibility of entry of these gases into cellars and underfloor voids through walls as well as through floors or oversites. Where remedial measures are undertaken, there will be a need to link the wall DPC with any radon measures.

Floors and flooring[30] explains the problems caused by radon in more detail.

Presence of deleterious materials

Building clients are becoming increasingly concerned that their buildings should not incorporate what have been termed 'deleterious materials'. Architects are often asked formally to undertake that no material is used from itemised lists submitted by the client. These lists usually divide broadly into two groups of materials – those which may affect the structural integrity of the building and those which may affect the health of occupants.

The concept of a deleterious material in this context is a contentious one. Any material or product, if used incorrectly, can lead to difficulties. Some materials may find wide application with very few limitations on use while for others there may be very significant limitations. In general, the concept of blanket lists of unacceptable materials should be resisted.

Examples of materials which may be relevant to walling and which regularly seem to cause concern for clients on grounds of structural integrity are as follows:

- high alumina cement. In effect this is not now approved for structural concrete although it may still be used for non-structural purposes
- calcium chloride admixtures for use in reinforced concrete. These are now not used as additives for reinforced concrete but are used up to certain very low levels in unreinforced concrete
- aggregates for use in reinforced concrete which do not comply with requirements of BS 882[64] and aggregates not complying with BS 8110[17]. Correct choice of aggregates is crucially important to the performance of concrete. Other important standards are BS 3797[65] and BS 1047[66]. There are also a number of additional considerations in relation to choice of aggregates which should receive particular attention in special cases

Clients' concerns on grounds of health and environmental grounds seem to relate in particular to:

- asbestos[67]
- man made mineral fibres
- timber preservatives
- lead paint
- urea formaldehyde (UF)
- chloroflourocarbons
- halons
- non-sustainable natural materials

In addition to the items listed above other items tend to appear on clients' lists for reasons which are not at all clear. Some seem to have been selected rather arbitrarily. To provide some guidance on these issues, a useful general text is *Hazardous building materials – a guide to the selection of alternatives*[68]. This guide is concerned with the detrimental effects which construction materials may have on health.

Volatile organic compounds

Some products used in the construction and maintenance of walls and their component parts, such as paints, wood based boards, and cleaners and polishes, emit volatile organic compounds (VOCs) which may be injurious to health such as solvent naphtha and related compounds from dampproof membrane materials, and other emissions from degraded PVC. Another common VOC is formaldehyde emitted from wood chipboard products and from UF foam cavity fills. To enable specifiers to select suitable products, information is available about the types and amounts of VOCs emitted. Tests are carried out according to a guideline drawn up by a working group under the auspices of the European Community Concerted Action of Indoor Air Quality and its Impact on Man.

So far as DPCs and DPMs are concerned the main sources of VOCs are those based on coal tar products, which may contain solvent naphtha.

Hygiene

Certain wall surfaces are required not to harbour dirt and germs, and to prevent water penetrating below the surface. The provision of such a surface can take several forms, for example:

- tiling and jointing with an appropriate grout
- sheet material with well-adhered joints
- flexible paints which accommodate cracking of the substrate

These finishes are dealt with in more detail in Chapter 10.

In hospitals and certain industrial process areas, if the skirtings are of the same material as the sheet flooring, then it may be possible to weld or otherwise join the upstand at the foot of the wall to provide virtually a tanked area. Integrity of the joint is of paramount importance.

Figure 1.61
Window security has been a problem for householders through the ages

Security

Buildings and their surroundings are exposed to risk from graffiti, vandalism, opportunist theft, professional burglars, arson, criminal damage, civil disturbance, espionage (both commercial and military) and terrorism. It is the external envelope which is under attack, with windows and doors the weakest links (Figure 1.61).

The risks are continually changing with increasing crime, and so the security performance of existing buildings must be continually reviewed. Estimates of the number of burglaries committed each year in the UK very between a half and one million, with, for example, dwellings in typical urban areas having a risk of burglary of 1–3% each year.

BS 8220[69] gives general prescriptive guidance on crime prevention measures. There are however no absolute criteria against which the performance of a building and its security measures can be judged. The patterns of crime continually change, and criminals move to new forms of crime or adapt their methods of attack to overcome the latest security technology.

The security performance of a building can only be judged subjectively against its ability to deter, detect, delay and possibly detain the perpetrators.

The risks have to be assessed in consultation with appropriate authorities or specialists to establish their likely probability of occurrence, the loss to the building owners and their insurers if they occur, and appropriate measures to reduce the risk.

For some building types and uses, known to be in high risk categories (eg banks, military bases etc) there are well established procedures for carrying out these assessments, and they will form part of the brief to the design team. However, for buildings subject to lesser risks, these procedures and methods of assessment are less well defined. The police authorities, via their Architectural Liaison Officers and Crime Prevention Officers, are, however, willing to provide advice on risks in their areas and suggest appropriate security measures. The police forces in England and Wales now offer the Secure by Design service for new housing.

Walls

Walls are only as secure as their weakest point. Therefore they should be designed and constructed to provide as uniform as possible a barrier to deter and delay an intruder. Windows, doors, ventilation openings, grilles, architectural features etc may provide weak points in the barrier and require special consideration.

All walls, even masonry walls, can be penetrated illegally by determined intruders. Apart from the use of explosives, vehicles have been used as battering rams to gain access. Even opportunist thieves have been known to break through a masonry wall rather than try and penetrate a security door.

In general:
- walls should be designed and built to withstand the likely level of attack
- openings should be kept to a minimum
- architectural features which provide concealment or provide access to the roof, and unprotected or unsecured windows etc should be avoided
- fixings, particularly of cladding, should be concealed, or, if exposed externally, be tamper-proof

Windows

Most domestic break-ins are unplanned and occur when a thief notices that a house is unoccupied and there is a chance to gain entry quickly without being observed. Many intruders will not risk noise or injury by breaking glass, but will try to lever windows open by breaking the catches and hinges. Windows with a single catch fastener are easier and quicker to break into, and therefore less secure, than windows with multi-point locks – that is, mechanisms that engage in the fixed frame at 3 or 4 points. Simple hinges on the outside can invite attack but some concealed or enclosed mechanisms delay burglars sufficiently to make them give up. Some hinge mechanisms have special security features that make it more difficult to lever open.

The type of glass and method of glazing can influence how easy it is to break in. Annealed glass is readily broken but can leave jagged edges that will put off an intruder. Toughened glass is much more difficult to break, except with a hard pointed tool, and it can withstand large deflections without breaking if a thief tries to lever a window open. Laminated glass is more expensive but it contains a clear middle layer that helps to prevent broken glass from falling out of the frame. This will delay a thief from gaining a quick entry and may cause them to give up. Sealed double glazing provides greater security because it is more difficult and takes longer to break than single glazing.

How the glazing is fitted into the frame is important. Some methods fix the glazing with clip-in beads on the outside of the window. In some designs these beads can be easily sprung off with a sharp tool and the glazing removed in one piece. Other external bead systems have special interlocking parts that prevent easy removal and therefore are more secure.

The British Standards Institution have developed a method for assessing security of windows, and windows that pass this assessment and the BS tests for weathertightness are awarded a Kitemark.

Insurance companies sometimes specify the fitting of security devices, for example bars, which can have rather unfortunate results, or they may take a completely commercial attitude, and simply raise premiums when the risk of breaking and entering is perceived to be greater.

Security provision

The following factors should be considered:
- doors, windows, locks and other associated hardware that are of appropriate strength and performance

Figure 1.62
Measuring the effectiveness of metallised glass in the window and copper mesh in the walls

- appropriate methods for fixing components (eg screws should be of appropriate size)
- tamper-proof fixings if they are accessible from the outside
- appropriate concealed cabling or conduits and any external cables protected if an intruder alarm system or supplementary external lighting is to be installed
- open trickle vents in windows presenting a high security risk
- security provisions restricting means of escape during a fire
- installation of appropriate electronic and electrical devices (eg intruder alarms)

Standards

Overall performance requirements for the security of walls and their component parts are not generally available. Prescriptive requirements are contained in BS 8220 and organisations concerned with high security buildings such as banks or the MOD have their own in-house standards. Because of the nature of the subject these requirements are not generally available.

Methods of testing and assessment for specific components, particularly glazing and locks, are covered by British and overseas standards. British Standards are also available for bullet-resistant, bandit-resistant and safety glazing[70,71,72].

Methods of assessment for other components for use in low risk areas are being developed by BRE and the Loss Prevention Council. Some European national standards exist and there is a possibility therefore of a European standard eventually.

Electromagnetic screening

Electromagnetic screening can be applied to walls, windows and doors in commercial and industrial buildings. It uses techniques that are unobtrusive and do not impair the quality of the working environment, while providing an acceptable level of performance at reasonable cost (Figure 1.62). In addition to the effectiveness of the shielding of different materials, such matters as the quality of joints around doors and windows are important[73].

Chapter 1.11   **Durability and maintenance**

Materials selected for walling and its ancillary components should be able to withstand the climatic agents described earlier without suffering structural damage or unacceptable cosmetic effects. Virtually the whole range of climatic agents are of concern, each to an extent depending on the specific material in question.

The environmental information required in a given situation depends very much on the nature of the material and structure concerned. Very rarely is only one parameter of unique importance to deterioration – more usually one parameter dominates but others contribute significantly, either independently or synergetically. For example some plastics and paint materials are sensitive to ultra-violet light, but high humidity, rainfall and certain pollutants also play a part. Frost damage to porous materials is a particularly good example of the combination of two agents – temperature and driving rainfall.

It is not possible here to give a comprehensive list of materials and their possible modes of failure but examples of some of the main concerns are listed in Table 1.6.

Survey methods

Finding out exactly what kind of wall construction has been used in a building may often be difficult, and some destruction in the interests of diagnosis of a problem may be inevitable. It will be a matter for professional judgement in balancing the severity of a problem with the consequences and costs of making good any damage.

One way of minimising damage is to use an optical probe. Optical probes may be useful in some circumstances, but BRE experience is that they need to be used with caution, since the field of view is restricted (Figure 1.64 on page 66).

Figure 1.63
Severe deterioration of stone in this capital from Wells Cathedral

Table 1.6	
Modes of failure for various materials used in walling	
Plastics and paints	Clouding, discolouration, embrittlement and cracking due mainly to the effects of overheating and ultraviolet light
Stone	Attack of acid pollutants (Figure 1.63), dissolving by rainwater, frost attack, crystallisation damage, staining by rainwater run-off.
Cements and concretes off,	Carbonation, corrosion of reinforcements, alkali aggregate reaction, staining by rainwater run-deterioration of high alumina concretes, formation of expansive thaumasite
Timber	Insect and fungal attack, moisture movements
Brickwork	Frost attack, salt crystallisation damage, sulfate attack on mortar, lime blowing, efflorescence, moisture movements, staining by lime, silica, iron, biological agents
Renders	Loss of adhesion, cracking and spalling, frost or sulfate attack at the render/substrate boundary layer
Metals	Corrosion, thermal movements
Laminates or composites	Delamination or loss of key due to differential moisture or thermal movements or chemical reactions

Figure 1.64
Using an optical probe. Inset is shown a typical field of view obtained with an optical probe. In this case a camera has been connected to the probe. The wire tie in the cavity is shown to be relatively uncontaminated with mortar droppings

Relative importance of factors

Wall finishes may be considered to provide for the following main functions:
- protection of the structural wall
- better appearance
- increased comfort and safety

The relative importance of these factors varies according to circumstances and budgets. However, one of the most important attributes is the maintenance of relevant functions over time; in other words, the finish must be durable. Because of its importance, durability has often been regarded as a basic property of a finish, but it represents only the length of time that the chosen properties persist, and depends as much on the conditions of use as on the properties of the finish.

Many factors influence the life of a wall finish:
- condensation, liquid water and other liquids
- industrial pollution
- indenting loads and impacts
- sunlight
- insects
- moulds and fungi
- high temperature

as well as the fundamental properties of the materials and adhesives used, and the compatibility of these with other parts of the structure and its behaviour in use. With so many different agents of destruction, it is fortunate that so many old buildings have survived (Figure 1.65).

For further discussion of the durability of particular finishes, refer to Chapters 9 and 10.

Compatibility problems

Solvents or other chemical constituents of organic compounds such as mastics, adhesives or paints, may migrate and stain porous masonry or affect other plastics materials such as cable insulation.

Electrochemical reactions can occur between dissimilar metals when they are placed in contact. Aluminium, zinc and carbon steels may be particularly vulnerable when in contact with each other or with

other commonly occurring metals such as brass and copper. Different types of steel such as carbon and stainless steels may themselves interact when in intimate contact.

These reactions can be prevented by separating the metals, for example with a non-conducting spacer. But this needs to be carefully done.

Corrosion

The long term performance of metallic components will be dictated to a large extent by their resistance to corrosion. Bimetallic corrosion of metals should be guarded against, both in fixings, and between fixings and claddings if the latter are of metal. Contact between two metals does not necessarily cause corrosion, but the wrong combination of metals under particular conditions (including the presence of moisture) will accelerate the corrosion of the less noble.

The Ministry of Agriculture publishes a map every five years illustrating the average atmospheric corrosivity rate for 10 km grid squares of the UK. This map is based on a zinc reference data base but corresponding corrosivity rates for other metals (steel, aluminium,

Figure 1.65
King's College Chapel, Cambridge, built in the late fifteenth century

copper and brass) can be calculated using Dose Responsive Function relationships. This may help to assess long term durability of a proposed solution in any particular location.

Metals perform best in a clean, dry environment. While it is never possible in walling to achieve these ideal conditions, the design should be such as to prevent, as far as possible, the lodgement of dirt, dust and moisture on the surface. This, in general, means the avoidance of horizontal or near-horizontal surfaces. And while it is not possible to avoid surfaces getting both wet and dirty, designs should be free-draining to reduce time of wetness. Ideally, cladding should be washed regularly. Fully exposed surfaces will benefit from washing by rain. However, the lee side of a building will require monitoring.

The conditions under which walling is required to perform become more onerous with increasing height. Great care will be needed to ensure that the walling is sufficiently weathertight such that joints, crevices etc in the design do not permit corrosion to occur unseen. It would be prudent, particularly at high levels, to assume that there will be some moisture ingress, to check for water entrapment, and to ensure that the frame material is adequately corrosion resistant or corrosion protected.

The presence of thermal insulation behind the walling could also complicate the situation as moisture could collect in some materials and be retained as it were within a poultice in contact with the frame. While it may seem that the major risk is from rain penetration, one should not rule out the possibility of condensation increasing the risk of corrosion, both to the frame as well as to the internal face of metal cladding. Ventilation of a cavity is the best way of reducing this risk, though of course there should be no inadvertent reduction of thermal insulation value.

Ferrous metals
Ferrous materials, except stainless steels, are normally in themselves insufficiently durable for use externally and require additional corrosion protection. This additional protection can either be in the form of a metallic coating (eg zinc) or an organic coating (eg PVC) or a combination of both (a duplex coating). The life of such ferrous metals is directly related to their protective coatings.

The most common metallic coating is zinc, applied by galvanising or sherardising, although other electro-plated or hot-dipped coatings (eg of aluminium zinc alloys) and aluminium are available.

The life of a zinc coating is dependent on the environment to which it is exposed and is proportional to the thickness of the zinc. The thickness of the zinc coating required to give protection depends upon many factors, but generally cladding sheeting is formed from pre-galvanised sheet; the methods of forming the sheet (after galvanising) generally restricts the total zinc coating weight to not more than 275 g/m^2 including both sides. With thicker coatings, the zinc may crack and spall on bending. The life of zinc coatings is reduced in contact with mortar and moisture, and thicker (or duplex) coatings are thus required for ties and lintels used in the outer leaf of masonry. Wire for this purpose is electrogalvanised and is ductile. See also Chapter 2.2.

One type of ferrous metal which can be used in certain circumstances without additional protection is a weathering steel. These steels have a low rate of corrosion, and can weather to an attractive colour. However, there is a major drawback as the run-off from such material is rust coloured and will cause staining to adjacent materials. Weathering steel cannot under any circumstances be used as part of a curtain walling system into which glass, particularly insulated glass units, will be inserted. Buildings with such cladding need provision for ensuring that run-off products can be dispersed effectively.

The only satisfactory steel for permanent use in and on external walls is austenitic (18–chrome 8–nickel) stainless steel. This is stable in all but very special circumstances, does not cause staining, and is widely used for fixings. It should be replaced by molybdenum/chrome/nickel steel where high chloride levels are expected. In very special conditions such as swimming pool interiors, stainless steel may not be suitable; nor ferritic stainless steel as it is prone to pitting corrosion. See also Chapter 4.5 in *Roofs and roofing*[24].

For reinforced concrete members and rendered claddings, it may be prudent to use protected reinforcement and stainless steel mesh in high rise buildings and in marine areas; otherwise, corrosion can be severe. But it can also be severe in relatively sheltered areas when the alkaline environment giving protection to steel reinforcement disappears (Figure 1.66 on page 68).

See also Chapter 3.1.

Aluminium
Aluminium is a suitable material for the external finish of a wall, as it has a low rate of corrosion, but it must be expected that both its appearance and its integrity will deteriorate with time. As the white corrosion product forms, the surface will become rough, and will entrap dirt and become unsightly. Pollutants and contaminants will also be collected, and there will be a risk of accelerated corrosion.

Depending upon the service conditions, aluminium may require protection (eg plastics coatings) as well as suitable alloy selection. Copper-bearing aluminium alloys should be rigorously avoided.

Anodising produces a layer of oxide, which can have added colour, on the surface of the aluminium which in practical terms delays the onset of corrosion. The corrosion product of aluminium is white; hence, if dark-coloured anodic coatings are employed, when deterioration occurs it is readily seen.

Figure 1.66
Severe corrosion has taken place in the reinforced concrete columns of this two storey house

Aluminium sheeting, pre-coated with organic coatings, is now available; the visually acceptable life of such material is essentially the life of its organic coating, but it depends on the type of coating, environment, thickness and bond. In aggressive environments it will be necessary to consider eventual repainting to restore appearance. Aluminium extrusions as well as sheet can be coated with organic paints, the more successful of these being the fluorocarbon and polyester based materials. Powder coating is a very common finish, applied by heating the powder to fuse it to the surface of the aluminium.

Expected life of surfacing materials

There is a British Standard on durability, BS 7543[74], which applies to walls. It gives general guidance on required and predicted service life, and how to present these requirements when preparing a design brief.

BS 8200[28] requires all panels and secondary framing to have durability equal to the life of the building. In practice not all parts of the system can be expected to achieve that figure, but special care must be taken to ensure adequate life for any components which are difficult to inspect and critical to safety.

Industrial pollution

It is known that in sufficient concentration, sulfur dioxide plays an important part in the deterioration of a number of building materials the most sensitive being calcareous stones and certain metals. There is evidence that the considerable reduction in emissions over the last few years has resulted in a reduction of weathering rates for steel.

There is little evidence to indicate pollution attack on brickwork. The effect of absorption of sulfur dioxide from the atmosphere will be significantly less than intrinsic sulfate attack on mortar by transfer from the brick itself. (Sulfates are only present in some types of clay brick.)

There is further discussion on pollutants in *Roofs and roofing*.

Resistance to impacts

This has been covered in Chapter 1.1

Resistance to ultraviolet and sunlight

Many walling materials fade on prolonged exposure to sunlight, including some woods, and some plastics yellow. Yet other materials are unaffected. Whether such colour changes are important depends on circumstances. Material specifications do change, and inspection of old samples is not necessarily a good guide to future performance. It may be possible to carry out laboratory assessments, though manufacturers' advice should also be sought.

Natural pollutants

Carbon dioxide, CO_2, is present in the atmosphere at a concentration of about 350 parts per million. It dissolves in pure water to give a slightly acid solution 'carbonic acid' with a pH of 5.6 (a neutral solution has a pH of 7.0). Dissolved in rainwater, CO_2 is able to react slowly with the carbonate component of calcareous stones and lime mortars leading to long term deterioration.

Carbon dioxide can also cause deterioration of concrete, in particular reinforced concrete. Steel reinforcement is protected against corrosion by the alkaline environment of fresh concrete. However, CO_2 gradually penetrates from the concrete surface inwards, neutralising the alkalinity as it progresses (carbonation). Further discussion of this phenomenon will be found in Chapter 2.4.

The chief source of carbon dioxide from human activities is the burning of fossil fuels. It has been suggested that, at present, peak urban concentrations reach 3000 ppm. The effect of CO_2 as a greenhouse gas is an issue beyond the scope of this publication. However, the implicit climatic changes would be likely to modify the value of several other agents affecting performance, most obviously temperature.

Chlorides

Systematic measurements of particulate chloride or gaseous hydrogen chloride over extended periods in the UK are limited. An 18-month survey in Leeds indicated almost equal contributions from marine and non-marine sources to total particulate chloride.

In a maritime climate such as that of the UK, rain almost anywhere may at times contain sufficient chloride content derived from sea salt to cause increased rates of corrosion of metals such as iron, copper, zinc and aluminium. Highest concentrations are found in sea spray from breaking waves (up to 3 mg/m³). Average values near to the sea but away from spray are about 0.1 mg/m³ and, well in land, 10–30 µg/m³. Figure 1.67 shows average rates of deposition as a function of distance from the tide mark for a large land mass. Note however that there are many other factors, other than air borne chlorides, which can significantly affect corrosion rates.

Ozone

Ozone is produced naturally by reactions at various levels in the atmosphere. The natural background level in the atmospheric boundary layer (the lowest 1–2 km) can be increased by photochemical reactions. Ozone does not affect inorganic materials directly, but it is a powerful oxidising agent and plays an important part in converting sulfur and nitrogen oxides to their respective acids, sulfuric and nitric, which can then affect inorganics.

Ozone is implicated as playing a role in the deterioration of polymer materials, including plastics and paints. In addition, its precise role is difficult to quantify since it can not be distinguished from other decay processes – in this case the action of heat and UV light.

Insects

Prevention of infestation of wall voids and passage of insects and other animals into the interior of buildings is usually achieved by restriction of external openings. However it is not practical, given ventilation requirements, to guarantee complete exclusion of all insects and other animals.

Mortar bees

From time to time BRE is asked about damage resulting from the burrowing of so-called masonry or mortar bees. The damage complained of occurs most frequently in soft mortar joints, especially those of older buildings, but also in comparatively new buildings in which the mortar is particularly weak (Figure 1.68). Soft bricks and stone have also been damaged on occasions.

A number of wild bee species whose normal habitat is earth banks and soft exposed rocks are capable of causing damage to buildings by burrowing. The most common species is *Osmia rufa*.

The damage is caused by the female bees boring into the material to form a system of galleries or tunnels in which to house the pupae. The gallery construction takes place during the early spring, and the burrowing and emergence activities are completed by early summer. Only a single brood generation is raised each year so, although the bees may be evident during the summer because of more frequent flights, the building fabric will not suffer further damage until the following spring.

The gallery system constructed by a single bee will not cause any significant deterioration of the building's fabric. However the brood raised in one year may overwinter within the galleries and in the following spring enlarge them or construct new ones. Over a period of a few years large numbers may become established in a small area. In these circumstances damage can become much more severe and in extreme cases has been sufficiently bad to require some rebuilding.

In cavity walls the bees may construct cells within the cavities and on occasions they have caused additional nuisance by invading the interiors of buildings.

The most effective method of preventing further damage is repointing of the walls in which the

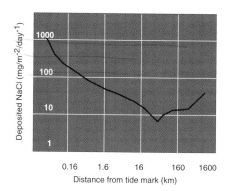

Figure 1.67
Deposition of sodium chloride aerosol as a function of distance from tide mark

Figure 1.68
The holes between the bricks are where mortar bees have tunnelled into the mortar

mortar joints are being attacked. The joints need to be raked out to a depth of at least 15 mm and then pointed with a mortar that is not too strong for the brick, but sufficiently hard to discourage the bees. For brickwork a 1:2:9 (by volume) cement:lime:sand, or 1:8 cement:sand with plasticiser, should be adequate provided it is used when frosts are unlikely. If there is a risk of frost a 1:1:6 cement:lime:sand, or 1:6 cement:sand with plasticiser should be used.

Pointing is generally best carried out during late summer or autumn. This avoids both frost and the activities of the bees. If the work can only be done in the spring, an insecticide solution should be injected into the gallery entrances, and the wall sprayed with the same solution after pointing to prevent attack on the mortar before it fully hardens. Emulsion or powder suspension insecticides approved under the Pesticide Safety Precautions Scheme are suitable. The Pest Control Officer of the local authority would be able to advise and in some areas may be prepared to carry out the spraying.

The use of insecticide spray or injection treatments are unlikely to achieve any lasting effect alone. In cases where the borings are in the actual bricks or stone a regular spray treatment in early spring may be the only effective method of control unless protecting the whole surface with an additional coating, such as a rendering, is an acceptable solution. If a rendering is used it should not be stronger than a 1:1:6 cement:lime:sand mix, and preferably a 1:2:9 mix.

Protection of openings from access by insects

Generally, mesh of 5–7 mm will exclude rats, mice and bats, and mesh of 3 mm or so will exclude most troublesome insects. Where the use or locality of the building justifies, mesh of these sizes should be applied to all external openings.

In the UK, intrusion of insects from outdoors is not regarded of sufficient significance to require screening of windows and doors

against them. However, to prevent harbourage and infestation problems, design of walls should avoid voids which are not tightly sealed off from the interior of the building.

Interior surfaces need to be considered in relation to building use. Where hygiene is of importance, finishes should not present crevices which could harbour insects such as silver fish, book lice or bed bugs.

Degrade of materials following infestation

Two types of problem occur in walls as a result of insect action: degradation of materials and infestation/harbourage.

Degrade of timber and some timber based board materials can result from the action of wood-boring beetles; Common furniture beetle, *Anobium punctatum*, is the most common species encountered in the UK. Though damage is potentially of structural significance, the risk of significant damage in timber framed structures is very low. In a survey carried out in 1993, BRE found that there had been a marked decrease in infestations of *A. punctatum* in UK properties up to 30 years old, though just under one third of buildings older than this can still be affected, with slightly greater numbers recorded for properties built before 1900.

House longhorn beetle, *Hylotrupes bajulus*, infestations can cause severe structural damage to softwood timbers, but are concentrated in the few areas to the south west of London, with the highest levels in properties built between 1920 and 1930. Infestations mainly affect roofs rather than walls. Areas of England where treatment is required are defined in the Building Regulations.

Guidance on recognising and treating insect damage in buildings is referenced at the end of the section, Moulds and fungus, on the next page.

Infestation by pest insects can occur in suitably sized voids in buildings. Where these voids have openings, for example between walls and floors, this can make pest control operations both complex and

expensive. Cockroaches and pharaoh's ants are usually only found in larger voids whereas book lice (psocids), bed bugs and silver fish utilise small crevices on internal surfaces.

Detailed requirements for the exclusion of some pests and general methods for reducing infestation by exclusion, reduction of habitat and food supply, and design and construction requirements for excluding pests and for facilitating the treatment of infestations are to be found in BRE Digest 418[75].

Animals and birds

The animals most likely to infest walls – usually cavity walls – are rats, mice and squirrels, gaining access through openings in the masonry; for example through broken airbricks, or through holes in fascias and soffits. They can be a considerable nuisance. In one dwelling examined by BRE investigators, which had been infested by grey squirrels, *Sciurus carolinensis*, many runs had been carved through the cavity fill, and over three sacks full of detritus, including several carcases, were removed. The removal of these animals is a matter for pest control agencies, though re-infestation needs to be prevented by adequate repairs[76].

Bats are known to colonise wall cavities, usually gaining access through openings at wall heads. Although bats are not regarded as having health implications their presence is often considered a nuisance. Remedial measures are problematical as bats are protected from any interference under the Wildlife and Countryside Act 1981.

Birds, particularly pigeons and starlings, commonly roost on external ledges in urban situations where roosting sites are at a premium. Fouling of wall surfaces by droppings is unsightly and may encourage subsequent surface growth by moulds and algae. Woodpeckers have been known to attack timber on buildings, and seagulls have been seen removing gaskets. Various parasitic mites carried by the birds may enter buildings through windows and other openings and cause dermatological problems for human occupants. Roosting prevention is usually by means of sprung wires placed just above ledges, or by spikes at suitable intervals. BRE Digest 418 gives further guidance.

Mould and fungus

There are two kinds of building fungi – those that cause wood rot, and those that do not. Wet rot occurs mainly at the bearings of timber joists in external walls; for example at the sole or head plates rather than in the studs. Wet rot decay of timber and timber based board materials can take place only where these are maintained in persistently damp conditions (Figure 1.69). Initiation of attack generally results from microscopic airborne spores, but can also occur where pre-infected timber has been used; in this situation, very rapid decay can occur in new construction. Under appropriate conditions damage may be rapid and severe, and therefore of structural significance. The dry rot fungus, *Serpula lacrymans*, can also occur, which is more devastating, though less common than wet rot.

Surface moulds be found on external and internal surfaces of walls and partitions, very often adjacent to ceilings and usually accompanied by persistent condensation or other form of wetting. Moulds are unsightly and

may also cause premature failure of paint films. On internal surfaces, as well as being unsightly, mould is thought to cause respiratory problems in susceptible individuals.

There is further discussion of rot and insect damage in *Roofs and roofing*. Guidance is available on the recognition of wood rot and insect attack in buildings[77] and on remedial treatments[78].

Joints

Joints are often the weakest points of walling, and it is usually the durability of infrequently occurring joints that receive least consideration by designers, fabricators and constructors. Joints between like components are more likely to receive adequate consideration than joints between unlike components from different sources. Extra attention therefore must be paid to joints between unlike components. For the same reasons, careful attention needs to be given to unusual or rare joints or those that occur at changes of plane.

Two-stage joints can be formed in both single and two-skin walling. The underlying principle for such design is that the water and air barriers are separated. Provided water is prevented (by geometry, and by having an air path past the water barrier) from reaching the air seal at the rear of such joints, performance can be reasonably assured (Figure 1.70). Two-stage

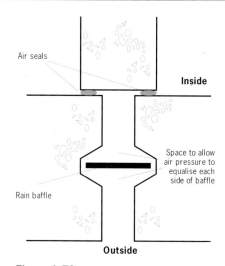

Figure 1.70
Separation of the functions of water and air exclusion

Figure 1.69
Fruit bodies of *Paxillus panudides* and decayed wood. This fungus prefers softwoods in very damp situations

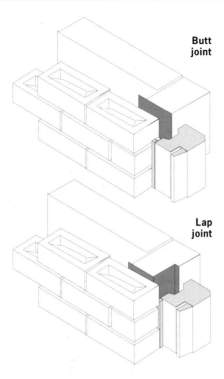

Figure 1.71
Face sealed joints should be lapped rather than butted

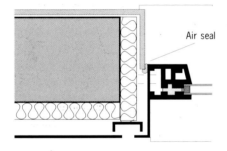

Figure 1.72
Two stage joint, with air seal on the inside, and an overlap on the outside to shed rainwater

joints are more likely to tolerate variations in workmanship than are single-stage joints.

Single stage joints, usually mastic or gasket sealed on the outer face, where a single mechanism provides protection against both air and water penetration, are usual in some types of curtain wall construction. Face-sealed joints in window-to-wall masonry situations should be designed to be lapped rather than to butt, wherever possible, since this will afford some protection to the seal (Figure 1.71).

Window-to-wall joints are usually better sealed at the back, and to have some kind of overlap or mechanical protection at the front, effectively creating a two-stage joint. This applies whether the wall is single or multi-skin (Figure 1.72).

Each joint design will have a maximum and a minimum dimension over which its performance can be assured. Such ranges will depend very much on the individual characteristics of edge profile of the joined components and the jointing product material and dimension. Figure 1.73 shows a typical Gaussian curve with such limits indicated. For those cases where corrections of positioning or other deviations are not feasible, the prudent specifier or contractor should have alternative designs of joint in mind for those conditions which lie outside the limits.

Fixings
Adjustability of fixings, for example for sheets used in overcladding, is normally needed in all three planes (Figure 1.74). It is not possible to give any universally applicable recommendations for the amount of adjustability needed since this will depend on circumstances such as the kind of construction, its characteristic accuracy, and the amount of control exercised. Nevertheless it will be possible from data on accuracy of the original construction and its movements in service, to calculate, for example, the size of slots within brackets and the size of holes for pegs.

All adjustable fixing mechanisms must be equipped with provision for the restoration of positive fixity following adjustment in order that accidental slippage under the maximum design load does not occur. Probably the most straightforward method of achieving this is by the compression of two serrated metal plates or brackets secured by bolting through (Figure 1.75).

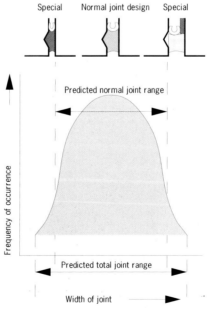

Figure 1.73
Normal (Gaussian) distribution indicating upper and lower limits for normal joint performance

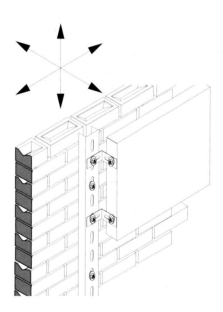

Figure 1.74
Adjustability of fixings is normally needed in all three planes

Structural surveys

It will be a matter of judgement when the building surveyor or architect needs to call in the assistance of a structural engineer to carry out a full structural survey.

Any preliminary examination and assessment of the wall or frame in order to determine whether or not to call in a structural engineer will need to consider at least the following:

● age of building
● use of building (in particular, access by the public)
● any features which make the building particularly vulnerable (eg span, local exposure, previous history of damage etc)
● obvious signs of distress
● degree of accessibility of the main structural members
● jointing techniques used for the main structure of the frame (eg welding, glueing etc) and any evidence of cracking or sloppy fit of bolts or snapped-off rivets
● condition of the walls or frame members
● evidence of rain penetrating the wall voids, and where it collected
● corrosion (in the case of metal structures) in extent or severity greater than surface corrosion
● surface damage, cracking or spalling (in the case of reinforced concrete structures)
● presence of fungal or insect attack (in the case of timber structures)

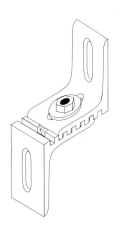

Figure 1.75
Principle of the restoration of positive fixity after adjustment

Where it is decided to call in an engineer to carry out an appraisal of a structure, it will normally be undertaken in accordance with the framework described in *Appraisal of existing structures*[79].

The report from the engineer to the client or his other professional advisers should include recommendations on at least the following points:

● items (numbers and quantities of members) requiring immediate repair or replacement for reasons of safety, stability or serviceability of the structure, either as a whole or substantial parts of it. (The assessments will need to be supported by calculations in appropriate cases)
● items likely to require attention within the short term (say two years)
● items requiring attention within the medium term (say five years)
● a suitable maintenance regime, if one is not already in place
● a time of next structural survey and appraisal

Cleaning

Few surfaces are truly self-cleaning under the action of rainwater since streams of run-off will follow certain routes rather than others. Some of the heavily textured or rendered surfaces will tend to dirty unevenly, and may be prone to algal growth. Early in the specification process, therefore, it will be necessary to consider whether uneven dirt adherence can be tolerated or whether periodic cleaning will be necessary, and whether or not it is likely to be carried out. Windows in particular should have some means of access to permit regular cleaning.

Trapezoidal profiled or sinusoidal corrugated sheet will self-clean much better if the corrugations run vertically than if they run horizontally. Anodised or mill-finish aluminium needs to be washed periodically in any case, to preserve its integrity.

The combination of marine salt and industrial pollution is recognised as a very aggressive agent for corrosion of metals. It can cause

premature failure of factory coated metals on windows, both aluminium and galvanised steel. Window frames can be quite vulnerable when compared with, say, roof sheeting or other items of cladding which is exposed and fully rain-washed. Window frames and associated components are often slightly recessed from the wall surface such that dust and salt deposits can build up if not regularly (ie at least every six months) removed by washing. Specifiers should see BS CP 153-2[80] and BRE Digest 280[81].

Ease of repair

Ease of repair is likely to be critical near to the ground or at access levels at any height of walling; it should be possible to remove and replace individual panels, for example, without removing and replacing a whole run. Special one-off replacement panels with unique surface profiles can be very costly to supply. If double glazing is installed from the inside, then no permanent obstructions to prevent reglazing should be fitted such as heater casings, partitions etc.

Some means of access by operatives to damaged walling will be necessary for replacement purposes; indeed the action of access in itself will increase the risk of damage. There may therefore be a case for the installation of a permanently available access system as an integral part of the cladding, and also some means of anchoring suspended cradles against sideways movements on high rise buildings.

Chapter 1.12

Anthropometrics and dynamics

Access

For access by wheelchairs, a clear width opening of doorways of 840 mm requires structural openings of 1000 mm. Doors should open a full 90°.

View out

In housing, window sill height should normally allow seated people to see out, and window heads should normally be above the eye level of people standing. Special consideration should be given to the needs of disabled people.

Reach

Maximum sideways reach for cleaning windows while standing on the floor adjacent to a window, and with minimum sill projection, is 850 mm (Figure 1.76). In the same situation, downward reach below an 1100 mm safety rail is a maximum of 550 mm.

A comfortable vertiocal reach over a 600 mm wide worktop is 1.7 m (Figure 1.77a) and, standing adjacent to the window, the maximum vertical reach is 2 m (Figure 1.77b).

Forces required to operate moving parts

The moving parts which are considered here include doors and opening windows, though controls on other manually operated devices such as blinds and shutters will in many cases not be very different.

Doors

The force required to operate a lever handle should not exceed 70N, or for a small knob 35N. For elderly or disabled people, lower values may be appropriate; for example the force required to open a door, or to close it, with one hand should be not more than 40N and preferably not more than 20N, and to hold it open not more than 7.5N. Young children may need special consideration. Handle height has little effect on capacity between 0.8 m and 1.1 m from the floor.

Windows

In the case of hinged, pivoting or horizontal sliding units, the force required to initiate movement should not exceed 80N and, in the case of vertical sliders, 120N. To sustain movement, the corresponding values are 65N and 100N. Elderly and disabled people may need special consideration and lower values may be appropriate[82].

Wear on moving parts

Maintenance to an agreed regime will be necessary for all openable windows, doors, adjustable ventilators and access equipment.

Rate of wear is usually thought to be a matter for agreement between purchaser, manufacturer and supplier, so long as the minimum required to achieve an agreed life is reached. Specification may be in the form of a performance test simulating 'n' years service on the part in question.

If a moving part fails in service, then it must fail in a safe condition.

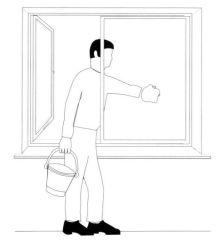

Figure 1.76
Maximum sideways reach

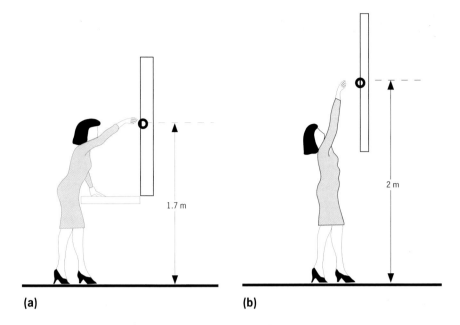

(a)

(b)

Figure 1.77
Comfortable and maximum vertical reach

Chapter 2   Loadbearing external walls and units

This chapter deals with masonry loadbearing walls. Given correct choice of materials, sound construction techniques and absence of abnormal destructive effects, either man-made or from natural causes, such walls can last for centuries (Figure 2.1).

Most frameless loadbearing external walls consist either of a structure made in the form of an assembly of prefabricated units – bricks, blocks or slabs of natural stone, fired clay, concrete or calcium silicate – or in the form of site-cast slabs of concrete. Generally units will be of regular and rectangular shape and will be bonded together with a mortar, but large units may be built without mortar. These walls are required to support their own dead weight plus the dead weight of any floors, partitions, ceilings, roofs and claddings which bear onto them or are fixed to them over the height of one or more storeys.

They will also be required to bear all the vertical applied load, lateral seismic load and lateral stacking applied load where relevant. Inner leaves of masonry cavity walls will be expected to share out-of-plane forces generated by wind loads on the outer leaves. All vertical and horizontal loads on these walls must ultimately be taken down to the ground via foundations. Vertical thermal and moisture movement is not normally important in a loadbearing structure and vertical differential movement in relation to an outer cladding is usually dealt with by design of the cladding and fixings. Horizontal

movement, especially shrinkage, can be a problem and is dealt with by vertical movement joints between panels. Where the wall must span horizontally to bear lateral loads some form of buttressing must be provided at the movement joint or slip ties must be used.

Devices will be needed to connect masonry to other parts of buildings including other walls, floors, beams, columns etc. All such devices need to be durable for an economic life, which needs to be taken into account when designing for all other functions.

Figure 2.1
The west front of Lincoln Cathedral, one of the sites used by BRE to monitor the effects of acid deposition on building materials

Chapter 2.1   **Brick, solid**

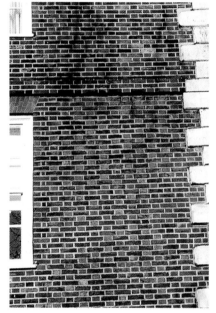

Figure 2.2
Double Flemish bond brickwork, Middle
Temple, London

This chapter deals with all kinds of fired clay facing materials, including brick and terracotta (but excluding tiles), normally encountered in solid walls. Concrete and calcium silicate bricks more usually found in the outer leaves of cavity walls are dealt with in Chapter 2.2.

Clay facing brickwork needs no words of introduction.

Terracotta is a fired clay material which can be modelled into units of different forms and patterns, and which may or not be glazed; the glazed material is also known as faience. There was much mystique about its manufacture, with some recipes being closely guarded secrets.

Although brickwork was used extensively in Roman times, it was not until the end of the thirteenth century that brickwork in England began to be used on a limited scale using bricks imported from the Low Countries. Tudor and Jacobean times, though, saw a massive increase in its use (Figure 2.2).

Since then it has maintained its popularity for buildings of all kinds. It is reported that in England, around one third of all dwellings still have solid masonry walls as their predominant structure. The older the property, the more likely it is to have solid masonry walls, with nearly two thirds of all dwellings built before 1918 having solid walls. (The EHCS[2] does not distinguish between brick and stone.) With respect to other building types, the change from solid to cavity construction took place over approximately the same time scale, though there is an increasing

possibility that some solid walls are thicker than one brick.

So far as Scotland is concerned, of all dwellings, nearly three quarters have cavity walls and the remainder solid, with brick-and-block two thirds, sandstone around 1 in 5, whin or granite 1 in 25 and most of the small remainder non-traditional[3].

Characteristic details

Basic structure
Bricks
The brick is designed as a unit which can be laid with one hand by the bricklayer, and at the same time laid in courses to produce walls of varying thicknesses. Bricks have been manufactured from an enormous variety of clay bodies, and the resulting properties of the fired product have been equally varied.

Clay types include the brick earths, for example, which produce the London Stocks; Tertiary and Cretaceous which produce the Gaults and the Wealdens; Jurassic which produce the Flettons; and Triassic (Keuper marl), Palaeozoic (Permian) and Carboniferous clays. A full description of the estimated 2,000 or more different brick types available at the height of brick production – some eight billion per annum – in the middle of the twentieth century from over 1,000 different brickworks, is to be found in *Clay building bricks of the United Kingdom* [83].

The range of manufactured types was equally vast: common, facing or engineering, extruded or wirecut, semi-dry or stiff plastic pressed with

Figure 2.3
A Warwickshire flour mill in locally made loadbearing blue engineering brickwork, sadly now demolished

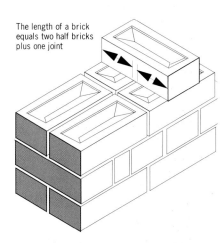

The length of a brick equals two half bricks plus one joint

Figure 2.4
The size criterion to ensure bond. English bond is shown

or without frog, horizontally or vertically perforated, sand or grog faced, shades from white through red to blue-black (Figure 2.3) – far too numerous to provide a complete listing here, let alone a description of properties.

Brick bonds
Bricks are bonded together in order to form a monolithic wall without serious continuous vertical joints. In order to bond successfully in a one brick wall, and so that the bricklayer can produce a wall with reasonably fair faces on both sides, the length of the solid brick used as a header needs to be twice its width plus one mortar joint (Figure 2.4). This rule is less critical with walls thicker than one brick, for differences can be taken up by varying the cross joint. It was also not necessarily rigidly applied with old bricks.

The actual size of the brick used in existing walls is not crucial until the time comes to cut out and replace bricks which have deteriorated, when there may be problems. In England, Tudor bricks tended to be around 58–60 mm thick and laid with thick mortar joints, whereas in later times, rather thicker bricks were used, laid with relatively thin joints, although size is not invariably a reliable guide to date. Between 1784 and 1803 a tax on bricks led to an increase in size, for large bricks paid the same duty as small[13]. Until the size of bricks was standardised in the 1950s, there were two common thicknesses in

the UK, the so-called northern brick averaging around 70–75 mm and the southern brick of some 10 mm less. Since the UK adopted the metric system, it will be found that the metric brick when used for replacement purposes needs wider joints than its predecessors.

Patterns of bonding have been developed over long periods of time. The early English bonds, in use from around 1430, consisted of alternate courses of stretchers and headers. Flemish bond, which consists of alternate stretchers and headers in the same course, was introduced into England in Stuart times (Figure 2.5). There are of course many variants of these bonds – doubles and singles, cross bond, garden wall bond, Sussex bond, monk bond etc. They mainly involve slight displacements of alternate courses, or doubling or tripling of headers, but all with the proviso that there should be no continuous vertical joints, or at least kept to a minimum.

It is sometimes considered that English bond gives a stronger wall than Flemish bond, but there is little scientific backing for this view. Of far greater significance is the effect of

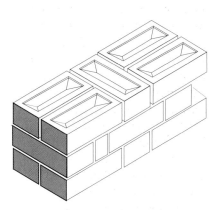

Figure 2.5
Flemish bond

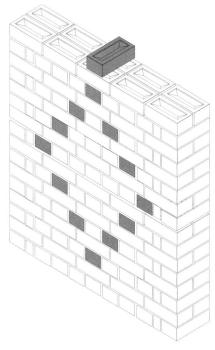

Figure 2.6
Diaper pattern formed by harder burnt headers

the mortar in the joints, and the presence of voids – in particular the lack of filling of bed joints, and also to a lesser extent, of perpends – and the laying of pressed bricks with their frogs facing down.

Some buildings with solid one brick external walls were built with the headers in a better quality or harder fired brick than the stretchers, giving a diaper appearance to the finished wall. This was done partly because it was thought, with some justification, that the lightly burnt through header provided an easier path for rainwater to penetrate the wall than the path interrupted by the transverse mortar joint between the stretchers, and partly for the decorative effect of the harder burnt or sometimes blue headers (Figure 2.6 on page 79). In any event, the weathertightness of the wall depends heavily on the condition of the mortar joints.

BRE investigators on site have occasionally found a wall built in two leaves in brick on edge, known as

'rat-trap bond' (Figure 2.7) Such walls are perfectly adequate structurally for two storey construction provided there are no obvious defects. They are perfect havens for vermin, however, which is how the bond gets its name.

Reinforced brickwork
Can take several different forms, with reinforcing bars threaded through vertical perforations in the bricks or blocks, fitted into gaps created by a special bonding pattern (eg quetta-bond), embedded in a grout occupying a cavity between two walls, embedded in concrete cast into pockets formed in the masonry, or simply laid in the horizontal joints to enable brickwork to span openings with minimal lintel support (Figure 2.8). Design practice is analogous to reinforced concrete and most forms can be created if required, including beams, columns, slabs, retaining walls etc. Hoop iron strips may be encountered in older work.

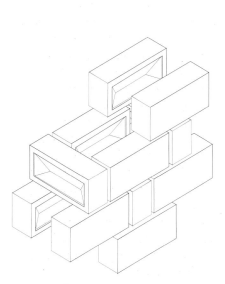

Figure 2.7
Rat-trap bond

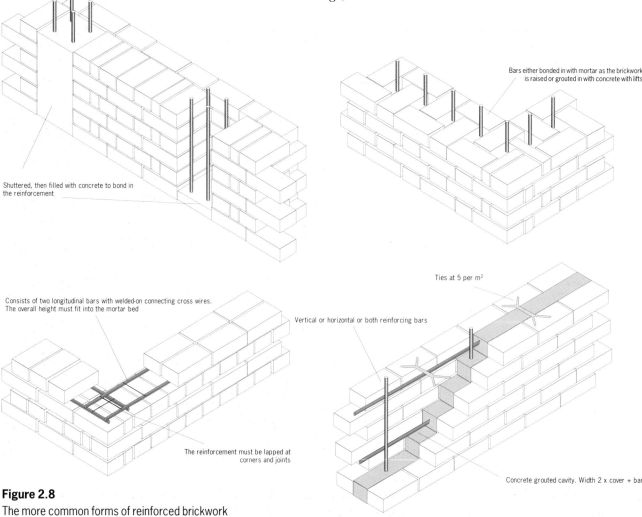

Bars either bonded in with mortar as the brickwork is raised or grouted in with concrete with lifts

Shuttered, then filled with concrete to bond in the reinforcement

Consists of two longitudinal bars with welded-on connecting cross wires. The overall height must fit into the mortar bed

Ties at 5 per m²

Vertical or horizontal or both reinforcing bars

The reinforcement must be lapped at corners and joints

Concrete grouted cavity. Width 2 x cover + bar

Figure 2.8
The more common forms of reinforced brickwork

Terracotta

The term terracotta tends to be used for the material rather than the unit, and the dividing line between what is a piece of terracotta and what is a specially moulded and fired brick can sometimes be a matter of opinion. Terracotta is unglazed, though partially vitrified; but when the clay body is glazed it is usually termed faience. Glazes may be either clear or pigmented.

Terracotta, or faience, was primarily used as a decorative material bonded to the external face of masonry, and is met with for the most part in heritage buildings. The material was first imported, mainly from Italy, and used in England in Tudor times (eg at Hampton Court Palace); English manufacture did not begin on a significant scale until the first quarter of the eighteenth century. The body is, of course, clay, with the addition of various quantities of other minerals such as sand and ground flints, depending on the desired colour and the need to control firing shrinkage[84]. It is the outer surface, or fire skin, which gives the durable qualities.

Terracotta was much favoured during the second half of the nineteenth century for use on public buildings where its decorative and durable properties could be exploited to the full. Frequently it was made to imitate limestone. Units fired to high temperatures under highly controlled conditions tend to be relatively uniform in appearance. If they are laid with narrow mortar joints the overall effect does not show weathering in appearance as does brickwork, and, for the most part, the material can retain a rather clinical and pristine appearance.

Mortars

The traditional mixes for mortars, before Portland cement was invented, were based on lime putty and sand. Such mixes, depending on the proportions, provided very workable and user-friendly properties with a long pot life, and developed their strengths slowly over time as the lime component hardened by absorbing carbon dioxide from the atmosphere and converting to calcium carbonate. However, they would always retain something of their initial resiliency when compared with mortars made from Portland cement and this is why the larger old buildings, which nowadays would be provided with movement joints, survive quite happily without them.

Between the wars black-ash mortars became popular in the industrial areas of the UK. These conform to no standard, but are thought to be compounded of ashes from local coal burning industries, possibly mixed with some lime and sand. These mixtures had very little hydraulic set and only limited long term set, but have survived, some with repointing, to the present day. Their main drawback is that they often contained acids which attack steel ties in cavity walls.

Fine (soft) sands, preferred for mortar preparation in the UK because they impart good plasticity and workability, are largely supplied from quarries unwashed. In this state they contain some silt and clay particles with the only requirement being a limit of 8% passing a 75 µm sieve for the general purpose (G) grade and a 5% limit for the structural (S) grade. Sands with excessive amounts of clay fines increase the amount of water required in the mix which can lead to lower strengths and higher shrinkage.

It is a common misconception that mortars always need to be as strong and as rigid as the bricks they join. Provided they are sufficiently strong to resist weathering, even relatively weak mixes will normally be strong enough for two storey work. It is only when large loads have to be carried, as for example in frameless loadbearing masonry multi-storey flats, that the question of matching mortar strengths to units arises. On the other hand, if there is insufficient lime or cement to bind the sand, durability problems will usually ensue (Figure 2.9).

Figure 2.9
This old half brick extension contrasts with the stretcher bond elsewhere. The mortar joints around the downpipe have eroded and repointing has been inadequate

Since the 1930s, mortars based on ordinary Portland cement (OPC) have become the norm. These would be batched on site from cement, sand and usually some lime to give added plasticity. Since the 1950s, a formulation plasticised by small bubbles of air beaten into the mix has been a popular alternative to lime plasticised mixes. To make these mixes more workable, an organic 'air-entraining agent' is added to the mix.

All site-batched mortars can suffer from batching problems which may affect their performance and durability. Occasionally too much cement is added which can create strong high bond, high shrinkage mortar which leads to cracking of masonry built with shrinkable units. More commonly, too little cement is added or too much water or air is incorporated giving weak mortars with poor durability.

Masonry cements are now covered by BS 5224[85] and BS EN 413-2[86]. Dry premixes of Portland cement, hydraulically inert fillers and a plasticising agent (air-entraining admixture) are supplied for mixing

with sand and water on site. They obviously eliminate some mixing and batching problems and thus provide a more precise control on quality than wholly site-batched mortars.

Retarded ready-to-use mortars perhaps offer the least room for error. The complete mixing process is carried out in the factory and mortar can be used as received. Since an OPC binder is used, a chemical retarder has to be added to delay the setting process. This is usually varied to control the set to a period of two to four days. This system obviously gives the highest level of quality control, but problems are still possible if further additions are made on site.

In view of difficulties which can be experienced on site in sorting out the various designations of mortars, there is perhaps something to be said for the use of a 'general use' mortar, which is able to resist all but the most severe exposure, and at the same time be able to accommodate minor movements. Such a mix would consist of 1:1:5 ½ ordinary Portland cement:hydrated lime:Type S or G sand plus an air-entraining agent.

Pointing technique, in addition to

altering the character of a brick or block wall, can also play havoc with its durability. Recessed joints, for example, could lead to reduced weathertightness, and should be avoided in all but the most sheltered locations.

Abutments

The most vulnerable situation for solid masonry is the parapet. Not only is it exposed on both faces, and therefore can be wetted from both faces as opposed to the wall on which it is carried, it does not have the benefit of heat gain from the building to help it to lose excess moisture and to keep it from freezing. Walls beneath parapets arguably should therefore have been protected with an effective DPC at roof level, as well as under the coping, otherwise they can become very damp (Figure 2.10). In certain cases, deterioration of the masonry in parapets can be such as to justify complete replacement; indeed this has had to be done on flat roofed industrial buildings constructed of London Stocks at BRE after a life of around fifty years. When replacement is undertaken it is worthwhile putting in an extra DPC.

Openings and joints

During rehabilitation a variety of lintels may be found, including steel, wrought iron, concrete, stone, timber and arched brick. Lintels may be found which are underdesigned; in some cases they may be absent altogether, reliance being placed on a wall plate or the door or window frame to support the wall above. In these situations considerable problems may be encountered during window replacement.

Main performance requirements and defects

Strength and stability

Most older brickwork was not 'designed' but built in accordance with traditional rules of thumb. Structural analysis of many existing walls would indicate that they are greatly over (or under) stressed.

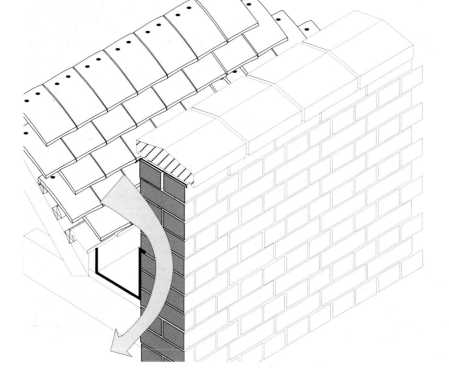

Figure 2.10
Parapets are the most vulnerable parts of external walls

However, whereas most will have stood the test of time and will not warrant remedial work, some walls will be found in a precarious condition. Very careful judgement is necessary to determine into which category each wall falls. Collapse of walls during rehabilitation does happen.

Walls are propped, stiffened and restrained by floors and roofs; also by buttresses in the form of returns, internal walls and chimney breasts; and even added features such as external buttresses and piers. Changes to any of these supporting elements may weaken a wall. Walls are also weakened by openings for doors and windows.

The current building regulations and codes require that building components providing lateral restraint (eg floors) must be tied or strapped to walls to give resistance to wind suction forces and to outward bowing due to cyclic movement. Ties are often absent in Edwardian and earlier structures (straps at the roof verge may be absent in quite modern buildings) but may be restored to a wall by providing through ties at floor level, either to the opposite wall or to a suitable anchor incorporated within the floor or roof structure. These ties often have decorative pattresses threaded onto the tie (Figure 2.11). Further information about remedial work is available in BRE Good Building Guide 29[87].

The bonding of bricks also determines the strength of a wall. Some housing was built of two leaves tied together with occasional brick headers and laid in a weak lime mortar. Under certain conditions, for example where the through headers fracture, the wythes can move apart, with potentially disastrous results. This can be dealt with easily by installing remedial shear ties as described later in Chapter 2.2.

Openings in walls – windows, doors and services apertures – create potential weaknesses in the form of narrow piers, thin spandrels and lintels. Any tendency for movement in a wall is likely to focus at these

openings. In the past it was common to provide relieving or discharging arches over windows, the theory being that the load coming onto the lintel was reduced, though the corbelling effect of brickwork arguably had much the same effect.

Narrow piers of brickwork between openings may be overstressed or offer inadequate bearing length for the lintels around the opening. Reliance on the wall plate, rather than a lintel, to span top floor windows is a common fault. Proprietary composite concrete gutter/lintels over top floor windows may be structurally unsound over long spans. Window and door frames may be the only support to the wall above.

Figure 2.11
Pattress threaded on the end of a remedial lateral restraint tie. It is still functioning, despite considerable deterioration of the masonry

Figure 2.12
This well-buttressed one brick thick wall to an outhouse has been destroyed over the comparatively short space of four years by a combination of thrust from the monopitch roof, which still stands, and vegetation. It is now in a very dangerous condition

Dimensional stability, deflections etc

Mortar strength and composition greatly influence a wall's resistance to cracking, and weaker mixes will be more tolerant of movement. Since one brick (225 mm thick) external walls are more likely to have been built in lime mortar, it will be unusual to find cracks which have resulted from movements of the materials within the wall itself, and it will also be rare to find movement joints. (For the required provision of movement joints, see Chapter 2.2.)

Minor cracking caused by differential settlement, expansion or contraction is common. Major cracking usually indicates differential settlement. Some apparently solid walls may incorporate rubble fill; this will result in reduced strength.

Timbers built into older walls as binders, grounds or plates may fail due to rot. A 'belly' in a wall might indicate lack of restraint, lack of buttressing or failure of poorly tied wythes under excessively eccentric loads. A leaning wall may indicate roof thrust, or inadequate thickness (Figure 2.12 on page 83).

Lintels installed on inadequate bearings may have settled or rotated. Arched brick lintels, especially shallow or flat arches, often slip as a result of quite minor movements in the structure due to the loss of abutments which provide the reaction to the arch force. Deflection is a common problem with timber or older steel lintels; stone lintels are more vulnerable to cracking.

For coefficients of linear thermal and moisture expansion etc, see Chapter 1.2.

Weathertightness, dampness and condensation
Penetrating damp

Solid one brick walls can be penetrated by rain, depending on the exposure and the type and condition of the bricks and mortar, though many walls of this thickness continue to perform reasonably well in the drier areas of the country or where exposure to driving rain is not severe. Before the widespread use of cavity walls, it was common, particularly in

exposed conditions, to render walls to improve weathertightness. If the render is too strong, it may crack and lead to rain penetration, so weaker renders often perform better (see also Chapter 9.2). Solid half brick (112 mm) walls cannot be regarded as weathertight.

As some of the alternative solutions which may be adopted, the Building Regulations for England and Wales call for external walls in situations of severe exposure to be at least 328 mm thickness of solid brickwork (equivalent to one and a half bricks), 250 mm of dense concrete blockwork, or 215 mm of autoclaved aerated concrete blockwork, all to be protected by a two-coat render at least 20 mm thick. See also BS 5628-3[19].

For older solid walls, which account for about 9 million dwellings in the UK, some of which may need remedial measures, there is usually no complete water barrier at any point through their thickness. Rain penetration therefore tends to be in a state of equilibrium with evaporative losses. The possibility of penetration is reduced by a number of traditional design features such as large overhangs, external rendering, string courses, which tend to throw off rainwater streams, coupled with the use of dense internal renders. Also, when the building is occupied, heat and ventilation inside help to prevent dampness appearing. If such buildings are unoccupied for long periods, dampness is likely to migrate further towards the indoor surface.

Concentrations of moisture can occur where barriers to downward migration exist, such as dense concrete lintels and DPCs.

Penetration through unit masonry invariably occurs preferentially through cracks at the brick/mortar interfaces. Generally speaking, the tooled joints perform best in resisting rain penetration, and raked or recessed joints worst.

Experimental work has shown that over 90% of leakages of running water occurs through such interface cracks and that these cracks may have an average width only of about 0.1 mm. This observation applies not

only to poorly built walls but also to walls which are well built with good standards of pointing for the external joints. Inevitably, some of these cracks are downward sloping channels, and the hydraulic head of water is sufficient to drive it through the wall.

The likelihood of a wall becoming completely saturated so that water begins to stream down the external face obviously depends on the type of brick or block, but nevertheless is thought to be relatively rare. At BRE, for example, only twice during working hours in a period of 25 years was it observed to occur on a particular London Stock brick wall some three storeys high.

The most satisfactory solution to rain penetration of a wall is to identify and correct the underlying defect. But in cases where dampness is widespread and correction is difficult it may be better to improve the water resistance of the outer surface. Available methods include repointing, application of a masonry waterproofer, painting, rendering, cladding and tile hanging which are dealt with in later chapters.

Water repellents
Water repellents are covered in Chapter 9.4.

Rising damp
The majority of solid walled houses constructed in the 75 years between 1875 and 1950, and some older dwellings, were built with a DPC of some description in the external masonry walls. Some of these DPCs may have deteriorated and failed but 'bridging' around the DPC is a far more common cause of rising damp. Buildings with a suspended timber ground floor will generally have a DPC located at, or above, the level of the wall plate supporting the floor joists (typically 175 mm below finished floor level); those with a solid floor will often have a DPC coincident with the top of the structural floor (typically 50 mm below finished floor level). This difference in level of DPCs can promote 'bridging' problems when solid floors are substituted for suspended floors, or where solid and

timber floors abut (eg solid floors in kitchens or sculleries and timber floors elsewhere).

Damp-bridging of DPCs is a common problem caused by external rendering, internal plastering, raised external ground, solid floors and bonded-in external walls such as screen or boundary walls. Previously-installed remedial DPC systems may be damaged or ineffective; and problems are commonly found with osmosis, evaporative tube and poorly installed chemical systems. Failure of physical DPCs, particularly those formed by engineering bricks or overlapping slates, may result from breakdown of the bedding mortar. Bitumen felt DPCs become very brittle with age and some can exude from the wall under pressure.

Ventilation

Ventilation of rooms required under the old Model Byelaws which preceded the building regulations was usually provided in the form of air bricks through external walls, controlled by hit or miss shielding on the inside of the wall. Some of the old airbricks may not have provided minimum background ventilation up to the standards now required. While building regulations are not retrospective, perhaps consideration ought to be given to whether such old airbricks are acceptable.

Trickle ventilators installed in window frames and glazed areas are dealt with in Chapter 4.1. If airbricks continue to be used, they should total 8000 mm² or 4000 mm², depending on whether the room in which they are situated is habitable or not. If the trickle ventilators are slotted, the minimum open dimension should be 5 mm and if they are holed 6 mm. It should be noted that these dimensions will not exclude insects. The majority of the more troublesome larger varieties of insects can be excluded by provision of a 3–4 mm mesh over ventilation slots. Care should be taken to ensure that any smaller mesh size introduced as a remedial measure does not reduce the ventilation opening size below that given by building regulations.

Rising damp in an old school building
A school building over a hundred years old had been built in a soft red brick, and dry linings had been installed on some walls in order to mask a very obvious rising damp problem. The samples obtained by drilling were tested for both moisture and hygroscopic moisture content. The wall was very wet with a moisture content in excess of 12% at 700 mm above internal floor level. There was a fall in moisture content to about 4% at a height of 1.3 m above internal floor level. Hygroscopicity was low, with values below 2% for most of the height of the wall. The graph (Figure 2.13) demonstrates the slight increase in salts at the level which rising damp has reached.

It was recommended to the client that a physical damp proof course was inserted into the wall; this was done and further moisture samples were drilled some six months after the insertion. It could be seen from the figures that the wall was now drying out and moisture content values were now below 4% for the full height of the area sampled. Values for hygroscopicity were very similar to the earlier values and for clarity are not plotted.

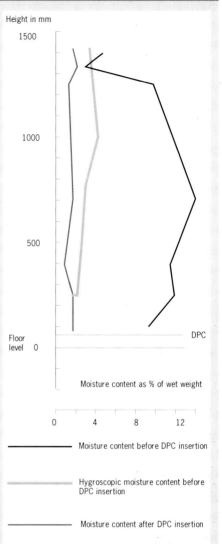

Figure 2.13
Moisture content readings before and after insertion of a DPC

Thermal properties
Standards of wall insulation as prescribed in the building regulations are far in excess of the insulation provided by a solid masonry wall. A one brick (225 mm) plastered brick wall will have a heat loss rate about 3.5 times greater, and a half brick (112 mm) wall 5 times greater than that currently permitted in new work. Further, the nominal insulation value of a solid wall may be reduced by saturation due to water absorption. In this respect, well rendered walls may be marginally better insulated than fair-faced brick walls. Similarly, claddings such as tile hanging provide only a further marginal improvement in thermal performance.

Rehabilitation work will provide an opportunity to improve insulation, either by adding insulation on the inside or outside of the wall. The method of adding insulation should be decided after careful consideration of structural characteristics, heating equipment, anticipated occupancy patterns and other possibly beneficial factors (for example improved external finish). A one brick thick wall plastered inside will give a U value of around 2.1 W/m²K. Substitution of lightweight plaster will give 2.0 W/m²K, while dry-lining will give 1.5 W/m²K. The addition of 40 mm of polyurethane to the dry-lining would give around 0.45 W/m²K.

Condensation
Solid 225 mm thick masonry walls do not generally suffer from interstitial condensation, but under certain prevailing temperature and internal environment conditions,

surface condensation can be a problem. Dampness in a solid wall from any source will further reduce even the minimum insulation it possesses and increase the risk of condensation on internal surfaces. Mould growth is often an unfortunate indication that surface condensation is occurring.

Reduced wall thicknesses (and hence reduced insulation) may exist in porches, previously bricked-up openings and where fireplaces have been removed. Half brick solid walls, often found in projecting back rooms to 'byelaw housing', are highly susceptible to surface condensation and mould growth (Figure 2.14).

Condensation can be a nuisance at window and door reveals, where there may be locations with reduced wall thickness. Corroding metal corner beads around window or door openings indicate that condensation has been a problem. Impervious external wall coatings may prevent or restrict drying out of water vapour from the wall and lead to dampness and an increased risk of condensation. Changes to heating, ventilation or occupancy patterns on the other hand can often lead to condensation and mould where previously no problems were encountered.

Fire

Masonry external walls, whatever their thickness, have an inherent fire resistance considerably in excess of the half hour or one hour normally required for low-rise buildings. Structural components within a wall and tested with it, such as lintels, require an equal fire resistance. In the case of conversion schemes (eg a three-storey house into multiple flats), an enhanced resistance may be required.

Lintels may have inadequate fire protection (typically only plaster); steel lintels being the most vulnerable. Loadbearing window or door frames will almost certainly have inadequate fire resistance (corner windows are most likely to be in this category).

Building Regulations 1991 Approved Document B[52] is the best reference for current requirements. In the case of proprietary components, for example lintels, information on fire resistance may be found in a relevant Agrément certificate. The BRE report, *Results of fire resistance tests on elements of building construction*[88] though now quite old, may contain relevant information. See also BRE Information Paper IP 21/84[89] and Chapter 1.8.

Noise and sound insulation

Figure 2.15 gives a rough guide to the insulation that can be expected from solid walls of different mass/m^2 in the mid frequency range; it does not take account of other factors, such as stiffness, air flow resistance and junctions with flanking walls, which may also be important for some materials.

The effects of windows

The effect on sound insulation of the insertion of windows into solid walls can be dramatic. If a solid wall has a performance, say, of 50–55 dB without windows, the insertion of openable single glazed windows, even if they are closed, will not usually give more than around 20 dB. Permanently sealed windows will enable the wall to achieve a higher performance, up to 30 dB, with

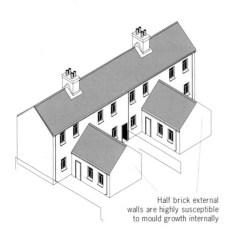

Half brick external walls are highly susceptible to mould growth internally

Figure 2.14
Projecting back rooms in 'byelaw' housing may have external walls only half a brick thick

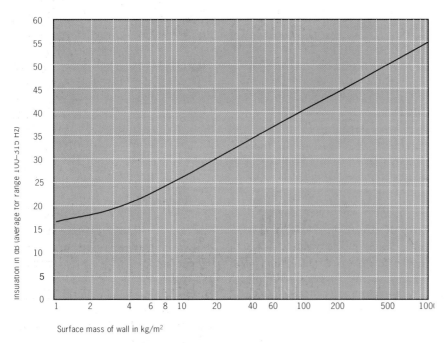

Surface mass of wall in kg/m^2

Insulation in dB (average for range 100–315 Hz)

Figure 2.15
Approximate relationship of sound insulation to surface mass for single leaf construction

thicker glass giving improved values. Sealed double windows will give up to around 43–45 dB.

Durability

Traditional walling materials of fired clay generally have a very long life expectancy (Figure 2.16). Some terracotta units dating from medieval times have survived in a superb state of preservation, though the glaze on glazed faience units sometimes crazes, giving rise to accumulation of dirt. However, some of the pale coloured bodies popular in late Victorian times have not proved durable, since they were underfired in order to develop the required shades[90]. Generally the higher the firing temperature, the more durable the material. Unfortunately, inconspicuous repair is not feasible.

So far as actual bricks are concerned, the durability of these has varied according to the type of clay and firing temperatures reached. Taking London Stocks as an example, their structure tended to be porous. Although first grade stocks were very durable, and second grade possibly so, milds and commons were not very durable. Many older, especially Victorian, buildings used mild stocks internally where they are perfectly serviceable, but if such bricks are re-used in external walling they can suffer rapid and severe frost attack. Gaults as a class were not very durable, but some particular specimens were. Some of the Keuper marl bricks were not very durable. Flettons, of course, are not suitable for the highest exposures, and the manufacturers offer advice on their specification.

Where the normal life of faced work has been exceeded or maintenance has been neglected, radical remedial action may be required. While bricks or blocks in a wall may be sound, the associated mortar, DPCs or render may be failing (Figure 2.17). It may be appropriate to limit further deterioration by providing extra protection to a wall (eg rendering or overcladding) rather than attempting to carry out a full restoration of the masonry.

However, this cannot usually be done with heritage buildings where careful cutting out and replacement, provided matching bricks are available, is preferred.

Brickwork may be deteriorating due to frost action[†] or sulfate attack, bricks below the DPC being more vulnerable. Salt crystallization may also be found. Mortar may be too strong, resulting in cracking, or too weak, resulting in washout by rain and enhanced frost damage. Bricks used for sills and below DPC level may not be of M or O quality to BS 3921[91] and therefore will be more susceptible to frost damage than the F quality.

Concrete elements are often found within external walls, for example as lintels, and cheeks to bays or canopies; these may often be of thin section and with inadequate thickness of concrete over the steel reinforcement. Precast lintels may also be found installed upside down. Rendering may have areas that are unbonded or cracked (see also Chapter 9.2). Timber lintels, bonding timbers and internal joinery may be decayed, and reconstructed stone lintels weak – problems which often come to light only when replacing windows.

White spirit used as a solvent for chemical injection DPCs can affect bitumen DPCs. (The possibility of this should be checked with the supplier.)

Certain metals, such as lead and aluminium can corrode when in contact with alkaline mortars; for example when they are used as flashings or when they are subjected to run-off from fresh mortars or concrete. (It is widely recommended that they are protected by bituminous paints where they are tucked into mortar joints but this is rarely done in practice.) Run-off from corrodable metals can stain masonry. It is a design responsibility to see that this does not occur, for remedial action is not always feasible.

Figure 2.16
Roman bricks re-used in the tower of St Alban's Abbey

Figure 2.17
A solid wall in a poor state of repair. While the bricks themselves may be in a reasonable condition, the paving is too high in relation to the slate DPC, the replacement pointing has failed, and there is evidence of movement

† A freeze thaw test has been devised by British Ceramic Research Ltd which is expected to become a CEN and later a British Standard.

Case study

Sulfate attack in parapets
Problems of spalling and cracking were investigated in the exposed, rendered parapet walls on top of a prestigious building in the south east of England some five years after completion. The walls had been constructed using N category bricks and problems of drying shrinkage within the render coats had caused very small cracks to open up, allowing water to penetrate to the brickwork and to become trapped. Drying shrinkage of the render had occurred because the second render coat had either been richer in cement or thicker than the first coat. It is probable that some of the water for the reaction was provided by the bricks being inadequately protected in winter and the render being applied over this wet substrate thus sealing in the moisture. Such wet conditions provide ideal conditions for sulfate attack. A white reaction product formed at the back of the render coats and within the adjoining bedding mortar between the bricks, causing expansion to occur.

A complete rebuilding of the parapet walls was necessary at a very high cost to the building owners. The rebuilding was undertaken using L category bricks, laid in air entrained mortar, which leads to improved durability of brickwork.

Sulfate attack
From 1931 onwards, the Building Research Station issued warnings about the susceptibility of the mortar in some brick walls to sulfate attack, particularly when the wall was rendered. A common factor to all was detailing that did not ensure that the brickwork stayed dry. And still failures occur!

BS 3921 defines two categories of clay bricks, normal (N) and low salts (L) types, on the basis of their content of water-soluble salts. Salts in clay bricks will be largely, but not exclusively, sulfates and are usually of sodium, potassium, calcium or magnesium. A brick that is classified as 'normal salts content' may release sufficient sulfates to cause deterioration of Portland cement based mortar or renderings in rain saturated walls and so is restricted in its application. The second (L) type is not so susceptible to problems associated with sulfate

attack as it contains low quantities of soluble salts.

Sulfate attack is mostly found in exposed brickwork (constructed using N category bricks) of parapet walls and in rendered brickwork where the render has become cracked and can no longer prevent water penetration. It is also particularly prevalent in exposed brickwork adjacent to flues (irrespective of brick type)[92]. However, it can also occur in conditions of high exposure to driving rain.

Efflorescence
Solid brickwork can suffer from migration of salts to the surface, with unfortunate effects on the appearance of the brickwork. Unless there has been a change in conditions affecting old solid brick walls, it is likely that any salts present will have been mobilised and brought to the surface long ago. Efflorescence is more likely to be encountered in relatively new brickwork in cavity construction; a full discussion of this phenomenon is to be found in Chapter 2.2.

Maintenance
Repointing should be the only maintenance required on a durable brick wall, and the frequency with which this should be carried out depends on the mortar, the finish of the joint, and the degree of exposure of the wall, though at least a 30 or 40 year life ought to be expected. Hand raking the old mortar wherever possible is preferred to the use of mechanical equipment. BRE investigators on site have seen the appalling damage caused by disc cutters.

Cleaning of terracotta must be done with great care, in particular the outer surface should not be mechanically abraded or acid etched to expose porous substrates, or durability will be compromised.

Work on site

Storage and handling of materials
Portland cements are the most vulnerable item. They need to be stored off the ground in covered accommodation.

Restrictions due to weather conditions
Bricklaying or repointing should not proceed under the following conditions:
● air temperature below 2 °C or when expected to fall below 2 °C
● when aggregates are frostbound
● in driving rain

Workmanship
Even the old building skills such as bricklaying need careful specification, and high levels of skill and supervision. Trying to solve an unexpected problem under the pressures of contract completion dates and the unpleasant conditions experienced on every site is highly inefficient.

To minimise any risk of structural movement or collapse, care is needed in the selection and positioning of temporary support for works to openings in external walls when replacing lintels or creating or enlarging openings (Figure 2.18). Before starting work, the size of load

Figure 2.18
Risks are occasionally taken when replacing lintels in solid masonry external walls. Here the old relieving arch has been cut away with just a soldier supporting the springing on one side

which will bear on the temporary support must be checked; a simplified method for assessing load is given in BRE Good Building Guide 10[93]. More detailed advice on loading and method of support will be needed if the existing construction is in poor condition[94].

In situ assessment of masonry walls

It is often of value to be able to measure the performance of existing walls, especially in cases having (or possibly having) legal consequences. Examples are where there are signs of deterioration, cracking etc; where a wall has partly or wholly collapsed; where the loading or usage of the walls may be changed; or where the imposed loading has changed due to external influences (eg an adjoining building has been removed).

A number of techniques have been developed for in situ assessment of masonry. The bond wrench was developed in the USA as a control test for the bonding of bricks to mortar in new work. This has been further developed by BRE into a site procedure which enables the bond of bricks to mortar to be assessed in built brick or blockwork, and is described in BRE Digest 360. In some cases it is also very useful to know the level of stress in a wall. This can sometimes point to overstressing or foundation movement, or give useful insight into movement related stressing. The test technique is based on the flat jack method and is described in BRE Digest 409.

Normally where mortar problems occur (eg of poor durability), they have been investigated by chemical analysis and the results may be compared with the specification which is normally by content of cement etc. The trend, however, is towards performance specification of mortars and strength then becomes the key parameter. A simple in situ test for mortar strength is described in BRE Digest 421[29].

Inspection

Although a rare occurrence (apart from boundary walls) there have been some collapses of masonry structures caused by the wind. Several of these failures have been investigated by BRE and the results have led to recommendations on appraising the stability of large masonry structures. Ways of appraising the stability of existing walls and techniques for strengthening them, and the design of new structures to ensure stability are discussed in *Safety of large masonry walls*[95].

All external walls should be inspected in detail and positions and directions of cracks noted; if more than a hairline, crack widths should be recorded and long term monitoring considered. If visual evidence of distress is found then a full plumbline survey should be undertaken. The possibility that the walls might have been originally built 'out of true' should not be discounted (Figure 2.19).

Solid brick walls should be inspected to determine any potential overstressed narrow piers such as the walling above each opening, internally and externally. Lintels may need to be examined to establish material, shape, size, condition and bearing width.

The position of the DPC relative to ground level all round the external perimeter of the building should be noted, together with the effect of abutments. Obvious signs of rising damp, for example the presence of salting and tidemarks, should be noted.

The external elevations should be examined for cracking and any features likely to affect the flow of rainwater down the building face such as the protection provided by the overhangs, leaks from rainwater goods or unsound renderings.

Any loadbearing door or window frames need to be identified, and checked that fire resistance is adequate. It may be useful to compare the extent of 'unprotected areas' with current regulations to see whether modifications are needed or justified.

Internal faces of external walls should be inspected for evidence of dampness such as mould growth.

The problems to look for in the main parts of walls are:
◊ narrow piers or returns overstressed
◊ defective bearings of long lintels
◊ spalling, cracking, oversailing, bulging or leaning
◊ dampness and algal growth
◊ inadequate or blocked air bricks

Figure 2.19
A bulging outer 'leaf' in a solid wall. Is this sulfate attack, or simply the parting of two wythes?

◊ pointing erosion
◊ slipping or spreading of brick arches
◊ sulfate attack
◊ lime, silica or iron staining
◊ lime popping

The problems to look for in respect of DPCs are:
◊ missing, damaged or deteriorated
◊ bridged by rendering
◊ pointed over
◊ below external ground level
◊ liquids injected into joints, not bricks
◊ plaster contaminated with hygroscopic salts

Chapter 2.2　　　**Brick and block, cavity**

Cavity walling as a means of resisting rain penetration is usually associated with modern buildings (ie post-1945). Naturally, the older the building, the less likely it is to have cavity walls. However, cavity walls became an established form of construction at different times in different regions. In parts of south east England, solid brick walls were still being built up to the late 1950s, but in more exposed regions, cavities were being used in the late nineteenth century.

It is reported that in England, just under two thirds of all dwellings have masonry cavity walls as their predominant structure, with the proportion of dwellings built before 1918 at around 1 in 22, and those post-1980 at 9 out of 10 (Figure 2.20). The EHCS[2] does not distinguish between brick and stone. Of the dwellings built before the end of the nineteenth century, only 3% had cavity walls. It has already been observed that 7 out of 10 dwellings in Scotland have cavity external walls.

Housing Association Property Mutual reported that, in the new dwellings which they inspected during 1991–94, about 1 in 10 had unfilled cavities, with the remainder split almost equally between partial fill and full fill of insulation material.

Characteristic details

Basic structure
External facings

There is an enormous variety of clay facing bricks which have been used over the years, and some aspects have already been covered in Chapter 2.1. Although not generally used in solid walls, calcium silicate bricks and concrete bricks are to be found in cavity walls – during the period 1991-94 it was reported that about 1 in 25 dwellings had external walls which used calcium silicate or concrete bricks in the external leaf.

Occasionally a surveyor may come across innovative external walls built from horizontally or vertically perforated through-the-wall units built with divided joints to maintain weathertightness.

Cast stone, although it may be intended to be used in substitution for a natural stone, and indeed made to look like it, is in effect a concrete block made with Portland cement and a suitable aggregate, sometimes pigmented. Cast stone is unlikely to be found in buildings dated before 1900, although the occasional one may be encountered. It is now widely used in areas with planning restrictions (eg Bath) which require the use of a 'stone' finish. The relevant standard for the material is BS 1217[96].

Generally dense concrete block faced walls are either left untreated, if the surface is relatively closed, or, if of a more open texture, finished externally with a two-coat rendering and roughcast or dry-dash finish; alternatively, a proprietary resin-based mineral-filled finish may

Figure 2.20
A post-1980 detached house. Around 9 out of 10 dwellings of this period are built in loadbearing masonry and the majority will have cavity masonry for all their external walls

sometimes be used. Lightweight blockwork similarly finished may also be used as a non-loadbearing cladding, for example on timber framed houses.

Brickwork as a replacement external skin

Brickwork as a replacement outer skin finds application mainly in low-rise construction, although it has been used in multi-storey construction where the structure can be adapted to carry the extra loads and the brickwork carried on shelf angles, for example, at each floor level. Since any concrete supporting structure has already undergone most of its shrinkage, the need for movement joints at each floor level is less exacting than with new-build; only thermal movements will be significant.

Any new brick skin will need to be tied back to the original structure, for example by resin-bonded anchors at the same centres as ties, as if one was designing new brickwork at that identical exposure.

Internal skins

Although often of brick, inner leaves may be of in situ concrete, clay hollow blocks, or breeze blocks, and there has been an increase in the use of ultra lightweight concrete blocks for the inner leaves of external walls since the building regulations prescribed improved thermal insulation standards.

Wall ties

Wall ties may be found made of many different kinds of materials: wrought iron, stainless steel, copper, copper alloy and mild steel coated with zinc, epoxy, bitumen and zinc with PVC. The design of the tie itself has also varied and in the past has included such variants as cranked bricks, bars with fish tails, and modern wire 'butterfly' ties (Figure 2.21). Butterfly ties have been available since the 1920s and were standardised from 1945 onwards, but stainless steel was only introduced in the 1960s.

Figure 2.21
Corroded examples of strip and wire wall ties

Openings and joints

Openings for windows, doors and services may create potential weaknesses in a wall, with narrow piers between openings, thin spandrels and lintels often acting as a focus for any tendency for movement within a wall (Figure 2.22). Lintels found during rehabilitation can be of steel, concrete, stone, timber or arched brick. Inadequately protected timber and steel lintels are particularly vulnerable and may be found in various stages of deterioration.

When cavity walls were first introduced, either a separate lintel or a brick arch was used for each leaf. With the increase in popularity of steel lintels since the 1950s, however, a number of combined lintels were introduced which supported the masonry in both leaves. The most usual were the common lintel (eg. the top hat or box lintel with shelf) and the boot lintel where the main structural lintel is in the inner leaf but has a simple metal or concrete boot unit (or toe) to support the outer leaf of masonry. Owing to a greater emphasis on preventing thermal bridges in the late 1990s, the separate lintels format is coming back into popular use.

Double lintels may be a source of problems. If a cavity tray is installed above the outer lintel, with weep

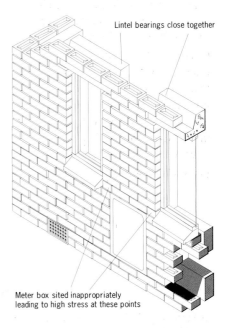

Lintel bearings close together

Meter box sited inappropriately leading to high stress at these points

Figure 2.22
Positions of high stress caused by proximity of lintel bearings, and openings for services

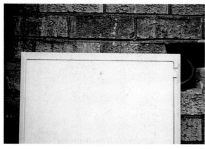

Figure 2.23
The brickwork over this meter box has collapsed because there was no lintel or other support

Figure 2.24
A poorly adhered sealant in a masonry wall movement joint. There are other joint defects too

holes above, it is too high to fully protect the lintel and window head. On the other hand, if it is installed to terminate below the outer lintel, weeps are not possible. Battening out the tray below the lintel is a possible solution.

Accommodation of services
One of the most common faults found on site inspections of dwellings has been absence of lintels over service entry points, and even brickwork above meter boxes can suffer from cracking if there is no adequate lintel (Figure 2.23).

Main performance requirements and defects

Strength and stability
The materials used in cavity walls of buildings up to three storeys will rarely be overstressed unless in narrow piers, but if constructed in accordance with Approved Document A1/2[20], the lowest storey has to be built of 7 N/mm^2 units, whereas 2.8 N/mm^2 units are allowed for the upper two storeys. Piers of brickwork between openings may be overstressed or offer inadequate bearing length for two lintels, or possibly three if an opening in an internal wall also abuts. Loadbearing masonry can be built to any height but above three storeys it has to be a calculated design to BS 5628-3[19].

Some form of structural linking (eg wall ties) between cavity wall leaves will have been included in most designs of cavity walls so that both leaves to some extent share loads, but their frequency and distribution will differ between walls. The current prescribed frequency is 2.5 ties/m^2 for most walls, and 5 ties/m^2 for walls thinner than 90 mm. Some metal types are liable to corrode. There is also some evidence that wall ties may be deficient, more especially in cavities which exceed 75 mm in width, and in narrower cavities at openings and unbonded edges.

Cavity walls are restrained by floors and roofs and propped by buttresses in the form of returns, internal walls and chimney breasts. Changes to any of these supporting elements may weaken a wall. Walls are also weakened by doors and windows and other openings.

Lintels may be absent over top floor windows and reliance placed on the wall plate to support the construction above. Proprietary concrete gutters/lintels over top floor windows may be structurally inadequate over large spans.

Accidental damage
Masonry can be remarkably robust in resisting collapse after impact damage. In a BRE experiment,

accidental-damage loadings to a domestic house were simulated by removing the masonry between lower window openings when a load was being applied at both first floor and upstairs ceiling level. The masonry was removed both from the front wall which supported the loaded floor and from the rear wall where the floor spanned parallel to it. In neither case was there any collapse due to the damage and there was evidence of a redistribution in the support to the load[97].

Wind pressures and suctions
Although damage to masonry walls is sometimes encountered after very strong winds, and collapses are not unknown, well built walls are remarkably strong[98].

In practice, most collapses of both leaves of cavity walls are a result of wind suction, since inward movement is resisted by floors and buttress walls. In the 1987 and 1990 gales, however, large numbers of houses lost their outer leaves due to suction forces on walls with corroded ties.

Dimensional stability, deflections etc
Minor cracking of walls caused by wall settlement or expansion, or by differential movement between the leaves is common. Major cracking of both leaves usually indicates differential settlement due to ground conditions, but some cracking may also be due to settlement of bearings.

Calcium silicate bricks and some blocks can shrink appreciably. A summary of the main points in relation to cracking is included here, and further information may be found in *Cracking in buildings*[99] and BRE Digest 360[100].

For coefficients of linear thermal and moisture expansion etc, see Chapter 1.2.

Regular horizontal cracks indicate expansion caused by advanced wall tie corrosion. A 'belly' in a wall might indicate lack of restraint, lack of buttressing, lack of wall ties or wall tie failure. A leaning wall may indicate roof thrust, or inadequate thickness or restraint.

Long lengths of masonry walling built in cement mortars will tend to crack unless they have been provided with movement joints (Figure 2.24). Many external cavity walls which should have movement joints do not, especially long walls with short returns.

Where walls have cracked and are to be repaired, movement joints can sometimes be cut into the existing wall without too much difficulty. The basic requirements are given below.

Clay brick walls

Typically, movements can be as much as 5 mm in 7.5 m of wall[101]. Provision is usually taken to be 1 mm per metre to cope with a combination of irreversible moisture expansion, and reversible thermal and moisture movements. The recommendations of BS 5628-3 for unreinforced masonry are not concerned exclusively with temperature size changes. However, the British Standard recommends that vertical joints in unreinforced fired clay brickwork should be positioned not more than 15 m apart to avoid cracking due to thermal contraction. Where wall lengths need to be subdivided, the joint spacing can be determined by working back from the ability of the chosen joint design to accommodate size changes.

BS 5628-3 gives the useful initial general guide that the width of joints in millimetres should be about 30% more than the movement joint spacing in metres. Closer spacing than 15 m is likely to be needed where comparatively little restraint is imposed on size changes – as in parapet walls, for example. The vertical joint nearest to a salient or re-entrant corner should be located not further from the corner than half the general joint spacing elsewhere. Additional joints (or, alternatively, reinforcement) may be needed at points of stress concentration – for example at openings. Movement joints in clay brickwork are to accommodate expansion, so must be able to close by prescribed amounts using compressible packing and non-setting mastics.

BS 5628-3 also draws attention to the possible need to maintain structural continuity at joints and the possible need to introduce slip planes to reduce shear stresses between parts that experience different size changes.

Calcium silicate brick walls

The material in these bricks usually shrinks after manufacture. BS 5628-3 recommends that, as a general rule, vertical joints should be located at intervals of between 7.5 m and 9 m, and that the ratio of length to height of each panel between joints should not exceed 3:1. As with clay brickwork, it may be necessary to introduce additional joints, or to incorporate reinforcement, at points of stress concentration. Although contraction is the predominant characteristic of calcium silicate bricks, their thermal coefficient is greater than that of fired clay bricks.

Concrete brick and block walls

Joints are specified to accommodate contraction of the wall, and the joints must be wide enough to prevent failure of the mastic in tension. BS 5628-3 recommends that, as a general rule, vertical joints should be at intervals of not more than 6 m. The standard also notes that the risk of cracking increases if the length of a panel exceeds twice its height. These recommendations apply in the main to half brick thick walls. Provided suitable precautions are taken, for thicker walls the distances between movement joints may be increased. At changes of section, similar considerations to those noted above for other masonry materials apply. (For examples of calculations, see *Cracking in Buildings*.)

Sometimes the requirement for a soft joint is overlooked until construction of that part of the wall is complete. A chase is then cut and filled with sealant giving the misleading appearance that the specification has been met (Figure 2.25).

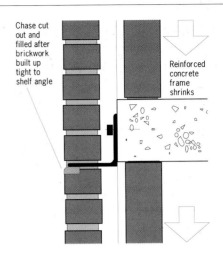

Chase cut out and filled after brickwork built up tight to shelf angle

Reinforced concrete frame shrinks

Figure 2.25
Has the specification been met?

Diagnosis of cracking

Cracks due to thermal contraction are usually of uniform width throughout their length and occur either at points of greatest restraint or at changes of section. Since these characteristics are shared by moisture-induced contraction also, diagnosis should consider first whether the materials concerned are susceptible to moisture-induced size change – either reversible or irreversible, or both.

Cracks due to thermal expansion almost always lead to displacement; in such cases identification of the mechanism by which a crack is produced is particularly important. The extent of displacement will show whether temperature size change, as calculated from the thermal coefficient and an assumption about the possible temperature range experienced, has been responsible. A ratchetting effect might also occur due to the accumulation of hard detritus in a crack.

Any displacement due to thermal expansion will be related to the largest expansion that has occurred in the building's history of cyclical size changes. Estimation of the amount of size change should, therefore, be based on the worst conditions likely to have been experienced by the building.

Cracking in lightweight concrete inner leaves in external walls of a factory

The BRE Advisory Service was asked to investigate the cracking to the inner leaves of some external walls to a factory.

The factory was constructed with a structural steel portal frame. The external facings to the walls were either of vinyl faced galvanised steel troughed sheets or of facing brickwork. The inner leaf to the external walls was 140 mm thick insulating concrete blockwork to varying heights from the ground floor slab. Above the working level the walls were finished with PVC faced, foil backed gypsum plasterboards.

The cavity between the blockwork and brick was partially filled with 25 mm expanded polystyrene. The cavity between the external metal sheets and the plasterboard contained 60 mm glass fibre quilt.

The cracking observed in the blockwork was of negligible consequence from a structural point of view. It had been due in the main to drying shrinkage of the blockwork, to thermal movements, and to wind loading movements of the structural portal frame. The problem was one of aesthetic acceptability. Filling and re-decoration would be suitable now that the shrinkage had taken place, but thermal movements would continue to occur. If the wall coverings did not hide the cracks, the cutting of movement joints would be necessary.

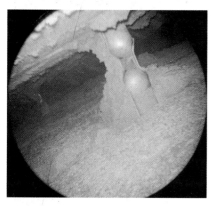

Figure 2.26
This photograph taken from underneath a strip tie through an optical probe shows a massive mortar deposit caught on the tie, leading to a high risk of water being conducted across the cavity

Typical displacements to be considered are:
● rotation of the ends of intersecting walls, return elevations and short returns
● oversailing where slip planes occur
● misalignment of planes at return elevations
● bowing

Weathertightness, dampness and condensation
Penetrating damp

A properly constructed cavity wall should be effective in preventing rain penetration. However, existing construction may well be imperfect in respect of the extent of separation between the two leaves (the cavity width), cleanliness of wall ties and the method of draining any rainwater which penetrates the outer leaf. Present information indicates that around 1 in 5 external walls have significant bridging of cavities, potentially leading to penetrating damp.

Where a wall is not built with full separation of the two leaves, some judgement will be required as to the need for modification. For instance, where the external and internal leaves are bonded together at window reveals with no vertical DPC: although this would not be

Figure 2.27
Solution of the dye fluorescein being applied to the previously wetted outer leaf of a wall to improve the visibility of water leakage paths through the cavity wall

acceptable in new work, it may not result in moisture penetrating as far as the inner face of the wall, perhaps because the building in question is relatively sheltered.

Currently the Building Regulations 1991 can be satisfied in respect of rain penetration (C4) by a cavity wall with leaves of any thickness, provided that the outer leaf is of masonry, the air cavity is at least 50 mm wide and bridged only by ties or by cavity trays, and with an inner leaf of masonry or a frame with a lining. Partially filled cavities should have a residual air cavity of at least 50 mm to ensure adequacy of rain resistance and to facilitate cleaning during construction.

Where the thermal insulation of a cavity wall is to be improved by cavity filling it will always be necessary to check the condition of the cavity to assess its suitability. The Building Regulations call for such filling to be in accordance with a current Agrément certificate, and with BS 5617[102], BS 5618[103], BS 6232-1 and BS 6232-2[104].

It is very common to find that bricklaying mortar has collected on wall ties, cavity trays and at the base of cavities resulting in potential routes for moisture penetration (Figure 2.26). Probably around 1 in 4 cavity walls inspected during 1991–94 having partial fill had inadequate provision for restraint of the insulation which could bridge the cavity.

Narrow cavity widths (less than 50 mm) increase the risk of rain penetration, as do wall ties which slope downwards towards the inner leaf of the wall, recessed pointing and inadequate projection of sills, copings, verges or eaves. Stepped DPC trays may lack stop ends at abutments, and may have unsealed laps or corners.

Site experiments on built cavity walls have shown that, after saturation, the outer leaf can quite commonly allow over 20% of incident rainfall to flow into the cavity. Rates of up to 80% have been observed in some wetting tests, especially with low absorbent bricks. (Figure 2.27).

Filled or partly filled cavities provide adequate resistance to rain penetration if proper attention is paid to design and workmanship and if certain constructions are avoided in areas of the country with the highest exposure to wind-driven rain. Guidance is given in the BRE report *Thermal insulation: avoiding risks*[36] and BS 5628-3. Cavity wall insulation, either as insulation batts built-in during construction, or blown into the cavity post-construction, should have been installed in accordance with a relevant British Board of Agrément Certificate or, for foam, with BS 5618.

A description of the methods for tracing the causes of dampness and to discuss the relative merits of the various remedial options is to be found in *Rain penetration through masonry walls: diagnosis and remedial measures.*

Aerated concrete blocks or panels clearly must be coated to resist rain penetration. A range of coating types has been investigated by BRE for water penetration, vapour resistance and durability. Results from test buildings show the influence of different coatings on moisture content gradients in the wall cross-section: in certain conditions persistent high levels or even penetration to the inner surface occurred[105].

Rising damp

Few buildings constructed with cavity external walls will have been constructed without a DPC of some kind. Where rising dampness is suspected, the cause is therefore likely to be breakdown or bridging of the DPC rather than its omission. Bridging is much more common than failure as most DPC materials will have a very long life in the protected inner leaf of a wall. The outer leaf should also have an effective DPC, but localised failure here is unlikely to cause dampness internally in the dwelling.

Where timber ground floors are used, the DPC in the wall would normally be below the level of any timbers but with solid floors the DPC may be at any level below the finished floor. With solid ground floors an overlapping link between the floor DPM and the wall DPC is a key detail in preventing rising damp; often there is no effective linkage and in older properties there may be no floor DPM.

Rising damp can also occur if a cavity wall DPC is bridged by mortar droppings which had accumulated during construction at the base of the wall (Figure 2.28).

Bridging of DPCs is commonly caused by external render, internal plaster, external ground, raised steps, solid floors and external works (such as screen or boundary walls) bonded to or abutting the cavity wall. Accumulation of dropped building mortar at the base of the wall cavity can cause bridging. Previously installed remedial DPC systems may be ineffective, particularly osmosis types, those relying on evaporative tubes, and poorly installed chemical systems.

Condensation

Masonry cavity walls are not usually affected by interstitial condensation, but problems could arise if impermeable materials are used on the wall. Any external wall can be affected by surface condensation, but cavity walls are most vulnerable where the cavity is bridged by materials with a high thermal conductivity. Where this thermal

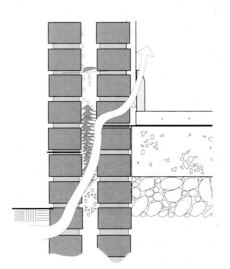

Figure 2.28 A common path for rising damp

Savings from cavity wall insulation

Twenty eight houses were insulated using blown mineral wool; the energy performance was then monitored for a year and compared with 27 'control' houses which had not yet been insulated. In ten of these trial houses an annual energy saving of 900 kWH or 9% of annual space heating consumption was realised, compared with the similar group of uninsulated 'control' houses. This was worth £19 pa to the average household at 1990 prices, and the insulated houses were warmer by 1.4 °C. It is estimated that it would have needed an extra 1700 kWh pa, or £34 pa, to achieve the higher temperatures if the houses had not been insulated.

Taking account of the higher temperatures and longer heating periods chosen by the occupants, the overall benefit of the package was worth about 2600 kWh pa or £53 per house. Very little disturbance was caused to the occupants during installation. No evidence of condensation on the walls could be found afterwards. Situated in an area of moderate exposure to driving rain, the houses showed no evidence of rain penetration or of any change in the rate of weathering of the external walls which might be attributable to the cavity fill.

Figure 2.29
Polyurethane board partially filling a cavity. In this case the lack of fit is unlikely to lead to a serious thermal bridge, but larger gaps are not uncommon

bridging is causing a surface condensation problem, the use of cavity fill insulation may not provide a solution, as it will not improve surface temperature at the critical locations. As with solid walls, mould growth is often an indication that surface condensation is occurring. (See also the discussion in Chapter 2.1)

Window reveals are vulnerable to condensation because of the reduced insulation value where the cavity is closed. Walls with poor pointing, bridged cavities, rising dampness or salts contamination are more likely to suffer from condensation. Rooms with external walls exposed on two or more sides or without direct sunlight or central heating are particularly at risk from condensation on the wall surfaces. There can also be problems with voids in cavity fill (Figure 2.29).

Thermal properties

Building Regulations require that new cavity walls have a U value of 0.45 W/m²K. A traditional cavity wall of twin brick leaves would have a U value in the order of 1.5 W/m²K. Where a clinker block inner leaf has been used, thermal performance will be slightly better; to improve to the current standard for new walls, at least 50 mm of thermal insulation will need to be added to the wall. The addition of 40 mm of polyurethane to a dry-lining would give around 0.45 W/m²K as would filling the cavity with mineral fibre.

Rehabilitation work will usually provide an opportunity to improve thermal insulation and cavity filling of walls will often be the most economic option, but where replastering or recladding is undertaken, it may be possible to increase thermal insulation as part of that operation.

Fire

The use of a combustible material in the cavity of an external wall is not thought to introduce an unacceptable increase in the fire hazard. Fire Research Station tests on walls of this type have indicated that the risk of fire spread is small

and that other considerations may dominate the choice of insulant. Fire performance is primarily affected by the degree of air access to the combustion zone and most insulants will provide only a minimum contribution to fire growth if the cavity is sealed. If perimeter sealing is not possible, careful selection of insulant is essential, considering both form and composition. Further guidance is available in *Fire risk from combustible cavity insulation*[106].

A fire test was carried out on three types of insulated brick-block walls with 100 mm wide cavities and one unfilled cavity was also tested. The fire, which reached a temperature of 1000 °C on the inner blockwork face over a period of about two hours, did not breach any of the walls or damage the stainless steel ties. Both the fully filled polystyrene bead insulation and the half filled polystyrene batts were destroyed by melting but did not ignite. The fibreglass batts were heat damaged but remained intact. The blockwork was severely damaged by the fire action and would need replacement and the brickwork bond was severely reduced despite only reaching around 100 °C. Some polypropylene wall ties had, unsurprisingly, melted.

Steel or brick lintels may have inadequate fire protection. Brickwork supported on window frames and lintels to bay windows or corner windows are particularly vulnerable in the event of fire.

See also the same section in Chapter 2.1.

Noise and sound insulation

Cavity walls may be marginally better than solid.

See also Chapter 1.9 and the same section in Chapter 2.1.

Durability

Existing wall ties may be corroded, the presence of black-ash mortar, in particular, tending to accelerate failure. Brickwork may be deteriorating due to frost action or sulfate attack. Mortar may be too strong causing cracking, or too weak resulting in accelerated weathering. Bricks used for sills and below the DPC may not be of 'special quality' and may therefore be more susceptible to frost action.

Concrete elements are often found within (or integral with) external walls as lintels and cheeks to bays or canopies. They may be of thin section and with inadequate concrete cover over any steel reinforcement.

A question frequently posed is whether the higher levels of thermal insulation in modern wall systems, especially cavity walls, are likely to increase the incidence of frost damage due to the lower wall temperatures and the likelihood of slower drying. This has been investigated, and average wall temperatures were calculated to be about a degree or so lower, but statistically significant field data for major problems has yet to become available. Cavity insulation also increases the range of temperature swings in the outer leaf and may exacerbate movement and cracking problems.

Clay Bricks
Clay bricks are specified in BS 3921[91]. The most durable is category FL signifying frost resistant and low soluble salt content and FN signifying frost resistant and normal soluble salt content; other categories of lesser durability include, for example, moderately frost resistant ML with normal or MN with low salt content, and not frost resistant OL or ON again with low or normal salt content.

Cast stone
Cast stone facings may not be homogeneous with other materials within the unit. If care has not been taken to match the movement characteristics of facing and backing, differential movement may lead to delamination of the face.

Lintels
Reinforced concrete lintels may be cracked (especially those of cast stone) or may have deflected. Steel lintels may be corroding. Lintels installed with inadequate bearings may have settled or rotated. Cast-in situ concrete lintels are prone to corrosion of reinforcement and to be of poorer quality concrete than those of precast concrete (which occasionally get installed upside down). Poor quality concrete in the toe of a boot lintel may result in deterioration, especially if there is a defective or missing cavity tray over. Boot lintels are prone to 'rotation', especially if they have no bearing on the outer leaf (Figure 2.30).

Sulfate attack
Sulfate attack on brickwork was discussed in Chapter 2.1. However, it can and does occur also in relatively newly built cavity brickwork as the illustrations below show (Figures 2.31 and 2.32).

Also, brick sills may bow upwards (Figure 2.33).

Concrete bricks and blocks
Concrete blocks used as the external leaf of a cavity wall without further protection can be adequately durable and frost resistant. However, as with all materials, severe exposure can take its toll. Where further assurance

Figure 2.32
Cracking in facing brickwork typical of sulfate attack

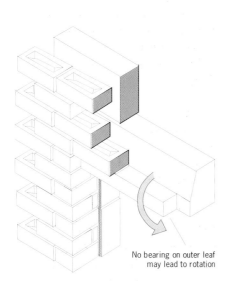

No bearing on outer leaf may lead to rotation

Figure 2.30
Boot lintels are prone to rotation if they have no bearing on the outer leaf

Figure 2.31
Sulfate attack has caused this facing brickwork to bow. Efflorescence is also present

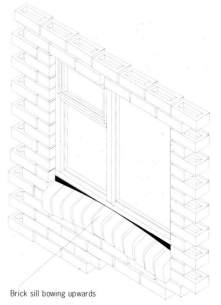

Brick sill bowing upwards

Figure 2.33
With sulfate attack in the outer leaf, a brick sill may bow upwards

Sulfate attack in the walls of a church

The BRE Advisory Service was asked to investigate the condition of the brickwork at a church situated near the coast. The external leaf of brickwork had bowed on the panels that faced west. One panel was checked with a plumb line and it was shown that the top of wall just below the parapet had moved backwards 34 mm from the vertical. At between 2 m and 2.4 m the brickwork had moved outwards 20 mm from the reference point at DPC level. Examination of another panel indicated a slight vertical bowing of the wall (5 mm on a 1.5 m height).

Samples of the mortar were drilled from the brickwork panel at one course above the DPC, then at 1 m intervals to the top of the wall. At each position the samples were collected both from the outer 50 mm and the remaining inner mortar. The outer drillings were collected as a reasonably dry powder, but further into the mortar it became very wet and was collected as a paste.

The sulfate content for mortar samples taken from the bowed panel was in excess of that normally present in a mortar mix, but other samples were not abnormal. Differential thermal analysis showed that there were significant amounts of ettringite present in the samples taken from the bowed panel.

The bowing of the brickwork panels could be explained by sulfate attack on mortar joints. The bricks were an ordinary quality facing brick with a sulfate content of 0.63%. It was possible that the bricks could have higher than recommended potassium content which made the brickwork more prone to sulfate attack. The brickwork was afforded little protection from the rain. The parapets did not provide sufficient overhang, and the brickwork would remain damp for long periods. Therefore, the three conditions for sulfate attack were present.

The bricks were not the correct choice of brick for the design and exposure of the church. The spalling of the face of the bricks could have been the result of frost damage, and there were examples of the faces of some bricks being forced off by ice. The bricks would appear to have been rather porous and readily absorbed moisture.

When replacing the brickwork, it would be important that a brick be chosen that would be low in sulfate content and had the required durability for the known exposure conditions and that the parapets be properly detailed; that is they would be protected by a coping with a good overhang, a damp proof course under the coping, and a cavity tray.

Crypto-efflorescence

Crypto-efflorescence was investigated by BRE in some sheltered brickwork walls in several blocks of flats. The problem had arisen in these areas because only a restricted amount of rainwater fell on the bricks which carried the dissolved magnesium sulfate salts present in the bricks towards the surface, drying relatively quickly through evaporation. The salts crystallising below the surface led to delamination and spalling of the outer skin of brickwork 2 mm thick.

Two main types of brick were used: one was an engineering brick and the other a much lighter coloured wire cut, extruded facing brick. It was the pale coloured facing bricks which had deteriorated in places as a result of the crypto-efflorescence. The problem had been ongoing for a number of years but it was unlikely that any reduction in structural strength had occurred.

A sample of failed brick was examined in the laboratory, using both optical and scanning electron microscopy. Most of the brick was composed of amorphous material containing elastic quartz grains. No conclusive evidence as to the origin of the salt could be determined. However the brick did possess a number of pores which appear to contain a magnesium-bearing white precipitate. The outer 2 mm of brick comprised a slightly more impermeable higher glass content layer which is normally found in extruded bricks of this type.

The salts responsible for the crypto-efflorescence problem were found to be the magnesium sulfate salts $MgSO_4.6H_2O$ and $MgSO_4.7H_2O$ which originate from the raw clay used for the brick manufacture. It was unlikely that the bedding mortar was involved in the process.

In a section of brick 100 mm thick, it takes only a background mobile sulfate level of 0.2% to build up and concentrate in a 1 mm thick crypto-efflorescence layer containing 20% sulfate. There was therefore plenty of potential for further problems to arise and although this strongly depended on the solubility of the sulfate compounds present within the brick interior, it was highly probable that further crypto-efflorescence would occur in the vulnerable areas if the problem was left untreated.

The possibilities for repair included:
● rendering the brickwork, using a metal lathing to support the render finish
● removing and replacing the outer skin of the affected brickwork

is required it may be available via an Agrément certificate.

Cast stone may be subject to crazing. As *Principles of modern building*[6] remarked, crazing of concrete products is a very variable phenomenon, varying in scale from wide cracks 100 mm apart to those which can be seen only under a microscope. The main cause of this crazing is differential moisture movement of the surface layer with that immediately underneath. The coarser mixes tend to craze less than the finer, and trowelling to a smooth surface tends to promote crazing.

If the published recommendations for external rendered finishes are complied with there need be no serious cracking of blockwork or its finishes, including rendering (see also Chapter 9.2). Shrinkage cracking is the main risk. The following aspects are particular important.
● Specification of mortar mixes: lean mortars deform more readily than rich ones and allow individual blocks to move within the mortar joint system without serious disruption
● Provision of vertical movement joints: simple butt joints formed during construction, or cut with a power saw within a few days after completion, at approximately 6 m centres and filled with a resilient sealant
● Provision of reinforcement in areas of high stress: galvanised or stainless steel bed-joint reinforcement located in mortar joints at window and door openings

Efflorescence and crypto-efflorescence

Clay bricks may contain water-soluble sulfate salts derived from the constituent materials produced during the firing process. Additionally, but to a lesser extent, sulfates can originate either in the mortar, or from atmospheric pollution or the ground. (Concrete bricks and blocks, and calcium silicate bricks are normally free from significant amounts of sulfates).

In a brickwork wall, it is possible for rainwater to mobilise these salts and transport them via capillaries towards the external surface of the wall (Figure 2.34). When evaporation takes place from these external surfaces, the salt concentration increases to oversaturation and the salts crystallise either at the surface (efflorescence) or below the surface (crypto-efflorescence), depending on the water supply. Efflorescence occurs when there is a plentiful water supply, but if the water supply is limited, the liquid front retreats inwards, resulting in crypto-efflorescence. Both processes can take place simultaneously or independently. Whereas efflorescence is merely unsightly, crypto-efflorescence causes delamination, spalling and cracking of the external brickwork surface as a result of the expansive crystallization of soluble salts within the brick pore structure.

Crypto-efflorescence is not unknown in the UK but our wet climate usually ensures that the salts come right to the surface where they form the less harmful efflorescence. Efflorescence is essentially a temporary condition. Once the salts have been brought to the surface and removed, it is unlikely to recur.

Crypto-efflorescence can occur if old bricks are re-used as they may be underfired 'place' bricks formerly only used in the protected internal walls of a building. If a brick is underfired, any unwanted sulfates present in the original clay will not have been driven off but will remain within the fired brick. If that brick also has a very impervious vitreous coating which will prevent the salts reaching the surface it will be even more susceptible to attack.

There is no chemical treatment for efflorescence which can be recommended. Dry brushing, which is labour intensive, can remove some of the material from masonry with a sound surface, but most deposits gradually reduce as wind and driving rain take their effect over a period of years.

Case study

Performance of concrete facing bricks

A group of factories were visited to investigate defects and to recommend appropriate remedial measures. The site had a driving rain index in excess of 7 – a 'severe' exposure grading.

The factories were single storey buildings, steel framed with gabled pitched roofs; the upper parts of the walls were faced externally with PVC coated steel sheeting, lower parts with concrete facing brickwork outer leaves to cavity walls. The inner leaves were of sandlime common bricks and the cavities contained 50 mm of thermal insulation.

Condition of the walling

Cracking, characteristic of drying shrinkage of the bricks and mortar, was observed in the concrete brickwork of the main areas of walling. This cracking was mainly confined to the mortar joints and had affected the full height of external brickwork but did not pass below DPCs near ground level. The cracks were no greater in incidence and severity than would be expected from normal shrinkage of the materials concerned. Their presence could be due to the omission of movement joints and restraint imposed by wrapping the brickwork closely around corner stanchions.

The condition of some of the bricks was poor, however, with flaking and erosion of brick faces in certain parts of the walls. A sample of some of the loose material was taken from one of the bricks. On chemical analysis this showed a Portland cement content of 2.32 % by weight (ie a cement:aggregate ratio of 1:42 by weight, which was equivalent to about 1:36 by volume). It was concluded that the bricks which had eroded were of defective manufacture.

Remedial work

The cracking that occurred to the main area of the building, although readily visible, was not conspicuous. While it might have been considered unsightly it was unlikely to have any real effect on any of the functions of the walls. Any attempt to repair the brickwork by raking and pointing cracks and, perhaps replacing cracked bricks by good ones would probably not be satisfactory. Incorporating a movement joint in the affected section might have been more successful, but, despite this, a crack could open between old and new walling rather than on the line of the new movement joint.

The inevitable blemish resulting from repair might well be more noticeable than the defect. It was recommended that fine cracks not visible at a distance of more than a few metres should be left alone.

Figure 2.34
Severe efflorescence is present on this stretcher bond brickwork

The mundic problem

Many domestic and small commercial properties in Cornwall and south Devon have been built during the twentieth century with concrete blocks using local sand and coarse aggregate available as waste materials from the region's metalliferous mining industry. Some aggregates have proved to be very unstable in service, sometimes necessitating demolition of properties. This is known locally as 'the mundic problem', mundic being the old Cornish–Celtic word for pyrite, which is abundant in many mining wastes and is central to the degradation process. Further information is available on the mundic problem[107].

Wall ties

Wall ties play a key part in ensuring the structural integrity of a wide range of masonry and masonry-clad buildings. Ineffectual bonding due either to using the wrong type of tie, insufficiently long ties or workmanship faults such as the widening of cavities beyond specification is a frequent problem in a wide range of building types. This has often resulted in the need to carry out very expensive remedial work and, in rare cases, collapses. The trend towards wider cavities to accommodate insulation has increased the risk of insufficiently long ties being specified or used. Further information is available[108].

Most reported wall tie failures have involved:
● inferior protective coating or none (Figure 2.21)
● inadequate thickness of galvanising
● aggressive (especially black-ash) mortars
● permeable (especially lime) mortars
● exposure to severe (eg marine) environments

However, the problem of wall tie corrosion could eventually affect virtually all cavity walled structures built with galvanised steel ties. Those built before 1981 are likely to fail

earlier because the standard specified a thinner layer of galvanizing than post-1981[†]. Furthermore, wall tie corrosion is not confined to cases of poorly-made ties, nor to aggressive mortars, nor to conditions of extreme exposure – it is just more rapid in those circumstances.

Austenitic stainless steel and duplex coated, zinc plus epoxy and zinc plus PVC coated mild steel ties in general suffer little deterioration. Copper alloy (bronze) ties may suffer minor dezincification which is indicated by a salmon pink colour on the surface of the tie. The only other material in wide use for ties is polypropylene (plastics) which is unlikely to suffer chemical deterioration provided light is excluded but might be subject to vermin attack.

BS 5628-3 requires the use of corrosion resistant materials for fixings (including wall ties) above three storeys. The specification of an appropriate austenitic stainless steel for wall ties effectively removes any risk of their corrosion. Site practices, other than the substitution of inferior wall ties for those specified, will then have little if any bearing on the risk of tie corrosion.

Diagnosis

Structures with galvanised ties are potentially subject to wall tie corrosion[109]. Cavity walls most at risk are those:
● built in black-ash mortars
● exposed to severe weather, especially in marine or industrial environments
● built between 1900 and 1940
● built with vertical twist ties during the shortages which followed the 1939–45 war, or during building booms, especially in the early 1970s
● built with galvanised vertical twist ties where the protective coating has exceeded its predicted life. This depends on its thickness, but in most cases will be of 35 years

● built with galvanised wire ties where the protective coating has exceeded its predicted life. This depends on its thickness, but in most cases will be of 20 years
● built with galvanised ties supporting the outer leaf of brick-clad timber framed construction where the protective coating has exceeded its predicted life. This depends on its thickness, but in most cases will be of 15 years

If the first two conditions listed above are combined with any of the remainder, surveyors should be particularly vigilant.

The surveyor's first need, of course, is to be certain that the wall concerned is of cavity construction. Next, the outer leaf should be examined for regular horizontal cracks at about 300–450 mm vertical spacing; they are usually more evident in the upper parts of the wall and likely to be more clearly delineated on a rendered wall. The surveyor should be aware of the possibility that cracks have been repointed or the wall re-rendered.

The cracks are distinguishable from the otherwise very similar cracks caused by sulfate attack on brickwork or rendered brickwork since they occur at vertical wall tie spacing intervals rather than in every bed joint. Corresponding cracks in the inner leaf are rarely found, but some may occur where cross walls are bonded to cavity walls. They may also be found in the inner leaves of walls enclosing unheated spaces.

If cracks indicative of wall tie corrosion are found in the external leaf, it is virtually certain that the ties are of the thicker vertical twist variety since wire ties have too little bulk to generate a significant volume of corrosion product unless the mortar joints are very thin. If wall tie corrosion is suspected, it is comparatively simple to locate a tie and to remove a brick, at that location, from the outer leaf, to allow direct inspection of the tie. A decision must be taken on the size of the sample of tie locations to be inspected. Guidance on sampling is contained in BRE Digest 401[110].

† It was announced in the House of Commons in January 1998 that BRE had identified a risk of accelerated corrosion in galvanised wire ties coated with a green film, when surrounded by moist urea-formaldehyde foam.

Metal detectors are useful to locate tie positions.

Alternatively an optical probe can be inserted through strategically drilled holes. However, it should be noted that corrosion occurs mainly on the part of the tie that is bedded in the outer leaf, particularly that part within the bed joint and close to the cavity face. If inspection by optical probe does not reveal corrosion of the part of the ties spanning the cavity, this should not be taken as firm evidence without some sampling and direct examination of parts bedded in the outer leaf. Other, generally more expensive, inspection techniques are described in BRE Digest 329[111].

In all cases of wall tie corrosion, consideration should be given to the possibility that the accompanying expansion has distorted the wall to an unacceptable extent or transferred to the outer leaf loads (eg roof loads) intended to be carried by the inner leaf. Surveyors should also be alert to the possibility that cavity ties are used where there is no cavity. For example, in some cross-wall housing, separating walls are projected beyond the face of the building and their ends cloaked by a half-brick skin carried up to the full height; stability of the brick skin, in these situations, can depend wholly on the integrity of the ties.

Location of ties
The most common technique is the use of a metal detector. These are specialised devices optimised to find metal within a range of about 100 mm – 'treasure locators' are not suitable. Also stainless steel demands specialised devices. Infrared thermography can be used for detecting wall ties, and the method is a smaller scale version of that used on whole buildings to detect heat losses. It is the action of ties as thermal bridges that is used in this technique. The test is totally non-destructive and theoretically can be used on any form of cavity wall construction that uses metal wall ties.

Replacement
Techniques have been developed for the reinstatement of cavity walls by insertion of new wall ties without recourse to demolition.

The range of techniques currently available are described in BRE Digest 329 which deals with:
● investigative test techniques
● the classification of the main problem (eg thick metal ties or wire ties)
● choosing appropriate remedial actions to deal with any existing structural deterioration and to prevent further deterioration (eg remove old ties or structurally isolate ties)
● choosing remedial tying systems appropriate to the structure, fire requirements, wall materials etc (eg public buildings require longer periods of fire resistance than domestic dwellings)

The above reference is supported by BRE Digest 401 which covers:
● surveying the building to establish the wall type and the density and layout of existing ties
● specifying tie sampling methods and rates for some typical situations
● classifying old ties by visual methods and by quantitative measurement of remaining zinc coatings; estimating remaining life and recommending remedial work in the short or long term
● the density and layout of the remedial ties; whether to specify by prescription in accordance with the Building Regulations and British Standards Codes of practice or by calculation
● the choice and validation of a suitable remedial tie system to ensure that the system can give an effective fixing in both inner and outer leaves of masonry
● sampling techniques and rates for the pre-contract testing of the chosen system
● quality control checks during installation by operatives, supervisors and independent professionals to ensure that the work is being carried out to specification. They include visual

checks, checks on the removal of debris and old ties, random proof tests of inner leaf connections, torque testing expanders and examination using optical probes and metal detectors
● final visual acceptance checks on the quality of repointing and making good, and random examination to ensure that the overall quality is satisfactory
● quality assurance documentation to verify the design basis of the installation

Using modern remedial tie technology, a complete cure of the problem can be effected. The cost is not excessive – from a few hundred pounds for a small centre terrace dwelling to perhaps £3,000 (at 1998 prices) for a large detached dwelling.

Maintenance
See the same section in Chapter 2.1.

Work on site

Storage and handling of materials
On arrival on site, all materials, especially lightweight blocks, should be stacked and protected from rain and snow so that they are kept as dry as practicable both before and during laying.

This recommendation is clearly important wherever blocks will dry out more or less completely (eg in inner leaves of external walls and in internal partitions) but research suggests that it may be less relevant to cracking in the wetter external leaf situations. Doubts have arisen because instances of minor cracking have frequently been observed in external leaves of lightweight blockwork in the early life of the buildings, occasionally even before the rendering is applied. Typically the cracks are fine, sometimes little more than micro-cracks though possibly as wide as 1–2 mm, and run vertically through blocks and perpends, extending through the full thickness of the blocks including any surface finish. On the coarser finishes they can only be detected by close inspection.

Workmanship

Workmanship is particularly important in construction of the outer leaf of a cavity wall, so as to reduce the quantity of water passing into the cavity, and in keeping the cavity clear of obstructions and other faults which can lead water across. If cavity insulation is to be used, good workmanship is also important for installation of the insulant, either during construction (for built-in board or batt materials) or after construction (for blown-in or injected fills)[112].

Inspection

All external walls should be inspected in detail and cracks noted. If these are more than a hairline, crack widths will need to be recorded and long term monitoring considered (Figure 2.35). If visual evidence of distress is found then a full plumbline survey may need to be undertaken. Also, walls might have been built 'out of true'. A metal detector may be used to locate wall ties if inadequate numbers or incorrect distribution are suspected.

Any potentially overstressed narrow piers should be identified, and the wall above each opening examined both internally and externally for signs of stress. Lintels may need to be exposed to establish material, shape, size, condition and width.

See also the same section in Chapter 2.1.

In addition to those listed in Chapter 2.1, the problems to look for are:

◊ inadequately protected or missing ties
◊ ties sloping the wrong way
◊ inadequately embedded of ties
◊ cracking and instability due to corrosion or absence of wall ties
◊ mortar on ties
◊ reduced weathertightness due to induced cracking
◊ damp penetrating to the inner leaf – cavity trays omitted or poorly detailed
◊ stop ends omitted
◊ rotation of concrete boot lintels
◊ weep holes absent or blocked

Figure 2.35
Cracking due to proximity of trees on a shrinkable clay site

◊ cavities too narrow
◊ cavities fill bridged
◊ cavity fill deteriorated
◊ thermal bridging and condensation
◊ corrosion of inadequately protected or damaged steel lintels
◊ presence of corroding hoop irons

Chapter 2.3

Stone rubble and ashlar

This chapter deals with natural stone as quarried, but used in thicknesses of more than 100 mm as a loadbearing element. When in thinner sections as non-loadbearing and carried on a separate frame it is dealt with in Chapter 3.1. Cast or reconstructed stone is a form of concrete, albeit made with carefully selected aggregates to match natural stones, and in most respects can be treated as a type of concrete. See Chapters 2.4 and 2.5.

Where stone is quarried locally, and there are several centuries of tradition in its use, there is no doubt that natural stone can form a durable and attractive external wall, weathering well over many years (Figure 2.36). However, there are certain basic points to bear in mind when repairing or maintaining stonework. This book, however, cannot possibly deal with detailed conservation issues, and reference should be made to *The weathering of natural building stones* [113].

During the period 1991–94 it was reported that about 1 in 12 of new dwellings had external walls which used natural or reconstructed stone in the external leaf.

Characteristic details

Basic structure
Rubble

Stone rubble walls can be found in many different forms, varying mainly on the amount of dressing given to the stones and whether or not they are coursed. Uncoursed stone as cut from the quarry used often to be laid selectively, either, depending on its nature, without any further dressing or, at best, a blow or two from the mason to get a closer fit. This is commonly termed common or random rubble. The quoins are often partly dressed. Where the wall is of a thickness greater than that of twice the average width of the stones, bonding stones, sometimes called through stones, are laid for the full width of the wall to prevent the wythes parting. In many areas of the country, mortar was simply clay or earth gathered from the site, though better class work will be found laid in lime mortars. This kind of wall has strong similarities with dry stone walling, except that to enhance weathertightness it was normally rendered or harled. It will not be found built much above one and a half storeys in height when built with clay mortars.

Where the rubble wall was intended to be left unfaced, that is to say with the stone exposed, the degree of dressing of the stones was often increased, and the stones laid to courses. Terms used, which reflect the degree of working and the sizes of stones incorporated, include uncoursed random rubble, irregular course or snecked random rubble, and coursed rubble (Figure 2.37 on page 104). Walls built to courses are much more consistent and stronger than those which are not, and are more amenable to calculations for strength and stability.

Figure 2.36
This Cotswold village has a remarkable consistency imparted by the limestone from which the majority of its buildings are constructed

Ashlar

The most highly worked stone, fully rectangular and coursed, is termed ashlar (Figure 2.38). Just how much stone is dressed in ashlar walls of course depends on the thickness of the wall; solid stone walls often consist of two wythes filled with rubble or undressed stone (Figure 2.39). Frequently too, stone walls were backed by brick masonry, sometimes with a waterproofing of painted bitumen incorporated between the two materials.

Mortars

Wherever possible, the fine aggregate for masons' mortars ought to consist of crushed stone of the same kind as the rest of the wall. This will ensure a reasonable match. The mix might best consist of 12 parts of finely crushed stone, to 3 parts of hydrated lime (or lime putty if available), to 1 part of Portland cement, though some mortars have successfully used less fine aggregate and more cement. In essence, the mortar should be less strong than the stone, so that it is sacrificial.

Figure 2.39
The wythes have parted in this solid stone wall, due mainly to the absence of through stones. Part of it has already collapsed, together with the hearting

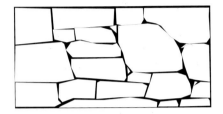

Random, or uncoursed, rubble

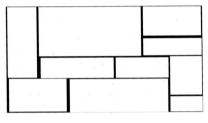

Irregular coursed, snecked or square random rubble

Coursed rubble

Figure 2.37
Common rubble walling (after Davey)

Main performance requirements and defects

Choice of materials for structure

Stone used for building purposes is mainly from sedimentary rocks, for example the limestones and sandstones. These can have pronounced bedding planes which originate from the way they were laid down, and these planes place a limit on the quarried sizes of useful stone blocks. Some building stones are from metamorphic or igneous formations, such as the granites, and do not have these bedding planes – size tends to be restricted mainly by handling and positioning criteria.

Stones may need to be accurately identified for replacement purposes, and specialist advice and access to reference collections may be needed. In choosing stone for replacement purposes, even if the original stone has been accurately identified, it is important to verify that the replacements are of acceptable standard since quality can vary, even within the confines of a small quarry[114].

Flint does not fit neatly into any formal classification of natural stone. Even in facing work, in Norfolk for example, it is sometimes found used straight from the chalk bed with thick mortar joints, but in better work, especially from the mid-sixteenth century onwards, is knapped into rough cubes and either roughly coursed or laid in a species of stack bond or in chequer pattern (Figure 2.40).

Figure 2.38
Part of Prior Park College, built by Ralph Allen in Combe Down stone

Figure 2.40
Coursed knapped flint with eroding pointing

Strength and stability

Assessments of the compressive strengths of building stones have rarely been needed in the past, and even the weakest stone can be expected to withstand the loads imposed upon it in normal use, although stress concentrations or accidents can cause failure of even the strongest (Figure 2.41).

One of the most vulnerable parts of a stone wall, both to weathering and to displacement, is the sloping coping on a gable wall. These copings for the most part rest on kneelers on the slope or footstones at the eaves, and are bedded in mortar up the slope. Those seen on site have mostly been in good condition, though reports have been received by BRE of attempts to emulate the detail, without the use of kneelers or footstones, where the copings have slid down the slope as a result.

Dimensional stability, deflections etc

For coefficients of linear thermal and moisture expansion etc, see Chapter 1.2.

Weathertightness, dampness and condensation

The mechanism of rain penetration through stone is similar in many ways to that of brick, particularly where the stones are absorbent, and relatively small in size. The softer stones, such as some of the

limestones, will absorb rainwater, to allow release by evaporation when conditions allow. The harder and less porous stones, such as granite, although relatively impervious as a material, tend to allow rainwater through imperfect joints. Flint depends almost entirely on the quality of the mortar joint.

Many ashlar faced walls have backings of brick with a coating of bitumen intended originally to reduce the likelihood of migration of soluble salts. This layer will help to reduce the passage of rainwater from the face of the stone only if it remains sound.

Thermal properties

The thermal insulation properties of a wall built in natural stone will differ very little from a wall built of brick or block of similar thicknesses and densities.

See also the same sections in Chapters 2.1 and 2.2.

Fire

See Chapter 1.8 and the same section in Chapter 2.1

Noise and sound insulation

See Chapter 1.9 and the same section in Chapter 2.1.

Durability

The UK has many examples of buildings built at various times throughout the last thousand years, which bear remarkable testimony to the longevity of stone as a building material, even if some of it does need attention from time to time (Figure 2.42 on page 106).

However, different natural stones have quite different modes of weathering, depending on their structure and chemical composition. Igneous rocks such as granite tend to endure extremely well, even retaining polish over many years. The siliceous sandstones, such as West Yorkshire and Darley Dale, are not attacked by water and atmospheric gases, and tend to be more durable than the calcareous sandstones such as Mansfield. The poorest sandstones from the point of view of durability are of the

Dampness in a heritage building
A historic stone built tower, part of the ancient fortifications for a south coast port, had been derelict for many years until the owners decided to put a roof over the building and convert it into a museum. Following its opening, the internal surfaces of the stone walls started spalling.

Monitoring was carried out by BRE to establish the internal conditions of temperature and relative humidity over a sixteen month period. Moisture content values were established for the walls and little evidence of rising damp was found. In order to check the influence of salts, hygroscopicity tests were carried out. Chemical analysis confirmed the suspicion that there were considerable quantities of chlorides in the stone, not surprising as the building was part of coastal defences.

The recorded relative humidities ranged from less than 30 to 76% with long periods in excess of 60%. The building owner was advised that if the relative humidity could be kept at around 50% or less and never be permitted to rise above 60%, then moisture pick up would be low with a consequent reduction in damage to the stone.

Figure 2.41
A gas explosion in this five storey Glasgow tenement has revealed the slender construction of the external stone wall

Figure 2.42
The Banqueting Room in the Palace of Whitehall, designed by Inigo Jones and built in 1619–21

where the wall is sheltered. The problems tend to be worse where the stone is bedded incorrectly, and the bedding planes run vertically instead of horizontally (Figure 2.44).

Fixings such as dowels and cramps have sometimes in the past been of ferrous metal (eg wrought iron), occasionally protected by bitumen. These fixings will corrode in time and should not be specified in replacement or new work. Slate dowels have been widely used in the past, but stainless steel or non-ferrous fixings are preferred.

Stones may also change colour on weathering, and newly inserted replacement pieces may take time to weather down to blend with the old.

Figure 2.43
This small country church in Worcestershire has no DPC, and centuries of weathering, rising damp and salt crystallisation have caused erosion, especially of the lower courses of sandstone. Some replacement has taken place

argillaceous kind with clay content which erode severely when wetted.

Limestones such as Bath, Clipsham, Portland and Ancaster which are cemented with calcium carbonate, and Magnesian limestones which contain magnesium carbonate in addition to calcium carbonate, are all susceptible to sulfur acids in the atmosphere.

Rainwater run-off from limestone can carry calcium salts which stain brickwork or other stones. If the run-off is onto sandstone, disruption of the sandstone may occur. If it happens, it is almost impossible to remedy, short of extensive rebuilding, and should have been provided for by the original design.

Some stones in the past have suffered decay from crystallisation of salts contained in the bedding mortars. A further source of salts is the ground from where they migrate upwards within a wall if there is a defective or non-existent DPC (Figure 2.43).

In the great majority of cases the deterioration is due to a build up of salts immediately below the surface of the stone – crypto-efflorescence – which forms a hard skin which then flakes off the surface. The accumulation of salts tends to be less where the surface is freely washed by rain, and is nearly always worst

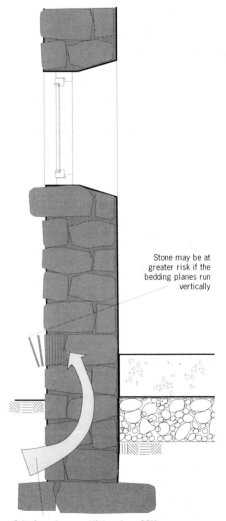

Stone may be at greater risk if the bedding planes run vertically

Salts from the ground if there is no DPM

Figure 2.44
Incorrectly laid stone; the bedding planes should be horizontal in this situation

Maintenance
The following are the basic principles of repair.

Plastic repair
This type of repair should be employed only for relatively unimportant areas which have decayed. It is essential that the area to be repaired is properly prepared and that the mortar mix is matched as closely as possible to the appearance of the original stone. Any decayed stone should be cut away and the edges of the are to be repaired should be undercut to provide a dovetail key. Feather edges to repairs should never be used. For larger areas a mechanical key in the form of ragged non-ferrous dowels or non-ferrous screws firmly fixed to the background should be employed.

Plastic repair should never continue across mortar joints. Rather, each block of stone should be treated independently from its neighbours and the repaired block pointed afterwards in the normal manner when the repairs have cured.

Mortar should be mixed to match the colour and texture of the stone being repaired and should never be stronger than the stone. Two layers of mortar will normally be used.

Repairs using new stone: 'piecing-in'
Repairs by this method can be quite small – down to 25 mm square on the face. The thickness of the repair will depend on the method of fixing and the size of the repair, but will not normally be less than 25 mm. Large pieces of stone can be fixed by normal masonry techniques using cramps and ties where appropriate. Smaller pieces can be grouted in, or fixed with nonferrous dowels and resin and grouted after the resin has set.

Repairing a cracked lintel
If a stone lintel has cracked across at an angle, and has dropped, it may be possible to jack the soffit to level, and to drill normal to the crack to insert stainless steel pins cemented in with a proprietary epoxy grout. This should prevent further slippage. Drilling from the back is preferred if access is available, otherwise holes drilled in the face will need pointing with a mortar made of crushed stone to match the original.

Preservation
It may be possible to treat certain stones to reduce the rate of erosion. However, specialist advice should always be sought.

Cleaning
Limestones may be cleaned by the use of copious amounts of water to which nothing should be added. Detergents are definitely to be avoided, as are caustic or peroxide solutions. Gentle brushing may be needed once the surface deposits have been loosened by the spray. Washing is likely to be less effective on siliceous sandstones.

Further information on the decay and conservation of stone masonry is available in BRE Digest 177[115].

Case study

Failed stonework in an office block
BRE was asked to inspect the stonework to an office building converted from a large mansion and to propose methods for its restoration. The stones which were to be repaired or replaced could be divided into categories:
- cracked but both pieces were still in-situ
- cracked and fallen off
- repaired with mortar
- not previously repaired, and showing varying degrees of decay

Stones in the first category were mainly lintels over window openings. It was proposed by the building owner's architects that, for those lintels with one central crack, the central part was to be cut out to allow a keystone to be inset. This would give the lintels a satisfactory visual appearance but would be unlikely to further prevent the movement that initially caused the crack. If there was a tendency for further movement, this would be accommodated at the centre of the lintel by cracks forming in the joints either side of the new keystone, but it would not prevent cracking occurring elsewhere in the lintel.

As to the remainder of the cases, whether or not a block should be completely replaced would be a matter for judgement. For minor defects it would probably be more economical to repair the block than to replace it. However where the amount of work required on a single block would be considerable, repair would no longer be the most economic option.

Mortar repairs to the upper surfaces of stones are more likely to fail in the long term than repairs to vertical faces, unless protected from the weather. Protection should prevent water getting to the repairs – best achieved with a covering of lead; the application of bitumen or silicone water repellents was not recommended.

Work on site

Workmanship
For those stones with bedding planes, probably the most important consideration is that the correct attitude of the stone in the wall is adopted. Stones should be laid as nearly as possible in the same attitude as they were originally formed, though bedding in tracery should be at right angles to the line of thrust.

See also the same section in Chapter 4.2.

Inspection

The problems to look for are:
◊ narrow piers or returns overstressed
◊ spalling, contour scaling, cracking, oversailing, bulging or leaning
◊ dampness and algal growth
◊ inadequate or blocked air bricks
◊ pointing erosion
◊ lintels deteriorating or cracking
◊ DPCs missing, damaged or deteriorated
◊ sulfate attack migrating from backing masonry
◊ inappropriate repairs
◊ face bedding of stone

Chapter 2.4   **Precast concrete**

This chapter deals with all kinds of precast concrete, including the large precast concrete panel system-built dwellings, flats and houses of the mid- twentieth century (Figure 2.45). It is reported that in England, around 284,000, or just over 1 in 100 of the total stock, have concrete panels as their predominant wall structure. As might be expected, none has been recorded as being built before 1919[2]. It is not possible in the EHCS to distinguish between the systems using large loadbearing panels and the smaller concrete panels carried on frames, to be described in Chapter 3.2.

Both low and high-rise construction is characterised by the great variety of external modelling to façades, and of the variety of finishes supplied to the precast panels. It is therefore not always easy to identify a particular large panel system (LPS) either by the modelling or by the surface finishes used.

Over 30 high-rise concrete systems have been designed of which the most widely used were Bison, Camus, Crudens, Jesperson, Reema, Skarne, Tracoba, Taylor Woodrow Anglian, and Wates. Some of the examples have by now been demolished, largely for social reasons, while others have been rehabilitated. For other systems, see the list in the full CRC catalogue[11].

To take one typical example, in the Reema Conclad system the external walls were of 7 inches thick precast concrete panels, with windows and door frames cast-in at the works[116] (Figure 2.46). Reema also built many low rise dwellings with small precast concrete cladding panels and these are dealt with in Chapter 3.2.

Several high-rise systems of large panel loadbearing construction were used also to construct low-rise dwellings; the most numerous included Balency , Bison, Bryant, Camus and Skarne. These systems normally have no capping at separating walls, and external corners formed by providing a surface finish on the returned edge of the façade panel.

Cast iron, low-rise, loadbearing panels

It is convenient to briefly mention cast iron panels here. The only housing system to have used cast iron panels was the Thornecliffe

Figure 2.45
A sixteen storey concrete panel block of flats

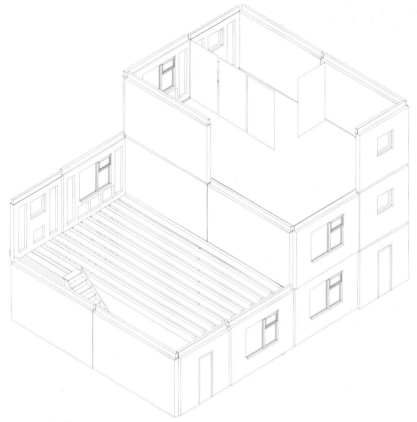

Figure 2.46
Schematic of the Reema Conclad system

(Figure 2.47), mostly in the Derby area, with a few in Atherstone, now demolished. Although the panels were smaller than most reinforced concrete ones, there are certain similarities in the way they are assembled, and, since they were rendered, they tended to look like reinforced concrete. Apart from losing their render, and decay of packing between some panels, they proved remarkably durable, though their thermal insulation was negligible.

Characteristic details

Basic structure

- Most of the large concrete panel systems were precast off site, but considerable numbers were also produced on site, mainly using the battery casting method where panels were cast vertically in multiples with subsequent panels being cast between the spaced apart earlier panels. A variety of finishes could be manufactured, mainly of the exposed aggregate or ribbed kinds, although

those panels cast horizontally could also be finished with tiles. L-shaped panels may also be found (Figure 2.48).

The structures obtained their stability from a combination of panels supporting each other, with floor slabs tying them together over comparatively short spans. Walls were commonly adjusted to level by means of threaded bolts set into the tops of the lower panels and the open joints packed with a dry mortar mix to achieve adequate sound and fire separation (Figure 2.49 on page 110).

Openings and joints

Windows were often cast into the concrete panels, so replacement is sometimes not straightforward.

Four basic kinds of joints were adopted in large panel construction (Figure 2.50 on page 110):
- Open drained joints (two stage, baffle and air seal)
- Face sealed joints (one stage, gasket or mastic)
- Cover strip joints (either one or two stage)
- Traditional (often mortar) joints

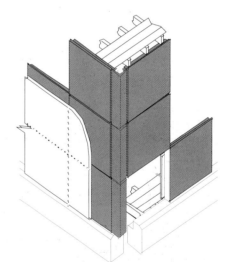

Figure 2.47
The Thornecliffe system

Figure 2.48
L shaped concrete panels were used in this development in South Wales

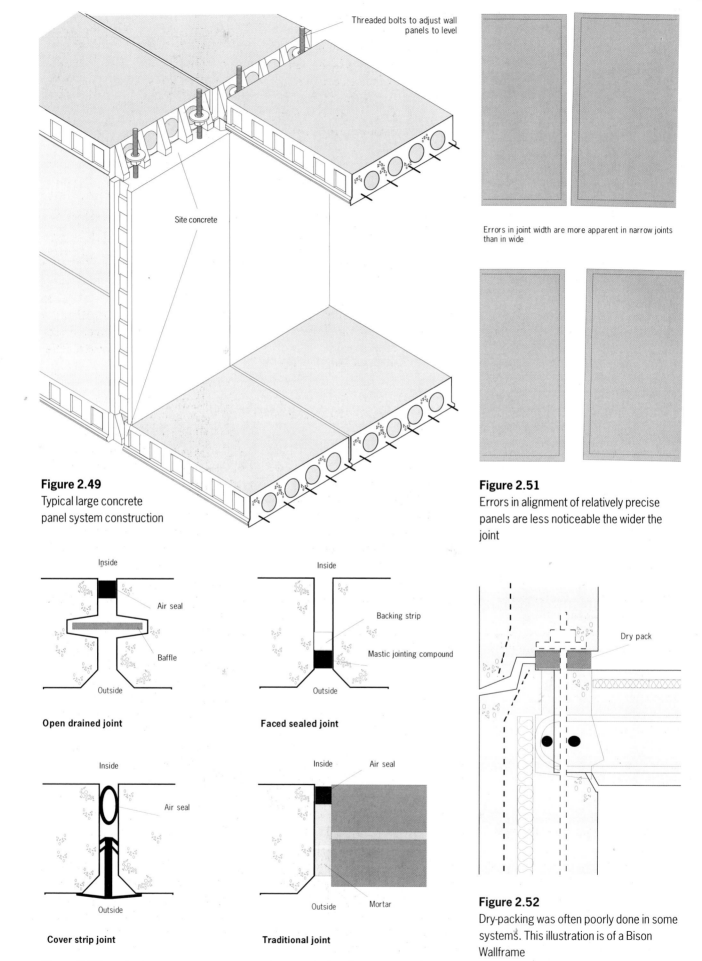

Threaded bolts to adjust wall panels to level

Site concrete

Figure 2.49
Typical large concrete panel system construction

Errors in joint width are more apparent in narrow joints than in wide

Figure 2.51
Errors in alignment of relatively precise panels are less noticeable the wider the joint

Inside

Air seal

Baffle

Outside

Open drained joint

Inside

Backing strip

Mastic jointing compound

Outside

Faced sealed joint

Inside

Air seal

Outside

Cover strip joint

Inside   Air seal

Outside   Mortar

Traditional joint

Figure 2.50
Four types of joints in large panel construction

Dry pack

Figure 2.52
Dry-packing was often poorly done in some systems. This illustration is of a Bison Wallframe

Of these the most popular initially was the open drained joint, with the face sealed second. Some open drained joints, though their profiles were unsuitable, were converted into single stage joints by attempts to seal the outer face, sometimes with doubtful results. The problem usually lay in the deterioration of the air seal at the inside of the joint rather than deterioration of the baffle.

The joints between the panels will have a considerable effect on the appearance of completed elevations, and designs should not have been finalised without taking the appearance of the joint into account. The wider the joint, other things being equal, the less noticeable will be errors in joint width (Figure 2.51).

Main performance requirements and defects

Choice of materials for structure
Perhaps the most serious problem with the reinforced concrete used in these systems was the use of accelerators which could have an effect on long term durability. Also of concern were the methods of providing connections (between the panels and slabs) which were sometimes inadequate. An extensive programme of modification had to be introduced following the gas explosion and partial destruction of a block of flats, Ronan Point, in 1968.

Strength and stability
Precast concrete components are almost invariably stronger than specified due to the need to obtain high mould optimisation. This sometimes results in difficulty when making fixings into the material.

Of the low-rise LPS systems, some, particularly Bryants and Bison Wallframe, had poor dry-packing (Figure 2.52).

With respect to high-rise construction, no major failure of an LPS building has been reported since the programme of strengthening such structures was carried out following Ronan Point. However, BRE recommends that all blocks required to serve longer than 25

years from date of construction should have a full appraisal for structural safety and durability, with inspections periodically thereafter.

Corrosion of reinforcement is unlikely to present a risk to the stability of LPS buildings unless it becomes substantial (eg following rain penetration).

Dimensional stability, deflections etc
All buildings move to a greater or lesser extent in service. Large panel buildings are no exception, and this movement needs to be taken into account when designing any overcladding. The original uncovered concrete panels, if they were light coloured, would have been expected to move approximately 3 mm total range per 3 m of panel height or width. If they were dark coloured, the movements would have been slightly greater, of the order of 4 mm.

When concrete panels are insulated and overclad, movements will be significantly reduced, but it will still be possible to estimate the expected range of movement given the method of overcladding proposed. Depending on the method of fixing the original concrete panels, the reduced movements on two adjacent panels could, however, still both occur at the common joint, and for overcladding design purposes this should be assumed to occur.

Weathertightness, dampness and condensation
High-rise blocks are vulnerable to increased exposure to driving rain. Problems have been reported with most systems. Rainwater leakage through the joints of large concrete panels seems to have been a major factor in prompting the consideration of overcladding. It is often also the case that repair or refurbishment of original designs to restore raintightness has been tried and has failed.

The actual rainwater load on a building is a product of its geometry as well as the incident rain and wind. The water does not necessarily run down in contact with the façade,

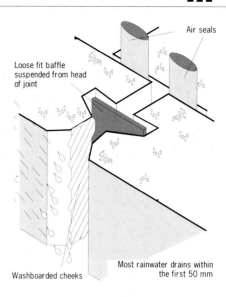

Loose fit baffle suspended from head of joint

Air seals

Washboarded cheeks

Most rainwater drains within the first 50 mm

Figure 2.53
Provided the air seal remains intact in concrete panel open-drained joints, most of the rainwater never even reaches the baffle

much falls as a curtain some 300–600 mm away. The wind will tend to drive run-off sideways. Water flows may be concentrated at vertical ribs, increasing run-off to many times the average.

Water load on joints
Rainwater drops carried in the air stream will enter the front of vertical joints in direct proportion to the open area of the joint. However, since most flows will be at an angle to the surface of the building (Figure 2.53), it follows that the depth of the joint, together with the topography of the opposing faces, will directly influence the amount of water reaching the back. This is why in concrete panel open-drained joints, if the air seal remains intact, most of the water never even reaches the baffle. On any kind of open joint, the greater the depth of the joint faces, therefore, the lower the water load on the interior of the joint. When considering any open joint design for overcladding systems therefore, it should be recalled that those with returned edges perform best, other things being equal. Sharp angles are better than rounded for encouraging water to flow down rather than across, but a minimum radius of 2 mm should normally be specified since very sharp angles encourage

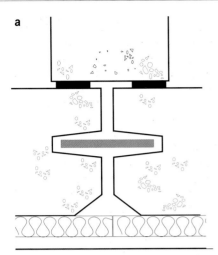

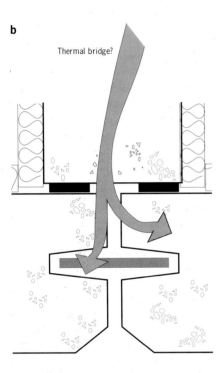

Figure 2.54
Thermal bridging at the junction of internal and external walls

thinning of applied finishes, with reduced durability.

When rain penetration through panel joints does occur, its source can be particularly troublesome to diagnose. This is because the water can percolate down through the many cavities in the external envelope before appearing on the inside of the building, perhaps some distance from the point or points of entry. Water trapped within the existing external envelope can cause serious damage to the fabric of the building, causing steel reinforcement to rust as well as saturating insulation and, hence, degrading its thermal properties.

In any face-sealed system of cladding the jointing products used in them (eg sealants and gaskets) will be subject to movements etc which must be accommodated. There are exacting raintightness requirements for such joints; they need to be virtually perfect. Rainwater will be drawn in by capillary action if any gap is less than a millimetre or so, and any gap greater than that will have water pumped through it by differential air pressure.

It may become necessary to provide cavity trays and weep pipes within cladding of this kind to allow rainwater penetrating the façade to escape without percolating to the interior of the building.

Thermal properties

Many large panel systems were designed and constructed before the 1973 fuel price increases, and by today's standards may have unacceptable levels of thermal insulation. Depending on the precise details of construction, an unrefurbished dense concrete panel wall with an internal lining of lightly insulated plasterboard might give a U value of around 0.85 W/m²K. Air leakage rates measured in LPS dwellings show, on average, only half the leakage rates of other dwellings, so this is not generally a problem, though it does affect the tendency to condensation[117].

There are main two options available for upgrading thermal insulation levels in the walls of LPS

construction: applying insulation to the inside or to the outside. External insulation is not always feasible (Figure 2.54a). Where there is a sound and weatherproof structure, or where the relatively simple replacement of weather seals in panel joints would ensure weathertightness, internal insulation may be achieved relatively simply and economically compared with external insulation and overcladding. However, it is rarely possible to provide for a complete elimination of thermal bridges at junctions of the external wall with internal walls and floors, since returning the insulation along the cross-wall may not be practical (Figure 2.54b).

Insulating the outside of the building enables the insulation to cover all such potential thermal bridges. It is expected that such insulation would be protected by overcladding.

BRE is sometimes asked whether it is reasonable to fill the cavities of LPS dwellings with thermal insulation. Although experience of the performance of such filled cavities is limited, nevertheless condensation on thermal bridges has been observed. A filled cavity also limits visibility if optical probe inspections on metal cleats are required. External insulation and overcladding is probably the best option, albeit more expensive (Figure 2.55).

Figure 2.55
Overcladding on a system built high-rise block

Fire

Fire safety has been thought to be a problem in high rise LPS buildings, but there have been no reports of failures in fire. Following the fire at Knowsley Heights, revised recommendations for the provision of cavity barriers in externally applied thermal insulation were introduced.

See also Chapter 9.2 and the same section in Chapter 2.1.

Noise and sound insulation

See Chapter 1.9 and the same section in Chapter 2.1.

Durability

Defects in concrete panel buildings do not appear to be system-dependent; that is to say, a wide range of defects can and does occur on all kinds of large panel systems. Many of the problems relating to LPS housing concern cracking, spalling concrete and falling surface finishes. Some of the reasons for these defects are inherent in the design, manufacture and construction of the building, and can be exacerbated by weathering. Many LPS dwellings have weathered badly and have a drab appearance.

Cracking and spalling of concrete panels has been found in a number of LPS blocks, especially where depth of cover has been inadequate. There is an increasing likelihood of debris falling from these structures unless adequate maintenance and repair programmes are carried out.

The main mechanisms of deterioration of the concrete members include carbonation of the concrete (a process whereby the concrete loses its natural alkalinity and so leaves the reinforcement at risk of corrosion – see feature panel aside) and chloride attack (having the same end result). The systems tending to have the poorer quality concrete, coupled with chloride attack, leading to enhanced risk of corrosion, include Bisons. Those with better quality concrete and no added chlorides include SNWs. Durability will be improved by overcladding the building, although carbonation may be slightly

---

Carbonation of concrete

In many large panel system and other concrete buildings, the concrete has spalled to some degree following corrosion of the steel reinforcing bars. This corrosion is usually brought about by the concrete losing its alkalinity over a period of time – a process known as carbonation. There is a simple field test for measuring its progress, provided a suitable fractured sample can be obtained (Figure 0.67).

BRE has examined carbonation in concrete of significantly different qualities by relating it to different methods of concrete production: prestressed precast, normally reinforced precast, and in situ concrete [118]. As might be expected, the 'higher quality' concrete produced the lowest level and spread of carbonation.

For the concretes in the BRE sample, there was an average of 80 years of protection to reinforcement with 20 mm of good quality concrete cover. However, in situ concrete cast around tie bars and fixings would have much less than 80 years of protection. Corrosion may be accelerated by the presence of chlorides and may even proceed, albeit at reduced rates, where the chlorides are present in relatively dry carbonated concrete.

Significant corrosion of reinforcing steel is usually indicated first by staining or marking, and then by spalling of the concrete cover. Surveyors should consider, however, the possibility that rust staining can sometimes originate with aggregates, and reinforcement not being to blame.

Although affected concrete can be cut out and a repair made, the long term durability of repairs is questionable, and the prospect of an increased incidence of spalling, with increasing carbonation over time, may be seen as supporting the case for overcladding

to exclude continued ingress of moisture to the original concrete cladding. Moreover, following repair, any patch will usually show, so overcladding has been adopted by some owners as a means of masking unsightly repairs. (See also Chapter 3.4).

A 30-year residual life would be a reasonable expectation for moderately carbonated concrete with good cover to the reinforcement. On the other hand, if the concrete contains significant amounts of chlorides, the residual life would be shorter than 30 years.

The question needs to be addressed of whether, if overcladding includes insulation, the warmer and drier environment of the concrete will adversely affect the rate of deterioration and, if so, whether this will be to an unacceptable degree.

The rate of carbonation in dry concrete is higher than that in wet concrete (because the latter is less permeable to carbon dioxide). Also, the rate will increase because of the higher temperature. This higher rate of carbonation clearly increases the risk of reinforcement corrosion; but the corrosion will only occur significantly in the presence of moisture, and this should be excluded by any overcladding. On balance, therefore, the corrosion risk should be reduced by overcladding or other surface coating, provided that the design minimises condensation within the concrete and genuinely excludes driving rain. However, if chlorides are present in the concrete, corrosion will continue in the protected carbonated concrete, so that overcladding in this situation cannot be relied upon to reduce deterioration. For further information, see *Carbonation of concrete and its effects on durability* [119].

---

accelerated in the drier conditions which then prevail.

As was implied earlier, concrete can be repaired, provided the results can be accepted visually. There are a number of techniques that can be used but, generally, these repairs are best left to specialist firms.

Ageing of the mastic or gasket is the most common fault on face-sealed joints. Failure in narrow joints, as would be expected, is more frequent than in wide joints. Replacement with a sealant or gasket with better performance (ie better movement accommodation factor) is required. Future performance depends on the thoroughness of

cleaning off the original material and priming of the jointing faces of the panels. Weep pipes must be preserved.

Cover strip joints, although vulnerable to vandalism, are amongst the most trouble-free with respect to weathertightness. If defective, the strips can easily be replaced. Traditional (often mortar) joints are usually to be found between the large panels and alternative forms of lightweight cladding. They invariably give trouble.

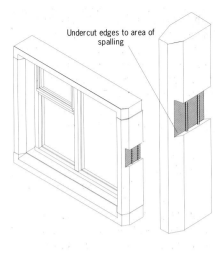

Figure 2.56
Patch repairs to precast concrete window surrounds

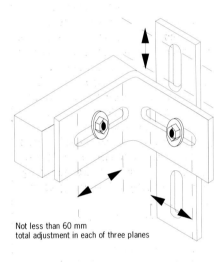

Figure 2.57
Fixings for overcladding support rails will need to tolerate rather large inaccuracies in the original building. The more practical designs are usually based on the twin slotted angle

The decision to refurbish using overcladding

Overcladding is not a panacea for all the ills of system building. On its own it will not reinstate the structural integrity of a building nor prevent further decay where inherent problems are to be found with the manufacture or assembly of the original components. Unless suitably designed and installed, it will almost certainly make it more difficult to identify any continuing deterioration. The decision to overclad must therefore be taken only after exhaustive consideration of the condition of the building. Reinforced concrete of the quality found in many LPS buildings must be recognised as a limited-life material, and provision made to monitor the performance of the buildings over time.

Maintenance

As already seen, there are normally two basic kinds of joints in the external envelope of LPS dwellings:
- two-stage (or open drained) joints
- one-stage (or face-sealed) joints

Both kinds have been found to be repairable in the majority of cases, though some have given rise to problems. In the case of two stage joints, renewal of the baffle may be awkward, though possible, but it is the air seal at the back of the joint which is crucial to performance. Even turning a two stage into a one stage by filling the exterior of the joint may not work. Repairing a single stage joint depends entirely on the integrity of the seal which, in turn, may depend on compatibility of old with new seals or adequacy of cleaning off the old.

Open drained joints, to be found for example on Taylor Woodrow Anglian, Bison, Jesperson, Reema and Wates, sometimes suffer from deterioration of the baffle material. Replacement should be straightforward. However, some joints have suffered breakdown of the air seal. This is not so easy to remedy, although it may be possible to insert the nozzle of a sealant gun into the rear of the joint to re-create an effective air seal. Alternatively, it may be possible to push a closed cell foam gasket through to the back of the joint if the baffle is first removed.

All spalled concrete should be cut out and repaired; there is certainly no advantage to be gained by painting the area with a bituminous paint system since it forms a vapour control layer in the middle of the cladding system. It is not desirable to leave any spalled areas unrepaired on the assumption that the overcladding will safely retain spalling which is already in progress.

BRE Digests 263[120], 264[121] and 265[122] deal with the mechanisms of corrosion, diagnosis and assessment, and repair of reinforced concrete.

With overcladding systems of the rain-screen kind, joints in the original external concrete panel wall should be made airtight before any overcladding work is begun, unless the proposed solution automatically includes an airtight barrier in itself.

This would involve:
- removing the baffle and gunning fresh sealant into the backs of all joints, or as far as can be reached into the back of horizontal joints, making sure that the verticals and horizontals interconnect
- sticking a new (could be self-adhesive) strip over the fronts of all vertical and horizontal joints
- injecting an expanding polyurethane foam into the joint to seal it.

Work on site

Workmanship

Precast concrete or reconstituted stone can be patch-repaired in situ – the so-called plastic repair method. The area to be repaired should be cut back to square undercut edges, preferably with a disc cutter (Figure 2.56). The mortar to be used should not be too strong for the original background. It may require a degree of experiment to obtain a suitable mix, but one of the order of 1:6 Portland cement:sand may well be strong enough. The mortar should be trowelled into place using coats of not more than 10 mm thickness. It

Before any decision can be taken on the range of refurbishment solutions which can be considered for an existing concrete clad building, it is essential to assess the condition of the whole building; in particular to ascertain the type, extent and causes of the deterioration of the cladding. There is no easy way of ascertaining this. The causes of many building defects are notoriously difficult to establish, and a very thorough examination of the building is necessary.

There are two main problems:
- whether the structure is safe (and whether it will it remain safe for a determinable period)
- whether the structure is sufficiently strong enough to carry any or all of the proposed range of options

The surveys should therefore give an idea of the remaining life of the building in its present state, even if no change in the system is to be made.

A preliminary survey can be carried out by means of powerful binoculars, but this will not be sufficient to reveal the condition of hidden parts and the condition of the concrete in depth. A procedure which permits closer inspection has been offered by some consultants who employ trained engineers to abseil down the face of the building. The visual inspection should be accompanied by spot checks of chloride content and carbonation (see feature panel on page 113).

Irrespective of the particular solution to be adopted for the building, it will be necessary to obtain an accurate survey of the condition of the whole of the existing external walling, particularly identifying the existence and condition of the fixings, whether any exposed aggregate or other surface finish is becoming detached, whether any corrosion of reinforcement and consequential spalling of the concrete surfaces has begun, and if the existing fixings are suitable to take the extra load of any extra materials – for example those in an overcladding system.

Of particular concern is whether the existing concrete panels will be strong enough to accept fixings for overcladding and will continue to do so for 30 years, and if so, of what kind. Some assessment will also need to be made of whether a proposed solution will actually introduce any side effects such as condensation within the concrete panels – it is conceivable that some potential solutions might actually reduce the building's remaining life.

With respect to the detailed condition of the panels the following will need to be determined:
- whether particular panels are loadbearing or non-loadbearing, and, so, differ in the thickness of the concrete, affecting their ability to accept fixings for overcladding
- whether the panels can withstand any or all of the potential fixing methods for overcladding, and whether, in turn, their fixings can take the extra loads
- chloride content
- rusting of reinforcement and spalling
- the extent of carbonation
- the accuracy of the existing building, in order to determine the extent of adjustability required in fixings and claddings; alignment of panels and windows to determine if claddings are required in non-standard sizes
- quality of the concrete and dry pack in the joints
- presence and condition of the ties between inner and outer leaves of sandwich panels
- position and size of any cracks and whether movement has stopped

Rusty streaks caused by iron-bearing aggregates should not be mistaken for corrosion of reinforcement or fixings; the latter is usually accompanied by cracking.

Where fixings are accessible, a check should be made to see that none is missing or insecure. BRE investigators have found concrete panels which could be moved under hand pressure. Where fixings are not generally accessible, it may be possible to inspect one or two samples at a particular point (eg a parapet coping which can be lifted). If corrosion of sample fixings is evident, a general inspection will be advisable.

It will be necessary to check that any levelling nuts have been backed off and that dry pack is adequate.

With purpose-made equipment which can be clamped to walls at reveals, it may be possible to take samples by drilling the external wall; generally, however, sampling by drilling or coring can be usually done only by using some form of platform and rarely from an unanchored cradle. A cover meter, the sensor head of which is usually light in weight, may be used with an extension arm via window openings. Wherever possible, the depth of cover found to reinforcement should be compared with the depth of carbonation to obtain an estimate of the future risk of corrosion to reinforcement; on this can be based decisions on the use of applied films intended to reduce further exposure to carbon dioxide. If carbonation depth is greater than or equal to cover depth, there is no point in trying to exclude carbon dioxide. (The risk should also be noted of introducing a vapour control layer in the wrong place if films are used).

The problems to look for are:
◊ carbonation
◊ spalling of concrete
◊ inadequate cover to reinforcement
◊ missing baffles
◊ missing air seals
◊ inadequate thermal insulation
◊ inappropriate painting of external surfaces
◊ missing dry pack
◊ rain penetration
◊ condensation

may be possible to cramp shuttering into place to assist in obtaining sharp arrises.

When undertaking overcladding work, it should be remembered that the original LPS building can have inaccuracies of the order of +55 mm over a typical storey, and even more over the height of the building; it will rarely be practicable to design a fixing for overcladding such that it can be adjusted to take out errors of this magnitude. Nevertheless, as much adjustability as possible and not less than ±30 mm should be sought, consistent with adequate strength. The more practical designs are usually based on the twin slotted angles (Figure 2.57).

Errors in panel alignment, other things being equal, may be more apparent against the rather precise and 'clinical' appearance of sheet systems compared with the more crude original. It should be remembered that errors in panel alignment will be less apparent the wider the joint, though, for filled joints, there will be a penalty in using more material.

Chapter 2.5   **In situ concrete**

In situ concrete, sometimes straight from the mould, or sometimes given a different finish (eg by hammering the dry surface), became a popular external walling material from around the middle of the twentieth century, even for prestigious buildings (Figure 2.58). Imperfections in casting were accepted, even welcomed, as a necessary expression of the manufacturing process. There have been disappointments, however, particularly in respect of appearance after several years of weathering. In consequence, fewer in situ concrete walls now seem to be built, and those that are, tend to be rendered or finished with applied cladding.

It is reported that in England there remain over a quarter of a million dwellings, or just over 1 in 100 of the total stock, that have in situ concrete as their predominant wall structure. A very few in situ concrete dwellings were built before 1919, but most were built between 1945 and 1980[2]. It is recorded that Sir Robert McAlpine had first used dense in situ reinforced concrete for housebuilding in Scotland as early as 1876.

Single leaf no-fines systems using removable shuttering were developed by, amongst others, Corolite, Unit Construction, and Wilson Lovatt; but the main developments came post 1939-45 war, with intensive exploitation by Wimpey and the Scottish Special Housing Association. Wimpey alone built around 300,000 dwellings using these techniques. BRE has published books on structural condition[123] and rehabilitation of no-fines[124], and it is therefore not proposed to go into detail with identification. Most surveyors should have no difficulty in identifying the normally rendered external elevations, pitched roofs with gabled ends to terraces (although this latter is by no means universal, since hipped semis were built in large numbers in parts of the UK). Most dwellings were two storeyed, but there were also bungalows and three storey flats and maisonettes. Some blocks of flats had flat roofs. Terraces of four, six or more are common, and the disciplines required on layout to make the most economical use of shuttering ensured that there were not many stepped or staggered layouts. Although most windows were conventional, and set in the shuttering, there were a few specially built dwellings with brick-built bay-windowed front elevations.

Early rendered in situ dense concrete dwellings are not always easy to identify, and one may have to find out more about the external wall construction to tell them apart. There were several thousand dwellings built between the wars of concrete single leaf walls poured between permanent shutters. The Universal system dwellings had a 7 inch wall between asbestos cement sheets, and the Fidler dwellings a 9 inch wall comprising two 2 inch clinker leaves with a 4 inch solid poured core, and rendered.

Other systems of in situ concrete include Incast and Forrester-Marsh.

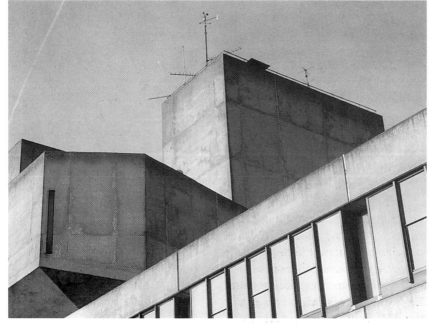

Figure 2.58
Exposed in situ concrete walling with no rain shedding details

The first Easiforms were built between the wars, and consisted of no-fines clinker concrete. Those built in the 1920s had cavity walls in which only the inner leaf was clinker. The Boswells of the late 1920s had precast concrete corner columns providing permanent profiles for the in situ poured cavity walls.

High-rise examples for the most part consisted of a concrete frame and no-fines panel infilling, rendered on the outside. They should present no problems in identification. The most numerous dwellings in this category are those by Wimpey.

Characteristic details

Basic structure

A typical loadbearing no-fines concrete shell is shown in Figure 2.59. External walls in low-rise dwellings built before 1951 were commonly 12 inches thick; those built 1951–64, 10 inches; and those after 1964, 8 inches. The walls were mostly built off a conventional brick masonry base on strip foundations. Reinforcement was incorporated, either in bar or mesh form. The first floor was supported off steel corbels cast into the loadbearing walls, or steel beams on padstones set into the no-fines; many dwellings had dense reinforced concrete eaves and ring beams. High-rise dwellings mostly had a reinforced concrete frame with infilling panels in no-fines.

However, there are other kinds of in situ construction, sometimes hybrids partly framed, such as the Universal system. This system consisted of solid external walls, cast in situ using clinker and gravel aggregate concrete. The walls were faced on internal and external surfaces with asbestos cement sheeting used as permanent shuttering. Pressed steel channel stanchions set at 4 foot centres temporarily supported the permanent shuttering and roof framing before the concrete walls were poured. Horizontal reinforcement in bar form was threaded through holes in the stanchion webs[125].

Occasionally too, the surveyor may well come across what at first sight would be described as a rendered wall, but which on closer inspection, together with the absence of day joints and other shuttering marks, and the lack of a very coarse aggregate, would be revealed as a pumped and sprayed in situ finish.

Main performance requirements and defects

Choice of materials for structure

No-fines concrete typically consisted of a single size ¾ inch coarse aggregate in a cement slurry. In practice the cement content ranged widely – samples measured by BRE investigators ranged from 5–20%.

Mixes for dense concrete will vary widely. See the note on accelerators in the section on durability.

Strength and stability

Minor cracking will be seen on a number of no-fines dwellings, caused by drying shrinkage of the no-fines material and subsequent changes in environmental conditions. It has no structural significance. Minor variations in original specifications are not considered to affect structural adequacy. Reinforcement generally has been found to be in good condition, with slight surface rusting only. Concrete quality achieved in the dense components was variable, and carbonation will be found to have penetrated to the reinforcement in many cases.

In a few cases where gable walls have an external leaf of brickwork, the wall ties may be corroding.

No instances of structural inadequacy were identified by BRE during the investigation of Easiform dwellings, although cracking in the external walls of some houses has occurred due to corrosion of the embedded reinforcement. In general this cracking is of a minor nature in post-1946 construction but is more substantial in a few of the oldest dwellings (50 years)[126].

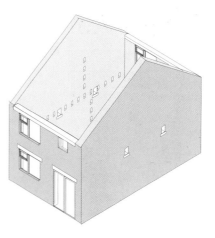

Figure 2.59
The shell of a typical no-fines house

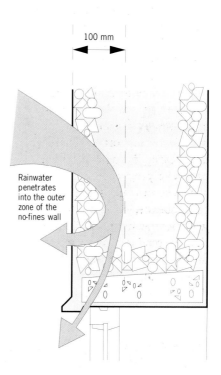

Figure 2.60
Rainwater penetrating cracks in the external render tends to find its way out again without penetrating through the wall thickness

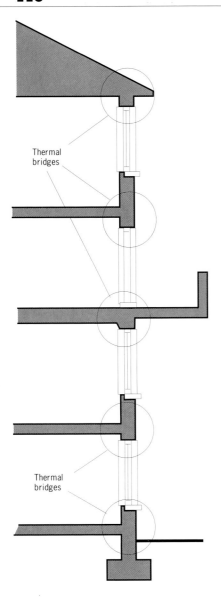

Figure 2.61
Locations of thermal bridges in a multistorey no-fines block

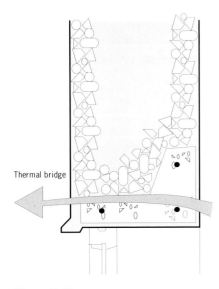

Figure 2.62
Thermal bridge at a lintel

Dimensional stability, deflections etc

For coefficients of linear thermal and moisture expansion etc, see Chapter 1.2.

Weathertightness, dampness and condensation

Monolithic dense reinforced concrete is vulnerable to shrinkage cracking; also cracks may occur at daywork joints. When rainwater runs down the outer face of such walls, even where they have been rendered, these cracks provide paths for rain penetration into the building.

On the other hand, no-fines concrete is not subject to cracking in quite the same way. Voids occur through the material, it is true, but rainwater penetrating through cracks in the rendering normally percolates down through the no-fines, close behind the render (Figure 2.60 on page 117). There is some evidence that the porosity of the aggregates used slightly affects the weathertightness. Where shrinkage of the render has been severe, particularly along the bellmouths over windows and doors, it has been necessary to strip the render and re-coat, in order to preserve weathertightness.

Thermal properties

Heat loss is one of the main deficiencies in external walls of this type, more particularly where the concrete is dense. A dense concrete wall, 200 mm thick, rendered outside and plastered inside, will give a U value of around 2.4 W/m²K. Substitution of lightweight plaster will give 2.3 W/m²K while dry-lining will give 1.7 W/m²K.

In the case of no-fines construction, some of the schemes were built dry-lined, and the airtightness of the wall depends heavily on the integrity of the external render coat. There were several places for thermal bridging within the wall thickness, usually dense concrete members such as lintels, balcony cantilevers and ring beams (Figures 2.61 and 2.62).

Where the render needs attention, the opportunity might be taken to use an insulating render

replacement. The addition of 40 mm of polyurethane, for example would give around 0.5 W/m²K.

Fire

No-fines concrete, rendered on the outside and plastered on the inside, does not present any significant fire risk. Provided cover to reinforcement is adequate, neither does in situ dense concrete.

See the same section in Chapter 2.1

Maintenance

Small shrinkage cracks in the external render should be repaired using a non-setting mastic rather than a cement:sand mortar.

Windows in no-fines construction were conventional and set in the shuttering before pouring. Consequently, replacement of rotted examples is not entirely straightforward.

Durability

Most dense concrete walls that have deteriorated have usually done so because of inadequate protection to any reinforcement incorporated. However, even if the concrete cover was adequate, deterioration of the steel could proceed as the carbonation front develops and the concrete loses its alkalinity, thereby putting the steel at risk. This process proceeds more rapidly if the concrete is of poor quality, and is relatively porous to carbon dioxide in the atmosphere.

Accelerators were permitted in steel reinforced dense concrete mixes in the 1950s and 1960s and maybe slightly later, but probably less frequently since then as the potential risks were identified. Levels allowed by the Code of practice of the time[127] were sometimes exceeded. For example, one case recorded over 2% of calcium chloride by weight of cement compared with 0.35% allowed[128]. Since one side effect of these additives is to further reduce the alkalinity and therefore the protection afforded to the reinforcement, their effects are gradual, and may show as corrosion of reinforcement and spalling of the surface at any time, depending on exposure. The standard was changed during the 1970s to prevent the use of calcium chloride in steel reinforced concrete.

Deterioration of this kind can usually be repaired, though some form of redecoration of the repaired surface will be necessary to disguise the repair. This type of repair is probably best undertaken in conjunction with upgrading the thermal insulation by means, for example, of an insulating render coat.

Work on site

Inspection

When checking the external walls for signs of rusting, the incidence of aggregate impurities causing rust-like streaking should not be forgotten.

In no-fines construction the steel corbels set into the external walls to carry the first floor joists should be examined where rain penetration has occurred through the wall.

Otherwise, the problems to look for are:

◊ rain penetration
◊ cracking, bulging or spalling of render coats
◊ inadequate cover to reinforcement
◊ inadequate thermal insulation
◊ inappropriate painting of external surfaces
◊ carbonation in dense concrete eaves beams

Chapter 2.6   **Earth, clay and chalk**

Well over half the world's population live in dwellings built from subsoils of varying types and using a variety of methods. Earth or mud was widely used in the British Isles, both in load bearing form as mass walls and as infill for timber framed walls (Figure 2.63). The latter category is also mentioned in Chapters 3.3 and 9.2.

There is much interest in the maintenance and repair of earth walled buildings and there is an increasing wealth of publications on the subject. The topic was one of the first to be examined by the Building Research Station and Special Report 5 was published in 1922[129]. A building constructed from stabilised soil blocks existed for

many years on the BRE natural exposure site, and was still in good condition when demolished.

There are distinctive techniques for construction dependent both on the characteristics of the locally available subsoils and local traditions.

● Cob, cleam, clob or clom – these walls are built of subsoil with a high clay content, mixed with straw and water (Figure 2.64), placed in situ on a stone plinth about 600 mm wide, heavily trodden down, and then pared back to an even line. Large numbers were built in the West Country up to about the end of the nineteenth century; some survive from the 1600s. Particularly good subsoil is found between Oxford and Aylesbury and is known locally as 'whitchert', meaning 'white mud'. Chalk cob is common in Hampshire, Dorset and as far north as Andover. Where chalk is the main constituent, slimmer walls are produced. Boundary walls around 300 mm thickness to a height of over 3 m have survived for well over 100 years

● Clay-lump (in Spanish: adobe) – mud bricks are formed from clayey subsoil mixed with chopped straw in a mould and then air dried. These walls are laid using a mortar of earth or clay. Buildings are found on the chalky boulder clays of Norfolk and Suffolk where the method was reintroduced from the Continent around the end of the eighteenth century, having formerly been used on a small scale in Roman times

Figure 2.63
Locations in Great Britain and Northern Ireland where earth buildings are situated

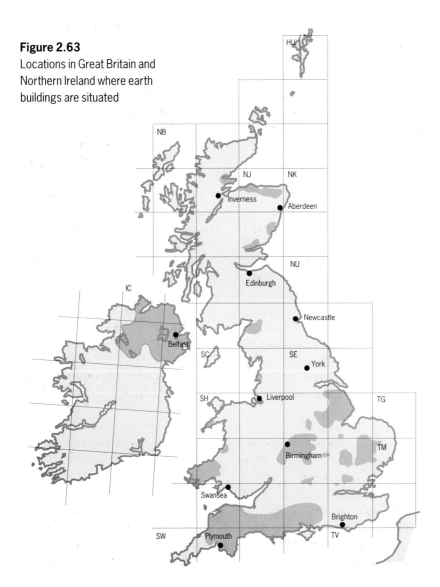

Figure 2.64
Carmarthenshire clom, from a farmhouse near Meidrim – probably eighteenth century

- Wattle and daub, mud and stud – a very common use of earth is as an infill to a timber frame construction. The wattle comprises staves driven into the frame with light twigs woven in between. The daub is then worked into the panel to cover the wattle both sides. The method has been used since Romano-British times. Its close relation, mud and stud, uses narrow oak planks set vertically into the structural timbers, again covered by the mud mix. Walls in this construction are most common in Lincolnshire
- Pisé-de-terre or pisé – built by ramming the clay mix between temporary shuttering, and struck later to allow the walls to dry. Techniques were developed in France during the second half of the eighteenth century and, although more common abroad, the method found particular favour in Scotland
- Stabilised soil blocks – blocks are hand formed in a mould and then tipped out to dry naturally. Increasing the density by moulding under pressure produces a better block and the Cinva-RAM method developed

in the 1950s in Chile was an early attempt to improve quality. BRE developed the method further under contract to the Overseas Development Administration with the hand operated 'Brepak' blockmaking machine. This exerts a compacting pressure of $10 \text{ N}/\text{mm}^2$ giving a block of dry density in excess of $2\,000 \text{ kg}/\text{m}^3$. The mix can be stabilised with hydrated lime or cement
- Turf sods – an easily won material that can be laid grass to grass and roots to roots in alternate layers. Bonding is with staggered joints as in traditional masonry. A variant is to alternate field boulders with the turf – a method widely used in Scotland until the mid-nineteenth century; a few walls of turf still survive, particularly in the Western Isles, but also on the mainland

The techniques of earth walling have been revived in the 1980s and 1990s, mainly to ensure authentic repairs and maintenance of existing buildings. A few extensions and small buildings have been erected and there is much interest in the wider use of the methods for new-

build. As eminently environmentally friendly systems they have great advantages; the materials are cheap and easily won; the existence of many farm ponds show the sites of material extraction. Straw or other vegetable fibre is freely available. Embodied energy requirements will be low, using mainly animal and manpower.

The BRE Advisory Service has continued to the present day to receive requests for information on the maintenance of earth buildings and has contributed to understanding the process by conducting research into moisture levels in earth walls.

Characteristic details

Basic structure
Cob is normally laid on a stone or brick plinth some 300 mm high and with a wall width of 450–600 mm. Plinths for chalk cob and pisé are usually narrower. Although not often found in older properties, new constructions must have a physical DPC laid in the plinth the usual 150 mm above ground level (Figure 2.65 on page 122).

Chalk cob recipes have traditionally consisted, for example, of one part of clay and straw mixture, well trodden, to which three parts of crushed chalk lump is added. Pisé on the other hand might consist of 1 part of clay to 2–2½ parts of sand. Small stones are occasionally added to both kinds of mix – up to half the total content.

A wide variety of mixes will perform well and local traditions will take account of varying soil types. As a mix example, in Suffolk in 1843, the Reverend Copinger Hill wrote: '*Clay for building should be clay-marl. If the clay is not good, chalk and road grit should be mixed with it ..... with moderate clay say seven-tenths clay, two tenths chalk, one tenth road grit.*'

The clay is mixed with chopped straw, well watered and then trodden by bullocks, horses or man. The wet cob is then forked onto the prepared stone plinth to a height of 500–600 mm with workmen on top

Figure 2.65
Old cob on the left, and new on the right. A house extension nearing completion

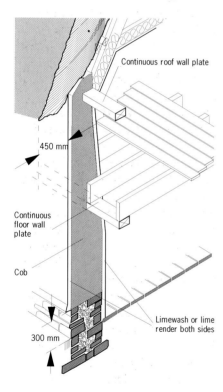

Continuous roof wall plate

450 mm

Continuous floor wall plate

Cob

Limewash or lime render both sides

300 mm

Figure 2.66
Typical section through a cob wall

of the wall treading it down. Surplus material is then pared down to finish up with a vertical wall. The waste material is recycled into the next mix. The first lift is allowed to dry naturally for a few days before the next lift is added.

Clay lump and stabilised soil block walls are laid in a similar manner to masonry with regular bonded courses using a clay or fresh earth mortar.

With wattle and daub walls the earth is non-loadbearing, the daub comprising subsoil which is clay based, well mixed with animal or vegetable fibre. The daub is laid against the lattice of sticks using plastering techniques. Mud and stud walls are generally thicker, with the mix beaten between the vertical laths and completely covering all the timbers.

Most areas of the UK with large numbers of earth walled buildings have local groups of enthusiasts with a specific interest in reviving the techniques and encouraging traditional repair methods[†].

Reinforcement may be found in the form of timber planks laid within the wall thickness. Sometimes chicken wire was used. Continuous

† The ICOMOS UK Earth Structures Committee can supply addresses of contacts.

wall plates are sometimes necessary for floor joist bearings and invariably for rafter bearings (Figure 2.66).

In some parts of the Western Isles, turf walls were laid just as the sods came from the fields, usually grass down.

Separating walls and partitions may be found constructed in pisé-de-terre, but partitions were not usually in cob because they would take up too much space.

Abutments
Chimney breasts and stacks can be constructed in their entirety from cob or from brick and stone.

External corners
External corners built in cob are occasionally found to be rounded, and this may well be intentional and not necessarily the result of exposure. It has been said that this reduces the tendency for the wall to crack. It is not usual to find masonry quoins in cob houses.

Openings and joints
Ancient practice was to saw out openings for windows and doors after the cob had dried, depending naturally on whether stones had been incorporated into the mix, and it may still be possible to do this when alterations are needed. However, more modern practice, certainly with pisé-de-terre, is to use formers strong enough to withstand the ramming, which are withdrawn when the material is dry.

Main performance requirements and defects

Strength and stability
The strength of a cob wall relies on its moisture content remaining stable. Wetting through a roof failure, or by moisture being held behind an inappropriate render can result in a build up of moisture at the base of the wall. The earth may then shear and a total collapse may follow. However the thickness ensures that loads are low, with compressive stresses normally being less than 0.1 N/mm^2.

Compression tests carried out in 1921 on samples of pisé made with London clay and breeze gave values ranging from 15–24 tons/ft^2, say 1.61–2.58 N/mm^{2} [129].

Dimensional stability, deflections etc

As earth materials are laid very wet, they shrink a considerable amount as drying occurs. The walls crack, particularly at corners and some patching may be necessary. Older properties may suffer from a variety of interactions between plinths, timber trusses and the earth walls.

Stone plinths may not be well bonded, and may be rubble filled. If the core has compacted, facing stones can be pushed outwards and the cob will be left only supported at the wall faces. A vertical split in the wall may result.

Early structures have timber trusses supported on the plinth, and damp decay to the wood will allow settlement. An 'A frame truss' may have a failure in the lower chord, resulting in the walls bulging at the top.

Rat burrows within the mass of the wall can result in total collapse if the structure is much weakened. Filling of the burrows with a mix containing broken glass is said to discourage further rat activity, although it is unlikely to restore the full strength of the wall.

Weathertightness, dampness and condensation

As already seen, the earth walls must not be allowed to become too wet otherwise there is a very real risk of collapse. Cob walls can be rendered with an earth and lime mix and finished with lime-wash. However, in some villages this was only done to the front elevation. Other elevations, and particularly barns, were just left as built. Weathering does occur and stones and straw will be seen proud of the earth, but material loss, even after centuries of exposure, is minimal.

The greatest dangers are from problems with the roof at the head of the wall, and from modern cement based renders used externally.

Although they may be laid over chicken wire, pinned to the surface, the cement and sand render invariably cracks. Rainwater enters through these cracks and concentrates at the base of the wall. Evaporation of moisture will be at a slow rate and in the wetter parts of the country the walls are at risk of collapse.

A similar problem arises from the 'improvement' of rural buildings by the erection of a single brick skin against the existing cob or clay lump wall. Again, the brick wall will not resist the passage of driving rain and there is a similar risk of failure.

Rising damp is not usually a serious problem with earth walls. It is feasible to install a remedial DPC in the stone or brick plinth, providing care is taken not to cause structural damage. Myths have circulated suggesting that cob should not be allowed to dry out too much, but recent research by the BRE Advisory Service has disproved this.

In new work, DPCs should be laid at the wall head as well as around all openings.

Thermal properties

Many commentators have recorded the views of inhabitants of earth walled buildings that they are comfortable to live in. P W Barnett recorded in 1922, after visiting 1880 built clay lump cottages at Harling (Norfolk), that *'the inhabitants of these dwellings say that, in spite of the fact that the walls are comparatively thin, their houses are cool in summer and warm in winter'.*

There is a shortage of information on the thermal properties of earth walls. The CIBSE Guide in Table A 3.22 gives values for crushed Brighton Chalk, mud and some soils at varying moisture contents and compactions. Gordon Pearson has produced a table of comparative heat losses through a range of walling: typically a chalk wall (10% moisture content) 450 mm thick, with 13 mm lightweight plaster and 13 mm external render will have a U value of 0.84 W/m^{2}K; a comparable clay wall (10% moisture content, 20% binder, 80% aggregate) 450 mm thick,

plastered and rendered, will have a U value of 0.83 W/m^{2}K.

Thermal admittance is also an important factor in providing comfort to occupants. Chalk and clay walls take up, store and release heat when the room temperature drops. A further application is the many miles of kitchen garden walls built of chalk and clay which provide warmth during the night to fruit trees.

Fire

Earth walls are non-combustible and tests carried out at the Australian Building Research Establishment showed that a 250 mm thick adobe wall achieved a fire resistance rating of four hours.

Durability

It is essential for newly built earth walling to dry before the finishing coats are applied both externally and internally. The material does not have to be bone dry, provided the finishes are permeable to water vapour, but any attempt to coat wet cob will lead to loss of adhesion of the coatings.

The longevity of earth and chalk walling depends on good protection from driving rain, splashup, and rising damp. The importance of a generous roof overhang cannot be over-emphasised – it should be at least 450 mm and preferably 600 mm for maximum durability.

Lime and bitumen have both been added to the mixtures in the past in order to try to improve durability. Soil stabilised by Portland cement has occasionally been used for experimental buildings, but this mixture should on no account be used in the maintenance of more traditional mixes.

Provided the wall is kept dry and is well protected, pisé-de-terre buildings can survive for upwards of several centuries, and even cob can provide occasional examples of similar longevity. Flooding is obviously a disastrous event leading to complete collapse.

Maintenance

All vegetable growth should be removed immediately. The condition of the roof protection is crucial to performance.

Renewal of the coatings, particularly the exterior coating, is important. The construction should be examined from time to time for rat runs, and the holes filled with a compatible mixture.

Work on site

Storage and handling of materials

Traditionally the mixing of raw material was carried out by bullocks in a pen (dung will be an added constituent but, contrary to popular conception, is not necessary for a quality cob). Recent works have used a mechanical excavator, both in winning the soil and in mixing it by running back and forth across the wet mix.

Restrictions due to weather conditions

Earth walling should not be laid in wet weather, and needs protection if rain intervenes during the building process. Cob is normally laid during spring and summer, which avoids frost conditions.

Workmanship

Cob working is a rural art. Essentially it is not highly skilled, but it does require experience. Pisé-de-terre is normally rammed between climbing formers (Figure 2.67), and wall thickness is self-evidently easier to control.

Inspection

The problems to look for are:
◊   deterioration or leaking of roofing
◊   inadequate roof overhang
◊   dampness
◊   incorrect external protection
◊   rotting of timber lintels
◊   external render cracking
◊   rat runs

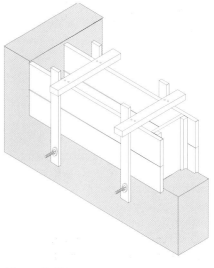

Figure 2.67
Pisé-de-terre is normally rammed between formers. The through bolts are necessary for stability; they were withdrawn from the wall subsequently and the holes pointed. (The drawing shows the construction of shutters used in the Building Research Station experiments of the 1920s.)

Chapter 3   External cladding on frames

One of the prime performance requirements of cladding is the separation of the uncontrolled external environment surrounding a building from the conditions required within the building in order that it can be used for its intended purpose. In practice this means that the cladding must provide a safe and durable weather resistant barrier with adequate sound and thermal insulation properties. Cladding is not normally required to make a significant contribution to a building's structural integrity, although it does have to transfer wind forces to the structural frame and often contributes to the overall shear resistance. This chapter deals with claddings in both masonry (Figure 3.1) and panel forms which are individually fixed either directly or via an intermediate frame, to the structural frame of the building. It is convenient for the purposes of this book to divide claddings into heavy and light.

Heavyweight units: these consist of a cladding made in the form of a wall of units – bricks, blocks or slabs of natural stone, fired clay, concrete or calcium silicate. Generally these will only be required to support their own dead weight but this is likely to be at least over the height of one full storey. Such cladding will not be expected to bear vertical applied loads but occasionally may bear lateral stacking loads.

Lightweight units: these normally consist either of a thin sheet material in subframes or a sheet material which has sufficient strength to span between fixings or studs by itself. Typical examples are glass fibre-reinforced plastics, metal sheeting, cement based composites, thin stone panels, and sheathing spanning between timber studs. Generally these too will only be required to support their own dead weight over the height of one sheet or panel and no applied load. Dead loads are taken to the structure via support fixings at the base of each sheet or frame, or ties acting in shear. They will also be required to bear wind loads by spanning between support and restraint fixings or to subframes. Thermal and moisture movement may be one of the principal actions so the fixings and the joints between panels must allow in-plane expansion or contraction to occur freely, otherwise excessive unintended forces can develop. In thicker materials, especially if highly insulating, out-of-plane movements can occur due to the differential movement between outer and inner layers, and very high and destructive shear stresses may develop in poorly designed systems.

Figure 3.1
Brick masonry cladding to a well known landmark in Battersea

Chapter 3.1

Masonry on steel or concrete frame

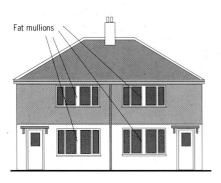

Fat mullions

Figure 3.2
A pair of Dennis Wild houses

Figure 3.3
The characteristic elevation of a Dorlonco house

Steel frames

Steel framing used within external walls first came into general use in the early years of the twentieth century, and many of the earliest examples were clad in brick masonry. In the main, the newly introduced techniques were exploited in public buildings, but it was not long before entrepreneurs saw possibilities to be exploited in the housing field too.

The majority of steel framed dwellings constructed in the years before the 1939-45 war were in systems which were designed to simulate, as nearly as possible, conventional loadbearing brick dwellings, and it is sometimes not easy to distinguish that a particular dwelling has a steel frame. One clue is to look at wide windows, where the steel column is sometimes incorporated within the frames, giving very fat mullions (Figure 3.2).

The following steel framed systems were normally clad in brickwork or rendered masonry: Birmingham, Crane, Cranwell, Cruden, Denis Poulton, Dennis Wild, Dorlonco, Nissen Petren, Presweld, and the two marks of Trusteel (Mark II and 3M).

Several of these systems do not warrant description. The Cranes were all bungalows in the Nottingham area. The Preswelds were few in number, and the Birminghams and Nissen Petrens even fewer. The Nissen Petrens were practically all either in Yeovil or Edinburgh, and in any event are the most easily distinguishable of all systems. Some at Yeovil are now listed buildings.

Trusteels are by far the most difficult to identify from outside, practically all being clad in brick, with some alternative tile hanging; but a glance in the roof space will reveal the latticed steel roof rafters of Mark II, and pressed steel channels of 3Ms. Roofs can be hipped, but the majority are gabled.

Dorloncos are characterised by their so-called double fronted elevation (Figure 3.3) with central front door. Claddings are mostly either brick or render on mesh (Figure 3.4).

Dennis Wilds (Figure 3.5) for the most part have hipped roofs, but the distinguishing feature is the roof truss with steel rod ties just above ceiling joist level.

Cranwells are normally rendered over very large clay block outer skins, and their windows and doors all have a very heavy precast concrete outer frame section (Figure 3.6).

Crudens are all in Scotland. They may have a variety of finishes: render, brick or concrete block.

Also included in this category are Homeville and Lowton Cubitt systems, since they normally have brick panels over separating walls and at gables, and could be mistaken for so-called rationalised traditional crosswalls. Some dwellings in these two systems are not easily distinguishable at first glance, although Homevilles are mostly in the southeast, and Lowtons in the north west.

The Trusteel system, one of the most widely used, employed a modular-sized steel frame (weighing less than 2 tons) for the loadbearing

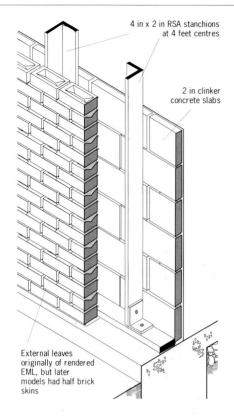

Figure 3.4
Detail of the construction of a Dorlonco house. Some had rendered block elevations

4 in x 2 in RSA stanchions at 4 feet centres

2 in clinker concrete slabs

External leaves originally of rendered EML, but later models had half brick skins

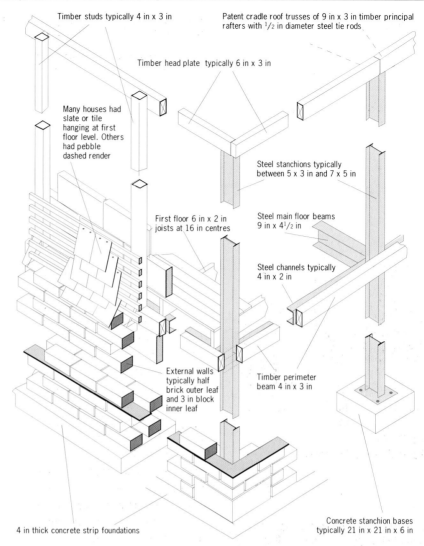

Timber studs typically 4 in x 3 in

Patent cradle roof trusses of 9 in x 3 in timber principal rafters with $1/2$ in diameter steel tie rods

Timber head plate typically 6 in x 3 in

Many houses had slate or tile hanging at first floor level. Others had pebble dashed render

Steel stanchions typically between 5 x 3 in and 7 x 5 in

First floor 6 in x 2 in joists at 16 in centres

Steel main floor beams 9 in x $4^{1}/_{2}$ in

Steel channels typically 4 in x 2 in

External walls typically half brick outer leaf and 3 in block inner leaf

Timber perimeter beam 4 in x 3 in

Concrete stanchion bases typically 21 in x 21 in x 6 in

4 in thick concrete strip foundations

Figure 3.5
Detail of the construction of a Dennis Wild house

members (walls, ceiling, roof) of small houses. Bricks were used only for the non-loadbearing external leaf; a saving of 50% in bricks and 50–90% in wood was claimed. The dry internal linings were laminated from two layers of plasterboard. Where a site was liable to subsidence the steel frame was combined with a reinforced concrete raft foundation, sometimes with jacking points for future use. The structural metal elements were made to very close tolerances[130].

Reinforced concrete frames
Masonry clad reinforced concrete frames were not used in low-rise housing, but they were used in medium and high-rise dwellings, and in many other building types.

Also covered in this chapter is non-loadbearing stone cladding in sheets which are normally thinner than 100 mm.

Characteristic details

Basic structure
Steel frames
Steel frames in public buildings dating from the early years of the twentieth century were commonly built up from simple rolled sections riveted together into sometimes quite complex box and cruciform shapes, stanchions as well as beams. The masonry covering the steel is usually carried on projecting angles at each storey, although cases have been discovered where masonry is unsupported over more than a single storey.

In low-rise dwellings, the steel was mainly in the form of rolled sections, although the most commonly used system, Trusteel, used sections fabricated from thin

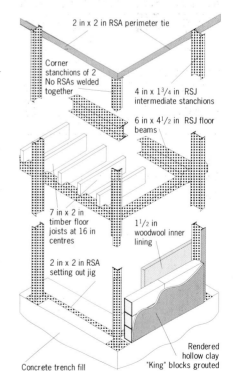

2 in x 2 in RSA perimeter tie

Corner stanchions of 2 No RSAs welded together

4 in x $1^{3}/_{4}$ in RSJ intermediate stanchions

6 in x $4^{1}/_{2}$ in RSJ floor beams

7 in x 2 in timber floor joists at 16 in centres

$1^{1}/_{2}$ in woodwool inner lining

2 in x 2 in RSA setting out jig

Rendered hollow clay "King" blocks grouted

Concrete trench fill

Figure 3.6
Detail of the construction of a Cranwell house

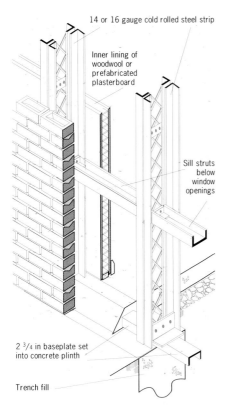

14 or 16 gauge cold rolled steel strip

Inner lining of woodwool or prefabricated plasterboard

Sill struts below window openings

2 ³/₄ in baseplate set into concrete plinth

Trench fill

Figure 3.7
Detail of the construction of a Trusteel Mk II house

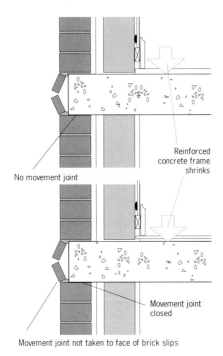

No movement joint

Reinforced concrete frame shrinks

Movement joint closed

Movement joint not taken to face of brick slips

Figure 3.8
Shrinkage of a reinforced concrete frame leading to brick slips detaching

sheet steel. In these cases the masonry would be separately founded with the steel frame used primarily for carrying floor and roof (Figure 3.7).

Reinforced concrete frames

Masonry on reinforced concrete frames frequently left the frame exposed, being carried on the horizontal beam with only a slight projection from the frame face. This practice almost inevitably leads to weatherproofing difficulties at the frame-to-masonry joint. Alternatively steel shelf angles are sometimes to be found fixed to beams or slabs, in which case the detailing can be similar to that for steel framing.

Sawn natural stone

Natural stone, sawn into sheets sometimes as thin as 50 mm, is increasingly being used for claddings. Research at BRE and elsewhere has shown that some types of stone undergo a significant loss of strength when subjected to environmental loads, in particular thermal cycles. More information is available in BRE Information Paper IP 6/97[131] and in BS 8298[132].

Main performance requirements and defects

Strength and stability

Brick cladding on steel frames, as already noted, is for high-rise construction mostly carried on shelf angles at suitable points in the structure, usually at floor levels where adequate bearings and soft joints are needed. For low-rise work, the masonry may well be separately founded on conventional footings.

One of the most important criteria for strength and stability is the integrity of fixings for the cladding, particularly in multi-storey buildings. A variety of tie patterns may be used to connect the walling to the frame. Some may be factory fixed to the frame, in which case it is essential that they permit vertical adjustment or are flexible.

Dimensional stability, deflections etc

A characteristic defect of this type of construction is shrinkage of concrete frames which squeezes brick cladding which has not been provided with horizontal soft joints at storey height intervals. This is exacerbated by the expansion of fired clay brickwork. It is unlikely to occur if concrete or calcium silicate units are used, as these also shrink in sympathy with the frame. The effects of these movements are exaggerated if the brickwork is set proud of the frame to allow the use of brick slips to mask the concrete support nibs. This happens because the eccentric loads cause bowing and the stresses have to be born by only two thirds of the bearing area of the bricks (Figure 3.8).

Both the concrete shrinkage and the brick expansion are more rapid at the outset and take several years to complete, but some can take decades before they start becoming apparent. Since this detail (with the brick slips) has now fallen out of use, the number of further failures will probably start dwindling.

Remedial work[133] entails the following procedure:
● checking the lateral stability of the masonry if the vertical spanning is interrupted by a soft joint
● checking that adequate tying exists between the outer panel and the columns, and between the outer panel and any inner infill masonry
● checking the condition of nibs, which may be damaged or sheared off, and repairing or replacing them with shelf angles or corbel studs
● cutting new movement joints and making sure that they are fully compressible
● installing additional ties if required for stability

For coefficients of linear thermal and moisture expansion etc, see Chapter 1.2.

Weathertightness

Single leaves of limited thickness can be expected to leak rainwater through to the cavity in conditions of driving rain unless protected externally with a rain resistant cladding. This water must be prevented from reaching the frame by suitable detailing of copings, DPCs, cavity trays and flashings. Defects characteristic of masonry on steel or concrete frame walls include rain penetration and draughts through defects in the original cladding, particularly in conditions of high exposure. Poor drainage will also take its toll (Figure 3.10).

Figure 3.10
Evidence of poor drainage in a brick clad reinforced concrete frame building

---

Fall of stone cladding

Some stone slabs fell from the south elevation of a building, injuring a number of people. The building had been erected around 20 years earlier, and comprised a reinforced concrete frame with windows and stone facings.

The damaged elevation above second floor level was divided into 12 bays by stone clad columns. Between the columns were windows with the spandrel panels beneath each window formed in stone. The stone slabs that fell came from what was essentially a string course at second floor level one stone slab deep which ran the whole length of the elevation. They fell from a position about one third of the length of the elevation from one corner of the building.

The slabs were natural stone about 890 mm x 470 mm and about 38 mm thick (Figure 3.9). They were fixed to the building with the long side horizontal. The top edge of

Figure 3.9
A piece of the fallen stone cladding

the stone was restrained by copper cramps inserted into two holes drilled into the top edge of each slab. These cramps were still in place on the building. Support for the slabs was provided by short (75 mm) angles (which appeared to be brass) the horizontal leg of which located in two slots cut in the back face of the slab near the lower edge and bedded in cement mortar. The angle was secured to the concrete backing by an expanding 6 mm proprietary fixing bolt. Each angle served the adjacent edges of two slabs, and some of the

bolts appeared to be loose and not to have been tightened to fully expand the fixings, and were easily removed from the sockets. The angle was in good condition and there was no sign of significant corrosion.

There were a number of fractured stone pieces, from fragments to quite sizeable pieces. All the pieces together would probably make up three or four full slabs. In addition, some of the slabs were virtually undamaged, but a number of these apparently had been removed by the fire brigade which first attended the scene. The blind over a shop front had been extended at the time the slabs fell and had probably absorbed some of the impact.

Measurements on the debris showed some variation in the thickness of the slab and in the dimensions of the slot. The full depth of the slot as cut, that is to say the maximum possible bearing, was 15–20 mm, whereas the actual bearing was in many cases less that this, down to 6 mm. Also some slots had indications that they had been damaged prior to fixing.

There was a hard mortar in the thin vertical joints between stones. There was no evidence of a soft joint in the whole length of the elevation in a vertical position, but there was some indication that a soft joint had been used at the head of the slabs, consisting of sheet polystyrene and a mastic pointing.

The fall had resulted from thermal movement in the long run of stonework, causing it to bow and lose its bearing. The thermal coefficient of the stone was of the order of $3\text{–}4 \times 10^{-6}/\,°C$. A temperature rise of as little as 18 °C would produce expansion in the 54 m length of the order of 8 mm. If completely restrained at the ends this could produce a very substantial bow of the order of 280 mm. Obviously the lateral movement had not been of this magnitude but there was every reason to suppose that repeated movement of the stone had caused some of the slabs to lose their bearing on the supporting angle.

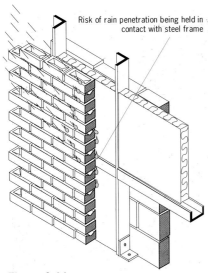

Risk of rain penetration being held in contact with steel frame

Figure 3.11
There are significant risks of deterioration when cavities of steel framed houses are filled

Figure 3.12
Cracked brickwork in a Dennis Wild house caused by rusting of the corner stanchion

Thermal properties

Lack of thermal insulation leading to high heating costs was common in many systems. Unheated bedrooms, combined with other factors, gives rise to mould growth following condensation.

It is difficult to give typical thermal insulation values, for no single construction can be said to be entirely typical. However, many systems were originally built just to better the 1965 Building Regulations target of 1.7 W/m²K or the 1976 target of 1.0 W/m²K.

Care must be taken in deciding whether and how to upgrade the thermal insulation of steel framed walls. Where the basic construction of the wall is little different from one of conventional construction – that is to say with masonry inner and outer leaves, and a cavity between – the main point of concern in considering the use of cavity insulation will be the possible poulticing effect of thermal insulation beads wetted by contact with the outer leaf against the steelwork contained in the cavity, or being drawn by capillary attraction into the interstices formed when certain kinds of insulation shrink (Figure 3.11). The condition of the surface protection to the steel is of crucial importance for durability. Other points to watch for include electric wiring in the cavity and frost damage to the external masonry leaf (cavity insulation may have the effect of slightly reducing the temperature of the outer leaf, making it more vulnerable to frost damage).

Fire

One general point to note is that if the structural frame is external to the external wall, it does not have to be fire-resisting, though the wall itself does. This may seem rather anomalous, but it stems from the fact that fire resistance of elements of structure is considered only in relation to fires generated within the building.

See also the same sections in Chapters 2.1 and 2.2.

Steel frames

Steel beams and columns may be partly embedded in floors and walls in buildings. These elements may also be used outside the building façade where, subject to their position in relation to windows through which heat may radiate and flames can jet, they can be found with no protection against fire. In both applications the elements are subjected to non-uniform heating such that large temperature differences across the steel section can be attained. These temperature differences cause differential expansion in the material and, when unrestrained, result in thermal bowing towards the fire[134].

Reinforced concrete frames

Reinforced concrete frames provide their own inbuilt protection.

Noise and sound insulation

For performance of masonry cladding, see the same sections in Chapters 2.1 and 2.2.

Durability
Steel frames

The greatest incidence of corrosion of frames in the steel systems examined has been observed in areas with high driving rain. Where dwellings also have a clear line of sight to open country, the problem may be exacerbated. In exposed locations, most steelwork paint protective systems were observed to be in process of deterioration, but this will not automatically mean that such coatings need renewal. It depends what service life is required from the dwelling.

BRE site observations of the corrosion rates of steel in low rise systems indicates typical rates of 3 mm in 20 years for steelwork in contact with wet cladding. The expansion caused by the rusting exerts considerable force on claddings, sufficient to crack brickwork (Figure 3.12).

Some Preswelds had fully galvanised frames, but these houses were in a minority.

BISF ground floor elevations, Dorloncos and Cranes are prone to cracked renderings following carbonation of the render and loss of galvanising to the mesh.

Although some rusting of rolled steel angle columns and beams has been observed by BRE in most steel framed systems, in no case has any dwelling needed to be taken out of service because of structural considerations. Pressed sheet steel sections have less steel content, and once the protective coating has disappeared, there is therefore less steel to survive. Trusteel dwellings, particularly the Mark IIs, should be particularly carefully examined. In most cases this will mean exposing a corner column. However, even nearly complete rusting away of a column at its base will not necessarily mean that failure of the complete frame is imminent. Normally there was some redundancy in the frame, and loads are shared with other parts of the structure. Corroded structural steel sections, other than lattice sections, are not difficult to cut away and replace, but it does mean opening up the structure (Figure 3.13).

Dorloncos have concrete floors and ceilings in which deterioration of reinforcement has been noted following carbonation of the concrete. Some Dennis Wilds external walls have bulged following corrosion of wall ties.

Reinforced concrete frames

See the same section in Chapter 2.4 which deals in particular with carbonation.

Maintenance

Maintenance of a steel frame is unlikely to be advantageous, unless the cladding can be stripped to allow corrosion protection to be restored. As was implied earlier, spalling concrete in the frames can be repaired, provided the results can be accepted visually, and there are a number of techniques that can be used. Generally, such repairs are best left to specialist firms.

For brick masonry, see the same section in Chapter 2.1.

Work on site

Inspection of a sample column of steelwork can often conveniently be made from the outside of the dwelling. The most exposed position should be chosen (Figure 3.14).

Inspection

In addition to those listed in Chapter 2.2, the problems to look for are:

Steel frames
◊ corrosion of steel frames
◊ corroded ties linking outer leaves with frames
◊ thermal insulation in cavities acting as wet poultices

Reinforced concrete frames
◊ spalling of concrete
◊ no soft joints at storey heights of panels in concrete frames

---

Figure 3.13
The internal leaf has been removed to reveal a badly corroded steel column. In this case there has been some load sharing, and the brick outer leaf is now supporting the column

Figure 3.14
Cutting away rendered masonry at a corner column. A vertical crack, signifying corrosion product, shows above the cold chisel

Chapter 3.2   # Precast concrete on steel or concrete frame

This chapter deals with a combination of materials which have either already been dealt with as a separate entity – the precast concrete panels in Chapter 2.4 – or as used in combination with masonry – the steel and concrete frames in Chapter 3.1. However, the examples described here are all clad with relatively small panels (Figure 3.15).

One of the main purposes of this chapter is to draw together information on the characteristics of certain systems used for housing, many examples of which remain in use. It is the external wall which provides the main distinguishing feature of these systems, and also the main source of problems. However, what is said is also for the most part relevant to other building types where the construction details are similar. Take, for example, the many systems developed for school building such as Hills (Figure 4.23 in *Roofs and roofing*[24]), CLASP, MACE, and SCOLA, where precast concrete cladding on steel frames was part of the vocabulary for the external walls.

Characteristic details

Where records have not survived, perhaps the most difficult task is to identify the system. A summary of the main distinguishing features which show mainly on the external walls is therefore given here.

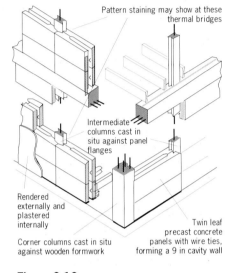

Pattern staining may show at these thermal bridges

Intermediate columns cast in situ against panel flanges

Rendered externally and plastered internally

Corner columns cast in situ against wooden formwork

Twin leaf precast concrete panels with wire ties, forming a 9 in cavity wall

Figure 3.16
The construction of the Underdown system external wall

With the early precast systems, those which give the most difficulty in identification include the Parkinson, Winget, Underdown (Figure 3.16) and Boot systems; in all of these, the concrete frame as well as panels were rendered over. All four had traditional timber roofs. Only the Boot system had continuous external wall cavities, and the other three may show pattern staining at the thermal bridges of the frame. Parkinsons normally have a section of rendered wall showing on elevation over the first floor windows, whereas the Winget and Underdowns do not.

The first 4,000 Aireys (so-called Duo-Slabs) built in the 1920s, and mostly in the Leeds and Edinburgh areas, had exposed piers at 4 foot centres with render on narrow slabs in between.

Figure 3.15
The archetypal precast concrete house – the Airey

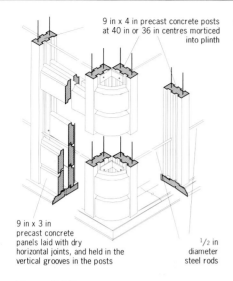

9 in x 4 in precast concrete posts
at 40 in or 36 in centres morticed
into plinth

9 in x 3 in
precast concrete
panels laid with dry
horizontal joints, and held in the
vertical grooves in the posts

1/2 in
diameter
steel rods

Figure 3.17
The construction of the Cornish Unit
system external wall

A very large number of systems exist where the frame and panel were left exposed externally. These may be conveniently divided into two groups – those with horizontally spanning panels and those with vertical spanning panels, usually of storey height.

In the first group of systems, having horizontally spanning panels, there is only one system which has the frame left proud of the wall surface: the Cornish Unit – Type 1 with narrow (less than 1 foot) infill panels on the ground floor only and a mansard first floor; and Type 2, having both ground and first floors with taller 2 foot high panels. The Type 1 is the easiest of all systems to identify at a glance (Figure 3.17).

Cornish Unit also built some three-storey blocks (Figure 3.18).

Of the other four systems in this group which have the frame enclosed – Mac-Girling, Unity, Woolaway (Figure 3.19) and Orlit – the Orlits with flat roofs are the easiest to distinguish, but unfortunately for the aspiring detective, some had pitched roofs.

Several hundred variants – Blackburn Orlits – were built in Scotland, and these had shallow pitched aluminium covered roofs. The Mac-Girlings have vertical flutes on the panels, which distinguish the system from Unity, while the Smiths are faced in brick slips. It is usually fairly easy to distinguish a Smiths from a traditional brick house since the mortar joints defining the panels are usually visibly different.

Reema low-rise dwellings are perhaps one of the most easily

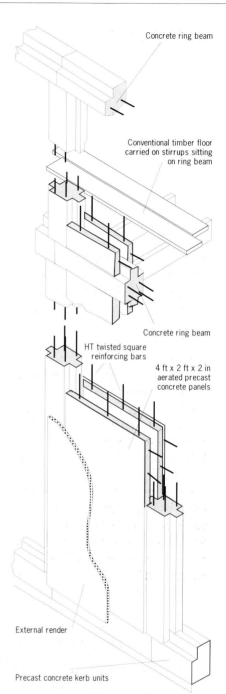

Concrete ring beam

Conventional timber floor
carried on stirrups sitting
on ring beam

Concrete ring beam

HT twisted square
reinforcing bars

4 ft x 2 ft x 2 in
aerated precast
concrete panels

External render

Precast concrete kerb units

Figure 3.19
Details of the Woolaway system

Figure 3.18
A Cornish Unit three storey block

distinguishable systems since they have storey height external corner and separating wall capping posts acting as permanent shuttering for the in situ frame (Figure 3.20).

The systems used exclusively for low-rise dwellings include Dorran (Tarran), Dyke, Myton, Stent, and Stonecrete. The Dorrans have a characteristic protruding band at first floor level formed by a precast concrete unit with a sloping external surface (Figure 3.21). The panels are 16 inches wide. Dykes have a similar projecting band, but are of post-and-panel at 2 foot centres. The other three systems in this group also have narrow bands at first floor level, but these do not project. The Stonecretes are post-and-panel at 2 foot 6 inch centres, while the other two are narrow panel with no posts. The Stents have one foot wide panels and the Mytons 16 inch.

Two other typical examples of low-rise systems employing precast concrete panels were the Firmcrete and Quickbuild systems. The Firmcrete system of constructing 'brickless houses' used chemically impregnated cement-bonded chipboard slabs. These sandwich slabs have a concrete core, and formed the separating and external

walls which rested on concrete footings and the foundation slab. The wall slabs were each 2 inches thick and the cavity was 5 inches wide.

In the Quickbuild system, external surfaces of wall panels sometimes were clad with brickbond tiling and ribbed aluminium, with precast concrete for the ground floor. Expanded polystyrene slabs were inserted between internal and external leaves to decrease the U value[135].

Steel, low-rise, concrete panel clad systems

This group includes the following systems: Bell Livett, Coventry, Cussins (Figure 3.22), Gateshead, Hitchins, Livett Cartwright, Open System and Steane.

The Bell Livett, Coventry, Gateshead, Hitchins, Open System and Steane systems were built in relatively small numbers, and are not dealt with here.

Cussins dwellings were clad in concrete panels, and, for the most part, faced with stack bonded clay slips simulating bricks (Figure 3.23). They can be distinguished from Smith system dwellings since the latter's clay slips are in stretcher bond carried across the panel joints.

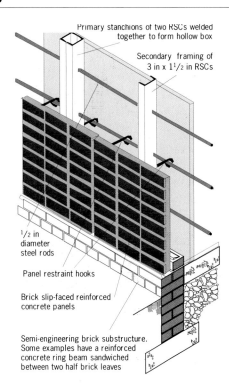

Figure 3.22
Details of the Cussins system

Livett Cartwrights are all in the Leeds area, and are characterised by the very tall, narrow (16 inches) cladding panels.

Basic structure

The steel frame of the steel framed systems is normally of small-section rolled steel. The Livett-Cartwright system is a typical example of the genre. The outer stanchions are mainly two storey, 5 inches × 3 inches, connected by horizontal 7 inches × 3 inches and 6 inches × 3 inches rolled steel channels clad with storey height reinforced concrete panels, the corner panels being L-shaped. The walls are lined with prefabricated plasterboard-faced timber panels, and the cavity contains a 1 inch thick, thermal insulation glass fibre quilt[136].

In the reinforced concrete columns (of those systems using concrete), because of the relatively small dimensions of the column, the cover to the reinforcement tended to be minimal. This led to corrosion rather more rapidly than was at first anticipated.

(Details of other systems are available in other reports in the series – see the full CRC catalogue[11]).

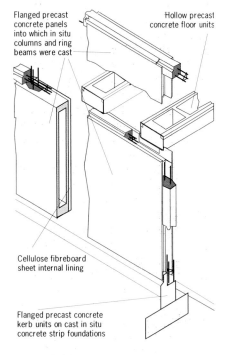

Figure 3.20
Details of the Reema low-rise system

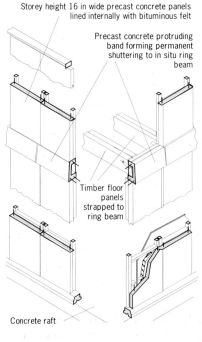

Figure 3.21
Details of the Dorran system

Figure 3.23
The deteriorating external wall of this Unity house is being replaced. Some temporary steelwork is in place, supporting the roof construction while work is undertaken

Main performance requirements and defects

Strength and stability
A degree of engineering judgement has to be used to determine when deterioration of the frame has progressed to the point that something needs to be done. Nevertheless, refurbishment of the structure is often possible, as Figure 3.23 shows.

See also the same section in Chapters 2.4 and 3.1.

Dimensional stability, deflections etc
It is in the nature of most concrete cladding supported clear of the main structure that restraints against its thermal contraction are generally small. The panels in these circumstances are relatively small and thin, and cracking is more often associated with expansive size changes in the cladding itself, rather than with contraction of the main structure. It follows that cracking is more likely to be due to compressive forces and is often accompanied by spalling.

BS 8200[28] provides data for many aspects of the design of concrete cladding, including those concerned

with temperature size changes. It provides thermal absorption coefficients for a range of cladding materials. In the absence of specific data, the following approximate coefficients may be used for building materials when dirty:
- light surfaces   0.5
- medium surfaces   0.8
- dark surfaces   0.9

The standard suggests that change of size should be calculated in two parts: firstly for temperatures above the temperature at installation, and secondly for temperatures below that at installation. In the absence of more exact data, the installation temperature could be taken to be 15 °C. This assumption for temperature must be treated with caution. If construction proceeded at temperatures below freezing, the assumption of a 15 °C installation temperature would then provide an appreciable underestimate of the expansive part of subsequent size changes.

For coefficients of linear thermal and moisture expansion etc, see Chapter 1.2.

Where the supporting frames are of reinforced concrete, progressive shrinkage of concrete frames combined with thermal/moisture cycling of infill panels may cause bowing of the panels.

Weathertightness, dampness and condensation
All cladding should resist the penetration of rain and snow to the inside of the building, and the cladding itself should not be damaged by rain or snow. In order to satisfy this requirement the cladding may either:
- be a lap-jointed moisture resisting outer layer (eg similar to tiles, see Chapter 9.1)
- be backed by a moisture resisting inner layer
- be backed by a clear cavity across which rain will not transfer

All joints, whether sealed or open, should be designed to prevent rain and snow reaching the inside of the building.

Thermal properties
See the same section in Chapters 2.4 and 3.1

Fire
See the same section in Chapters 2.1 and 3.1.

Noise and sound insulation
See the same section in Chapters 2.4 and 3.1.

Durability
Deterioration of concrete structural components in most systems of the type described in this chapter may lead eventually to unsafe conditions. It was this fact that prompted the UK Government to designate a number

Figure 3.24
Severe cracking has been revealed at the foot of this reinforced concrete column

of systems which would enable certain dwellings built using these systems to be put into mortagageable condition.

The following systems were designated under the Housing Defects Act 1984: Airey, Boot, Butterley, Cornish Unit, Dorran, Dyke, Gregory, Hawksley, Myton, Newland, Orlit, Parkinson, Reema Hollow Panel, Schindler, Stent, Stonecrete, Tarran, Underdown, Unity, Waller, Wates, Wessex, Winget and Woolaway.

It is evident that the causes of cracking of the structural concrete components should be investigated, if necessary by opening up the construction (Figure 3.24). Although repairs to external concrete components are feasible, they are rarely economic and rarely acceptable visually, and this has led to the practice of removing or making redundant the defective components in a number of low-rise systems. External cladding may be completely replaced and roof loads transferred to new external walls. PRC Homes Ltd has licenced contractors offering refurbishment packages.

Spalling of exposed concrete surfaces due to carbonation of the concrete and use of chlorides in the original mix has been common. Many untreated softwood windows have rotted, and by now must have been replaced with more durable alternatives.

The reinforced concrete door and window surrounds in Cranwells have by now carbonated, and rusting of the reinforcement has led to spalling of surfaces. Repairs described earlier for concrete systems are relevant.

There has been some detachment and corrosion of restraining hooks on Cussins cladding panels, leading to failure in high winds. Corrosion of steel spacers between Cussins cladding panels causes the joints to open and leads to rain penetration. The spacers need to be removed and replaced with a non-corroding alternative pack.

So far as the steel frame is concerned, some corrosion has been found, particularly on those sites exposed to driving rain[138].

Maintenance
See the same section in Chapter 3.1

Work on site

Storage and handling of materials
Precast concrete components distort, or break, if not stored correctly. Proper provision for supporting any replacement components must therefore be made. Proper lifting equipment must be used to prevent unnecessary stress on the component.

Inspection
In addition to those listed in Chapter 3.1 for steel and concrete frames, the problems to look for in cladding panels are:
◊   corrosion of ties of panels to frames
◊   carbonation of concrete
◊   bowing of panels
◊   disruption of joints

Chapter 3.3   Timber frame

This chapter covers both old and new designs of timber frames – that is to say, a brief note on the half-timbered structures with wattle and daub or brick nogging infill which date from medieval times, but more information on newer forms which are totally different in character. The maintenance and conservation of old structures is not dealt with in detail in this book, and information concentrates on the mostly imported systems of the interwar years and just after (Figure 3.25), and the brick veneered systems of the period since the 1960s. There is further discussion of timber cladding in Chapter 9.3.

Although the chapter concentrates on domestic construction, it should

be remembered that many systems were developed for the school building programme between the years 1950 and 1980 – Eliot, Punt (see Figure 4.22 in *Roofs and roofing*[24]) and Derwent, to name but three. What is said here in relation to domestic construction applies also to a considerable extent in other building types.

During the twentieth century there have been between half and two thirds of a million dwellings built in timber frame in many systems, the vast majority since the mid-1970s. More than 100 different systems have been identified. The systems developed before the 1939–45 war were mostly timber-clad and are readily recognisable as

being of timber frame construction. Many postwar systems – particularly those built since the 1960s – have brick claddings and are not always easily distinguishable from loadbearing brick cavity walled construction. In most cases, there are particular clues that indicate the hidden timber frame construction.

The claddings used for timber frame dwellings include brickwork, tile hanging, horizontal or vertical timber boarding, and rendering. One cladding system can be applied throughout the building, or – as on many of the constructions built since the 1970s – different systems can be used in combination. With brick cladding, the windows are usually set back in the reveals, and there may be small gaps or soft packing beneath the window frames, under the eaves and at the top of verges in order to accommodate any differential movement between the timber frame and the cladding (Figure 3.26 on page 138).

Internally, the external walls are dry-lined, usually with plasterboard nailed directly to the timber frame. However, many systems built soon after the 1939–45 war used fibreboard linings, possibly fixed over timber boarding.

In the interwar period, six systems were in use, none particularly numerous, examples being LCC (Figure 3.27), Solid Cedar (Figure 3.28), and SSHA. A few were built in systems at Aberdeen, Lanark, and Newcastle. All these systems were directly clad with horizontal boarding except the Newcastles, which had storey height vertical boards and a horizontal band at floor

Figure 3.25
Imported timber frame and timber clad houses

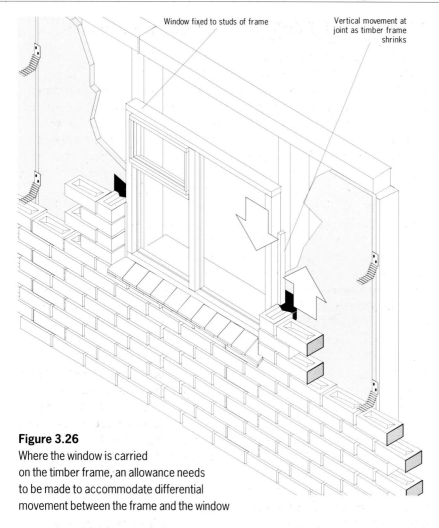

Window fixed to studs of frame

Vertical movement at joint as timber frame shrinks

Figure 3.26
Where the window is carried
on the timber frame, an allowance needs
to be made to accommodate differential
movement between the frame and the window

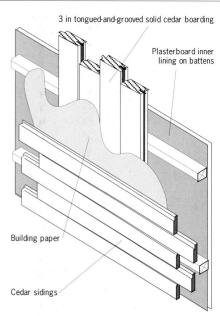

3 in tongued-and-grooved solid cedar boarding

Plasterboard inner lining on battens

Building paper

Cedar sidings

Figure 3.28
The planking system in Solid Cedar
dwellings

Figure 3.27
Houses built for the former London County Council

Figure 3.29
The first lift of brick cladding in place on the
breather membrane-clad, timber structural
walls

level. Solid Cedar, and some SSHA and Aberdeen houses had solid planked (or staved) walls, and the rest stud walls.

Timber, low-rise, 1945–65

During the period 1945–65 some 20,000 timber frame dwellings were built in eight systems. The most numerous were Calder, Scottwood, Swedish, Spooners and Weir Timber. A few were built in Canadian, and Simms Son and Cooke. Most of these systems had cladding fixed directly to the frame – such cladding included timber boards, plywood, and fibre cement sheets.

Identification of the original Calders is straightforward since they had shallow pitched roofs which included monopitches. Most were clad with fibre cement panels, and had a dark coloured horizontal band at first floor level. The Scottwoods mostly had rendered plywood external walls, with a self coloured projecting band at first floor level. A few examples had brickwork ground floor elevations. This system is the most difficult to identify, since it is possible to confuse examples with other systems.

The Spooners were built mostly in short terraces or semis. They were commonly clad in brick, although a few early examples had the upper storeys clad in sheet metal somewhat similar in appearance to BISFs. Windows were mostly timber, although a few early examples were steel.

Timber low-rise, post-1965

The more prolific of those current in the first ten years of the period since 1965 include Anchor, Anvil, Engineered Homes, Eurodean, Facta, Frameform, Guildway, Hallam, Laing, Metatrim, NBA Silksworth, Purpose Built, Quickbuild, Rileyform, Rowcon, TRADA, and Weir Multicom.

There are many more systems current in the period since 1975. It is not possible in the space of this book to give any identification notes for particular systems, save to say that practically all these examples have been clad in brickwork (Figure 3.29). Some earlier examples were clad in timber boarding over a cavity.

Normally the windows are fixed to, and move with the frame, and a good clue to this class of timber frame systems is to look first to see whether there is a deep window reveal, and 'soft' joints between windows and cladding and between cladding and soffits. Another clue is to tap the inner wall lining over a window. Timber frame walls usually have timber lintels, which give a more solid sound than dry-lining on masonry. Failing this, identification usually means lifting a switch plate to look into the cavity. Gable walls are usually of unlined stud framing.

During the 1980s and 1990s there have been considerable improvements in the techniques of timber frame construction. For example, in quality control of components fabricated off site, and speed and accuracy of erection on site which in turn helps the speed with which the building can be enclosed and waterproofed.

Characteristic details

Basic structure

Early medieval timber frames consisted of storey height posts set into wall plates at around 600 mm centres, and these frames were frequently jettied or cantilevered above a bresummer bracketted from the storey below (Figure 3.30). Since the infilling could be weak and unable to resist lozenging of the frame, diagonal wind braces were used at end bays. Occasionally, purely for decorative reasons, the frame was embellished with curved members. And occasionally it may be found that some window and door frames are structural, and integrated with the remainder of the loadbearing frame.

When oak, the preferred material for the frames, was needed in enormous quantities for building warships of the line, economies were introduced into building work, and the most obvious effect of this was that the centres of studs was increased, without real detriment to the strength of the wall.

Four main forms of modern timber frame dwelling construction have been identified:

Figure 3.30
Thaxted Market Hall

- Balloon frame: two-storey height or eaves-height external wall panels with the studs continuous from ground floor to roof (Figure 3.31a)
- Platform frame: storey-height external wall panels which are erected upon platforms formed by the ground and upper floor construction (Figure 3.31b). This is by far the most common form in the UK
- Post-and-beam: a structural frame of widely spaced timber posts and beams. Planked, joisted or panelled floor and roof units span between the beams, and non-loadbearing infill panels span between the posts to form the external wall claddings
- Volumetric box: assembled from room-sized prefabricated boxed accommodation units

The external walls of timber frame dwellings built between the wars are typically either timber framed panels with large-section studs, or panels of virtually solid timber planking. Both alternatives are directly clad with timber boarding. The timber was rarely preservative-treated, excepting the occasional brush application of creosote. Generally there was no insulation between the studs (Figure 3.32).

The dwellings built between 1945 and 1965 were clad wholly or mainly in single leaf brickwork, with the brickwork separated from the timber frame by a cavity and connected to the frame by metal wall ties.

Most systems used between 1966 and 1975 had an external sheathing on studs and a separate cladding of brick. Some systems were directly clad with timber boarding or plywood claddings fixed directly to the frame or tile-hung panels on the front and rear walls.

To meet the revised thermal insulation requirements introduced in 1966, most timber frame systems incorporated thermal insulation between the frame studs of the external wall panels. The increased temperature gradients meant that there was more chance of condensation within the external walls. To counter this threat, vapour control layers were incorporated into many of the timber frame dwellings built during this period. The layer was usually of aluminium foil-backed plasterboard, never very effective, or was a separate layer of polyethylene film positioned on the warm (internal) side of the frame. The breather membrane was either of building paper or of light bituminous felt and was intended to project below the sole plate to protect it, as well as performing its main function of keeping liquid water from reaching the panel, and releasing internal water vapour (Figure 3.33). At this time there was no insulation under the floor finish, and there was consequently a risk of a thermal bridge at the perimeter under the sole plate.

Towards the end of the period, the timber sole plates of some systems were treated with preservative. There were also a few instances of the timber frame itself being treated.

Most separately clad systems have single leaf brickwork cladding connected to the timber frame with metal wall ties through a nominal 2 inch cavity. Other separate claddings, such as horizontal or vertical timber boarding, are fixed to

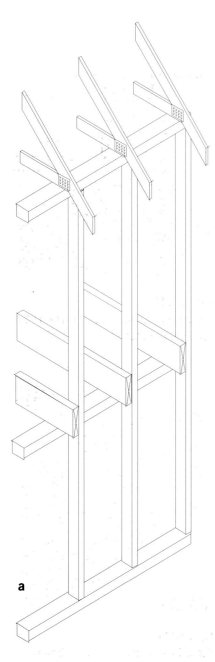
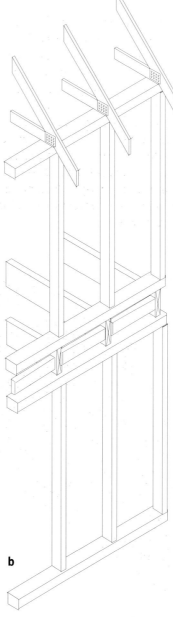

Figure 3.31
Balloon frame (a) and platform frame (b)

timber battens over the frame. There is usually a breather membrane fixed to the external face of the plywood sheathing. In some cases there is no sheathing; alternatively, the sheathing is of fibreboard, and timber bracing, plywood corner gussets or metal strapping is used to resist racking loads. For some systems, only the corner panels in a wall may be externally sheathed. The provision of ventilation to the wall cavity varies, but the cavity is usually fire stopped at separating wall positions.

The external walls of directly clad systems are clad for most of their area (particularly at ground floor level) with timber boarding or rendered plywood. The cladding is fixed over a breather membrane either directly to the timber stud frame or through the external sheathing. The linings used in separately and directly clad systems are usually of plasterboard fixed directly to the stud frame over a vapour control layer.

Some of the timber frame systems have rationalised traditional or conventional variants. A minority of systems were of post-and-beam construction, with the vertical timber posts carrying the floor and roof loads. The posts were infilled with timber frame panels[139].

Abutments

A timber sole plate is usually fixed to brick foundation walls or to concrete raft foundations. It provides a level and accurately dimensioned base to which the timber frame wall panels (or, with some systems, the suspended ground floor joists or floor panels) are fixed. A DPC is located under the sole plate to prevent rising damp reaching the timber components.

The sole plate is fixed by metal straps or brackets that are cast into the concrete raft or tied into the brick foundation wall with masonry nails or anchors. The sole plate is generally above floor screed level, but in some systems the floor screed is finished level with the top of the sole plate. For some systems the bottom rail of the timber wall panels

also acts as the sole plate, and is located directly on the DPC on the foundation wall or raft.

Accommodation of services

Soft joints will need to be provided round service entry points to allow shrinkage of the timber frame.

Main performance requirements and defects

Choice of materials for structure

Oak for the frames of buildings came into general use around the sixteenth century whereas the occasional use of other species, such as elm, may be encountered in earlier work which has survived.

Structural members of timber frames of small dwellings built since the 1939–45 war are almost certain to be softwood of a variety of species nominally classified as non-durable, clad with plywood sheathing[140]. Some dwellings may be found sheathed with bitumen impregnated cellulose fibre boards and braced with steel straps (Figure 3.34). Rather confusingly, at least one of the timber frame systems was clad with steel sheets.

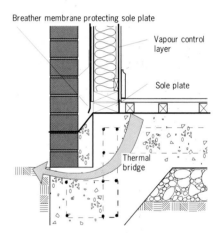

Figure 3.33
The breather membrane projects below the sole plate to protect it

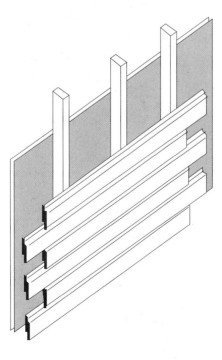

Figure 3.32
Typical simple timber frame construction from between the wars

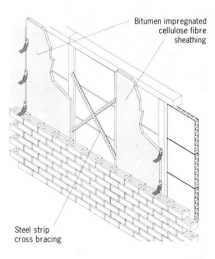

Figure 3.34
Some timber framed houses will be found with cellulose board sheathing and steel bracing instead of plywood sheathing

Strength and stability

Timber frames in heritage buildings were designed on a rule of thumb basis founded on long experience. Nearly all features were overdesigned, promoting longevity, and safety factors in consequence were large though undetermined.

In modern timber frame, to a considerable degree, frame and cladding act together to provide the required strength and stability to the walls. Exploratory tests on single frame elements and small areas of brickwork indicate a substantial strength contribution even from unreturned brickwork[141].

Based on the evidence of the BRE studies, most system built timber frame dwellings constructed between 1920 and 1975 have no serious structural faults. In dwellings built between 1945 and 1965, no instances of the frames of dwellings being compromised on structural stability grounds have been observed by BRE, although some Calders have been subject to differential settlement. Decay in the structural frames of Scottwoods could compromise structural stability if nothing is done about the decay observed. Most timber in the structural frame is potentially vulnerable to decay if the moisture content remains above 22% for comparatively long periods, and

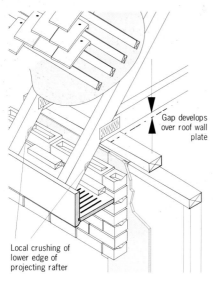

Local crushing of lower edge of projecting rafter

Gap develops over roof wall plate

Figure 3.35
Roof loads transferred to the non-loadbearing outer leaf

some softwood species are at risk at 20%.

In other systems the most common defect affecting structure is nailing error, observed in several systems. Where this occurs to a significant proportion of fixings in any particular area, integrity may be reduced. Misplacing is not easy to spot on inspection unless suspicions are aroused because of some visually apparent distress in the structure; for example following high winds.

Structural rigidity of the frame in timber, low-rise post-1965 systems is normally provided by the plywood or fibreboard sheathing. There have been some cases of decay of this sheathing, particularly in Frameform dwellings.

By far the commonest fault in this category of dwellings has been the lack of adequate provision for differential movement between frame and brick cladding. Although structural performance is rarely compromised, there have been cases where transfer of roof loads has taken place from frames to what should have been non-loadbearing outer leaves (Figure 3.35).

There have been cases of missing wall ties, particularly at gables, and some cases of walls, especially gable peaks, being damaged by wind suction. Spooners suffer particularly in this respect, showing as bulging wall panels. Replacement of wall ties will be needed where inspection (eg by optical probe) reveals this deficiency.

The structural use of wood based panels, for example for sheathing purposes, is discussed in BRE Digest 423[142].

Dimensional stability, deflections etc

Infilling to the early oak frames with brick nogging, in either herringbone, stretcher or stack bond, was better able to resist distortion of the frame than wattle, but considerable movements could still take place without threat of collapse (Figure 3.36). Sometimes the wood was left to weather naturally, and sometimes was given a coat of tar, largely for cosmetic reasons.

Cracking in the brick veneer of timber framed housing

Following some disruption in the external cladding of some timber framed houses, the BRE Advisory Service was asked to investigate. The front and rear elevations were either brick or rendered blockwork clad to first floor height, above which was vertical tile hanging, with an aluminium cover to the top of the brick and blockwork. The gable ends were clad with brickwork or rendered blockwork to the full height.

Cracking was reported some three years after the dwellings were occupied. Also there had been a considerable deterioration to the brickwork, ie erosion of the mortar and cracking in the mortar joints. There had been cracking of the rendering on the blockwork and an outward movement of the gable end, such that the overhanging verge on some dwellings was no longer present.

Some of the cracking was seen to be typical of sulfate attack, which was confirmed by laboratory analysis. However, other movements in the external leaves could not be explained by this mechanism alone, and more investigation on site revealed two further problems – a lack of sufficient wall ties, and differential movement between the timber frame, which had shrunk, and the brick cladding. This movement had not been allowed for by using soft joints.

The gable end brickwork to one property was removed to expose the wall ties, building paper and plywood sheathing. The first three layers of wall ties from the DPC were of the chevron type fixed to the timber frame studwork. Above this level, butterfly ties had been used. They were fixed to the studwork with one or two nails bent over the wire section of the tie, which was then bent to lay on the mortar joint. These ties were quite inadequate for their intended use. From the number of ties actually counted, compared with what would normally have been specified at 600 mm horizontal centres and 450 mm centres vertically, only 30% of the required ties were actually present.

Installation of extra ties was required to restore the integrity of the structure. Since the differential movement had all taken place by the time of the inspection, re-sealing the disrupted joints would be sufficient to restore weathertightness.

Figure 3.36
'Considerable movements could still take place without threat of collapse.' A farmhouse near Tenbury Wells

Timber-framed, masonry-clad walls

Differential movement between the timber structure and the separate brickwork is a potential source of problems. This movement is mainly the result of shrinkage across the grain of horizontal timber members such as floor joists, rails and plates. Most of the shrinkage occurs after construction and during initial occupation, as the building dries out. If insufficient allowance is made for differential movement, the components fixed to the timber frame – such as windows, rafters and direct claddings – can bear on the brick cladding (Figure 3.38). The rule of thumb recommendation is to provide 6 mm soft joints in the ground storey, 12 mm in the next storey, and so on.

Window sills, tilted or rotated if of timber, or broken if of tile, are among the problems associated with differential movement (Figure 3.39).

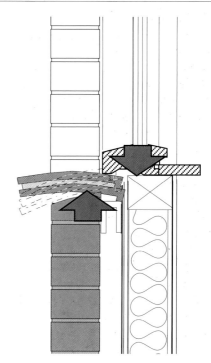

Figure 3.39
Rotation of a sill caused by the shrinkage of a timber frame

Figure 3.37
Here a complicated strap ties together beam and external wall plate to the head of a post (Photograph by permission of G C R Hughes)

Where movements have occurred in the past, for example where bearings of bressumers have pulled off their supports, it is not uncommon to see blacksmith-made straps spiked into the timber and bent over and built into the offending wall. Restoration of stability is of course still feasible by similar, albeit ad hoc, methods (Figure 3.37).

For coefficients of linear thermal and moisture expansion etc, see Chapter 1.2.

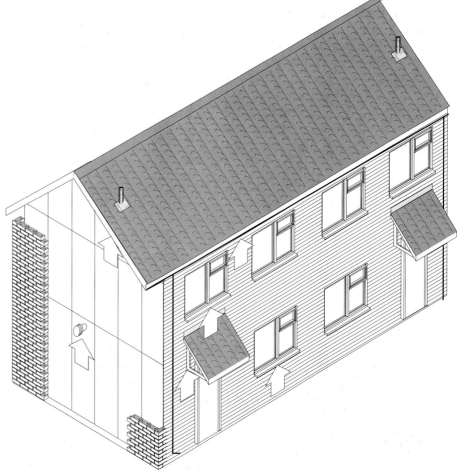

Figure 3.38
Differential movement between frame and cladding needs to be allowed for at a number of points

Another typical movement related defect is a reverse slope on the horizontal flashings at the joint between the first floor cladding, which is carried on the frame, and the masonry ground floor cladding, which is not. Leakage from disrupted flashings may contribute to any sheathing decay. Cracked sealant between the masonry and frame-borne claddings may also occur as a result of differential movement. Some sealant failures at these junctions may be the result of movement between the timber structure and the masonry cladding caused by wind action where the wall ties are either inadequately fixed or insufficient.

Weathertightness, dampness and condensation

The joint between the frame and nogging in old structures is vulnerable to rain penetration. Clay daub would of course swell on being wetted, to give a reasonably tight joint, but a lime mortar joint would not do likewise. Ad hoc solutions may sometimes be encountered (Figure 3.40).

The bituminous felt membranes used between the wars and in some post-1945 systems have minimal water vapour transmittance. Any vapour transmission occurs mainly through the lap joints. However, inspections have shown little evidence of high moisture content or decay in the timber structure of systems where bituminous felt has been used. These dwellings generally had minimal insulation and indoor ventilation rates were relatively high. An efficient breather membrane was therefore less essential.

In some dwellings where there is no sheathing, the bituminous felt has disintegrated at the foot of the external wall. Disintegration of the felt is particularly likely where mortar droppings have accumulated in the wall cavity. Breather membranes have shown localised decay where a defective detail has caused persistent rainwater leakage.

Inspections indicate that in many older dwellings the aluminium foil which used to be used as a vapour control layer has corroded and become ineffective as a vapour control layer. However, there is little evidence of resulting high timber moisture contents. The polyethylene vapour control layers used in some post-1966 systems were generally in good condition when inspected, although on occasion the layers showed some loss of flexibility.

Carelessness in site work also may play a role in lack of weathertightness (Figure 3.41).

There will be an obvious risk to timber frames built in flood susceptible areas.

Thermal properties

There was little if any additional thermal insulation installed in any of the interwar systems initially. U values can be expected to have been little better than those common in solid masonry at that time, especially if the walls were not airtight.

While many dwellings may be

Figure 3.40
Timber hood mould protecting a vulnerable horizontal joint

Figure 3.41
Damage caused to breather membrane by panels being dragged across rough ground. Repairs are rarely done effectively. The DPC has also suffered, and the sole plate inevitably will be at risk of rising damp from the torn DPC

prime candidates for increased thermal insulation, this is not easy to achieve. Perhaps the best way, where there is room, is to put an extra lining internally after installing a vapour control layer, making sure there is no existing vapour control layer since this could trap moisture (Figure 3.42). Removal of the original outer cladding and re-cladding is a more expensive solution, and, moreover, one which destroys the original appearance of a dwelling.

A small amount of thermal insulation was incorporated into some later dwellings. Where systems have stout board linings over studs, it may be possible to inject thermal insulation into the closed cavity between each pair of studs, although care must be taken to avoid cavities carrying electric cables. The cavities between the sheathing and the brick cladding should not be filled.

Fire

Most timber frame construction has been faced with an external leaf of brick masonry, and therefore there has been little difficulty in complying with requirements. However, fire can spread rapidly in timber frame dwellings once the wall cavity is breached, particularly where the ceiling or wall linings are flammable. Some Swedish Timber Houses and Scottwoods had fibreboard linings with poor spread of flame characteristics. Since 1965, there have been statutory requirements for the standard fire performance of linings of wood based materials. Where these requirements were not achieved by the original linings used, many building owners have replaced the linings.

A timber framed wall consisting of an external skin of at least 100 mm of brickwork or blockwork, covering studs at least 37 mm wide spaced at not more than 600 mm centres, either sheathed or not sheathed, and lined internally with 12.5 mm plasterboard skimmed with 10 mm lightweight plaster will give half an hour's fire resistance.

Spread of fire within wall cavities and into wall cavities of adjoining houses was identified as a risk in site inspections of timber frame housing under construction. Defects included cavity barriers omitted or removed during construction, not positioned or not fixed properly, not fully closing the cavity, not tightly lapped or butted (Figure 3.43), and imperfections of fit not made good by fire stopping. Examination by optical probe may be required, or a stiff wire can be used to check the existence of paths between the wall cavities of adjacent dwellings.

Noise and sound insulation
See Chapter 1.9 and the same section in Chapter 2.1.

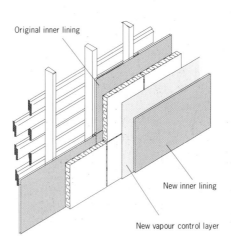

Original inner lining

New inner lining

New vapour control layer

Figure 3.42
Adding thermal insulation to an existing timber frame dwelling

Durability
General comments
Timber decay is the principal potential danger to older timber frame dwellings, and sapwood even in allegedly durable hardwoods may deteriorate (Figure 3.44).

However, where the timber has been kept dry, long life is normal. Timber in any structure is potentially vulnerable to decay if the moisture content remains high (that is to say, more than 22%) for sustained periods and the timber species is non-durable. Sapwood is

Figure 3.44
Decay in a medieval sole plate

Figure 3.43
Cavity barriers being installed incorrectly. They should be tight butted, not lapped. Lapping as shown leaves gaps

always vulnerable at this moisture content.

Timber based sheet materials may be at risk from moisture-induced decay, and some may be subject to a permanent loss of integrity if wetted for sustained periods. Unpreserved timber can also be subject to attack by wood-boring insect species such as the common furniture beetle, *Anobium punctatum.*

Although systems have been designed to prevent the timber from becoming damp, wetting can occur from four main sources:

● rainwater penetration through the external envelope
● condensation generated by water vapour from within the building
● rising damp, which occurs if the DPC is ineffective
● leaking plumbing or domestic appliances

The most common positions for decay are in the sole plate or bottom panel rail and in the sheathing at, or just above, these members (Figure 3.45). Decay in the sheathing and framing higher up the wall is, however, rare. Isolated instances of wetness and decay may occur under windows. Wetting under windows may occur because the window frames are badly fitted or poorly maintained. A flashing, situated over the horizontal frame rail below the window, is needed to prevent water entry.

Water is able to cross the cavity if the wall ties slope downwards towards the timber frame (this slope being caused, perhaps, by differential movement).

British Standard BS 5268-5[143] now recommends – taking into account the risk of decay, cost of replacement and safety – that the sole plates of timber framed structural panels are preservative treated and, depending on the timber species and sapwood content, that the framing members also are treated. However, the sole plate and framing in many of the timber frame systems used in the period up to 1975 were not treated.

In system built timber frame dwellings constructed between 1920

and 1975, the incidence of timber decay is slight. However, some timbers contain sufficient inherent acid content to attack metal fixings and coverings. Certain preservatives (eg copper-chromium-arsenic – CCA) and fire retardent treatments can have a similar effect in damp conditions. Most corrosion problems associated with CCA arose because salt based formulations were being used. The risk is significantly lower with oxide based formulations.

The metal fixings used to connect the timber components – such as screws, bolts, nails, wall ties and metal plate connectors – are generally unlikely to show much deterioration apart from surface corrosion. On the other hand, instances have been seen of failed wall ties having no protective coating other than paint. Localised cracking of brickwork cladding has occurred in dwellings of some separately clad systems, generally attributable to expansive corrosion of metal wall ties or lintels.

The incidence of timber decay in the timber frame systems inspected by BRE was generally very low and, with a few exceptions, the decay was localised and superficial. In cases of decay in timber structures, the principal causes appeared to be water penetration, either through faults in the cladding or through the junctions of cladding with windows and doors, and the occasional plumbing leak. These defects in the cladding were caused by poor construction or lack of maintenance.

Replacement of areas of decayed structural timber in timber frame buildings is relatively easy, and in many cases the replacement of components, or even wall panels, would be possible. However, it is important that the cause of any condensation or water penetration is determined and the problem rectified before components are replaced.

Where timber frame external walls are to be thermally upgraded, the suitability of any existing vapour control layers should be carefully considered. It may be necessary to install new ones.

Case study

DIY alterations to a timber frame wall
On one site inspection, investigators came across a case where a DIY enthusiast had inadvertently created a number of potentially disastrous defects. The plasterboard lining to the external wall of one of the rooms had been cut away, and studs supporting the floor above together with a portion of the thermal insulation blanket and the plywood sheathing had been completely removed above work top height.

Into the space thus created, a tank containing tropical fish had been inserted. Not only was the structure vulnerable (it was fortunate that load sharing had taken place) but the humidity from the heated tank and the house could migrate into the cavity, thereby wetting the structure.

Figure 3.45
The cavity tray here is not tucked under the breather paper, and any water running down it will be fed to the sole plate. (There are other defects too - the barriers are too short and too small to fill the cavity

Comments on particular timber framed systems

In early timber clad low-rise systems, in which the cladding was fixed directly to the studs or staves with no provision for ventilation to remove rain penetration, there were some instances of localised decay, particularly in the LCC examples. These instances were, for the most part, concentrated in areas of severest exposure. The LCCs had some problems with dry rot earlier on, but this has now been eradicated. Some Solid Cedars have had sections of external claddings replaced because of decay.

Vapour control layers were introduced into some of the systems used between 1945 and 1965, the exceptions being Scottwoods and Lanarkshires. Scottwoods can suffer considerably from decay due to rain penetration via defective window detailing. In contrast, the Simms Son and Cooke dwellings have had remarkably few problems during 40 years service.

Problems which may be encountered are very high levels of moisture in sole plates, or in lower panel rails where there is no separate sole plate. Moisture meter readings taken in CCA preservative treated sole plates may be subject to considerable error, and care should be taken to obtain an accurate assessment. Some Spooners may have unprotected wall ties. Calders had asbestos cement claddings, not now considered acceptable if they are deteriorating.

BRE have observed instances of sulfate attack of the outer brick skin in some examples. This has mainly been concentrated in examples where there have been defective cappings to brick panels, for example in Frameform, allowing outer leaves to become saturated for long periods. Spooners suffer particularly from corrosion of steel lintels. There also have been examples of rotation of lintels, causing cracking of the brick outer leaf.

Where aluminium foil backing to plasterboard has been used as a vapour control layer, this will by now have corroded, and will have ceased to perform its function. There may be consequential risk of interstitial condensation within the stud wall if thermal insulation thicknesses are increased. Polythene membranes began to be introduced around 1966, but were not in general use until some years later.

Maintenance

There are a few dwellings of some system types where decay has occurred in particular parts of the construction. However, the decayed timber is usually localised and therefore relatively easy to cut away and replace; for example some timber sole plates, not preservative treated, have been subject to decay and will require local replacement.

Missing or blocked vents to suspended timber ground floors may be encountered; as in traditional construction, their presence is essential.

Work on site

The inspection and assessment of system built timber frame dwellings is usually straightforward, particularly if the system and any potential vulnerabilities associated with it can be identified in advance. Inspections should be undertaken every five years or so. Only in exceptional cases should it be necessary to open up the construction to investigate areas of high moisture content.

The procedures for carrying out surveys of timber framed houses built before 1975 have been published by BRE [139] and, for those built since then, by TRADA [144] and TBIC [145].

The BRE procedure can be used on its own or as a supplement to the TRADA procedure where the latter leaves the condition of the dwelling in doubt.

Where surveys indicate deficiencies related to structural stability, durability or fire protection, and hence the need for further investigation, guidance on re-examination of the building exterior and interior, and the wall cavity and frame, is to be found in BRE Good Building Guide 11 [146]. Guidance on the interpretation of information collected is given in Good Building Guide 12 [147].

The problems to look for are:
◊ rot
◊ absence of soft joints to permit shrinkage of frames
◊ rotation of lintels
◊ high moisture levels in frames
◊ sulfate attack in claddings
◊ fractured pointing at window/wall joints
◊ missing or misplaced cavity barriers
◊ missing wall ties
◊ corroded wall ties

Chapter 3.4   **Sheet cladding over frames**

This chapter deals with all kinds of opaque lightweight sheet cladding, metal sandwiches and composites, either used as part of the original building or as overcladding. Sidings of wood or plastics are dealt with in Chapter 9.3. Overcladding is defined as any additional thermal insulation and weatherproof skin installed over the original external wall[148]. For the purposes of this book transparent or translucent cladding is treated as curtain walling and will be found in Chapter 4.3.

Sheet cladding over framing can take many forms, and employ many different kinds of materials. It is to be found in most building types, though industrial type buildings probably provide the majority of examples (Figure 3.46). Systems were also developed for other building types – the Oxford Regional Hospital Board system for health buildings for example. However, many of the features and problems of this kind of construction are typified by those seen on housing, both low and high-

rise, and the chapter therefore concentrates on this building type.

In England around 51,000, or just over 1 in 330 of the total stock of dwellings, have metal sheeting to their predominant wall structure with very few indeed being recorded as built before 1919[2].

The most relevant standard is BS 8200[28]. This standard includes a checklist for design and a reasonably comprehensive discussion of performance requirements, and design and production criteria.

Timber, low-rise, interwar, steel clad

The two systems in this category are Weir Douglas (Figure 3.47) and Cowieson. Practically all these are situated in Scotland. The steel sheets had plain lapped horizontal joints and cover moulds over the vertical joints. The finish was paint or harling.

Steel, low-rise, steel clad

Included in this category are Atholl, BISF and Telford systems, together with a few Rileys. The earlier examples of Atholls have strips of steel covering the vertical joints in the steel panelling (Figure 3.48) or the steel stanchion exposed (Figure 3.49); the postwar examples all have butt joints.

BISF system built dwellings are the commonest of all the steel systems, characterised by ribbed steel sheeting to the first floor elevations (Figure 3.50) above render-on-mesh ground floor elevations. They present no problems in identification in their original form, although many have by now been refurbished and

Figure 3.46
Sheet cladding to the new Burn Hall , a laboratory for conducting fire experiments, erected at BRE in 1996

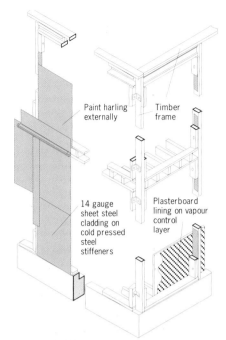

Paint harling externally

Timber frame

14 gauge sheet steel cladding on cold pressed steel stiffeners

Plasterboard lining on vapour control layer

Figure 3.47
Detail of construction of the Weir Douglas system external wall

Figure 3.48
An interwar Atholl system external wall with steel strips covering the edges of the sheets (showing some distortion)

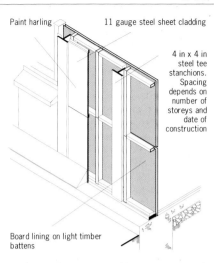

Paint harling

11 gauge steel sheet cladding

4 in x 4 in steel tee stanchions. Spacing depends on number of storeys and date of construction

Board lining on light timber battens

Figure 3.49
Detail of the early Atholl system external wall with steel tees housing the edges of the sheets

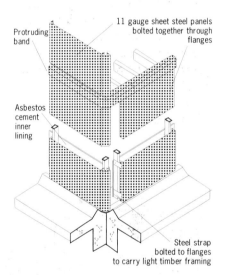

11 gauge sheet steel panels bolted together through flanges

Protruding band

Asbestos cement inner lining

Steel strap bolted to flanges to carry light timber framing

Figure 3.51
Detail of the Telford system external wall

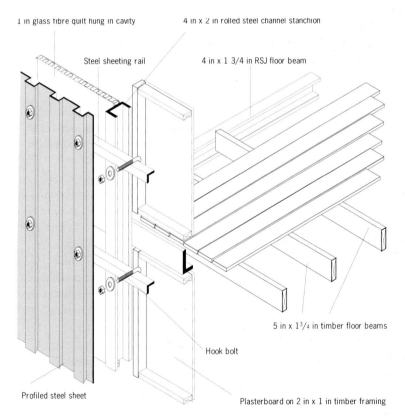

1 in glass fibre quilt hung in cavity

4 in x 2 in rolled steel channel stanchion

Steel sheeting rail

4 in x 1 3/4 in RSJ floor beam

5 in x 1³/₄ in timber floor beams

Hook bolt

Plasterboard on 2 in x 1 in timber framing

Profiled steel sheet

Figure 3.50
Detail of the BISF system steel sheeted external wall at first floor level

present marked changes in appearance. The few Rileys in Manchester have a superficial similarity to the BISFs built adjacently, with similar cladding to the first floor elevations.

Telfords were relatively few in number. They are characterised by a protruding band course at first floor level (Figure 3.51).

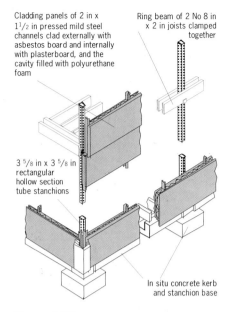

Cladding panels of 2 in x 1¹/₂ in pressed mild steel channels clad externally with asbestos board and internally with plasterboard, and the cavity filled with polyurethane foam

Ring beam of 2 No 8 in x 2 in joists clamped together

3 ⁵/₈ in x 3 ⁵/₈ in rectangular hollow section tube stanchions

In situ concrete kerb and stanchion base

Figure 3.52
Detail of the Hawthorne Leslie system external wall

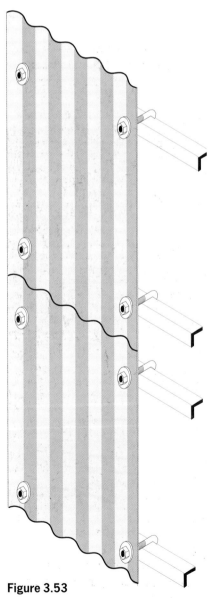

Figure 3.53
Asbestos cement sheets used as cladding

Steel, low-rise, light panel clad

The following systems are included in this category: Arcal, Arrowhead, British Housing, Hawthorne Leslie (Figure 3.52), Howard, Roften and 5M. British Housing, Rileys and Roftens were built in small numbers and are not dealt with here except to record, as noted earlier, that some Rileys had aluminium panels on the first floor.

Most examples of Arcal dwellings are rendered and dashed over mesh on timber framed panels. Arrowheads are normally tile hung, and all have a very deep panel over first floor windows below the eaves. Hawthorne Leslies have asbestos sheet covered, foam filled panels with a resin based fine aggregate coating. Timber cover strips are fitted at separating walls and external corners. Bungalow versions can be found.

Howards all have concrete panels below ground floor window sill height, and asbestos cement sheet cladding above. They also have asbestos cement cased chimney stacks. A significant number of Howards have already been insulated externally.

5Ms all originally had flat roofs, although some by now have been re-roofed with pitched. Much tongued-and-grooved boarding was used on the ground floors, with tile hanging at first floor level and plywood fascias situated between the different claddings, and also just below eaves. Flank walls were clad with concrete panels.

Characteristic details

Basic structure

The basic structure of lightweight cladding, whether it is original or overcladding, is essentially a thin sheeting of inert material, board or protected metal carried on a timber or, more usually, metal framing which in turn is carried on the main structure of the building.

Boards

Asbestos cement boards made with Portland cement and asbestos fibres, and formerly used for cladding, were normally stiffened with some form of corrugation, with joints formed by simple overlaps rarely taped. The sheets were available in a variety of profiles, and are usually found fixed with hooks or bolts to sheeting rails (Figure 3.53). Since they were vulnerable to impact damage, they were not often used at ground floor level on buildings vulnerable to vandalism or accident.

It is almost universally accepted that asbestos fibres pose serious hazards to health[149] and therefore asbestos cement is being phased out of production in many countries. Sheets reinforced with cellulose or PVA fibres have now largely replaced those formerly reinforced with asbestos fibres. Nevertheless many industrial buildings remain which are still sheathed in asbestos cement. Occasionally they may still be found on housing, for example temporary bungalows built soon after the 1939–45 war.

The newer sheets are supplied with a wide variety of decorative finishes including enamelling in a range of artificial colours or fine aggregate in natural colours. The standard boards are normally 5–9 mm thick, with the aggregate finished sheets around 13 mm. They are fixed either to timber or metal battens, which are frequently spaced away from the primary structure (Figure 3.54), or can be substituted for glazing in light cladding.

Boards of sandwiched materials of various kinds are available, and examples of the insulating kind include expanded and extruded polystyrene, polyurethane, polyisocyanurate, foamed glass, phenolic foam and mineral fibres. They are available in various thicknesses, and some are supplied with permeable but water resistant coatings. Most, however, will normally require some form of outer skin. There is no difficulty in achieving any particular U value found to be economic.

Where used for overcladding, boards can be fixed directly to the old structure. If of sufficient strength, boards can be held away from walls by timber or metal battens or cleats.

Plain boards are available in fibre reinforced cement or calcium silicate, or with resin binders. Some boards have smooth polymer finishes. Fibre reinforced, predecorated cement based boards consist of compressed and autoclaved sheets of Portland cement reinforced with natural and synthetic fibres and fillers, with surface coatings permanently fused or bonded to the base sheet. (Chapter 3.5 deals with these materials).

Boards can usually be found with the following properties.

Dimensions: 4 mm and 7.5 mm thickness with aggregate finish typically 6.5 mm; overall size 3070×1240 mm and 2520×1240 mm.

Weights: 4 mm, 7.4 kg/m²; 7.5 mm, 13.1 kg/m²; with aggregate finish, 20 kg/m² depending on type.

Flat boards should have a distortion-free surface and are obtainable in a wide range of colours. Surface textured boards have a lightly dimpled, sculptured surface. The structure, size and colour of aggregate finished boards is determined by the nature of the quarried and crushed stone.

Sheet metals
Sheet metals are usually of galvanised steel or aluminium. Galvanised steel is available with polyester coatings, silicone polyester coatings and vinyl coatings. Aluminium is available with polyester coatings, PVDF and modified alkyd coated sheets.

Profiled steel and aluminium sheets are not usable below 1.5 m height above access level, since, in their commonly available thicknesses, they will not meet the impact damage test. Steel greater than 0.8 mm is satisfactory above 1.5 m, but aluminium would need to be thicker. The manufacturers should be asked for test results.

Where long lengths of steel sheet are to be used, the manufacturer should specifically be asked whether any curvature due to the 'roll memory' has been taken into account.

Steel or aluminium sheets can be pressed into panels with returned edges, or shallow drawn

deformations can be pressed in the centres of the panels, or they can be used as corrugated sheets. Although sheets are available in lengths of up to 20 m, care must be taken to provide for thermal movements in the lengthways dimension of the sheet; movements are automatically compensated for in the width direction because of the inherent flexibility given by the corrugations.

Sheet metals coated before forming should not have sharp arrises or sharp radius bends. The quality of the cover on the back of panels or sheets should be no less consistent than that of the front, especially for those components to be used in a rain-screen design where the backs will be wet for long periods. It is not possible, however, to maintain the same cover on cut edges, which are likely to show deterioration earlier than the rest of the sheet. Machine sheared cut edges are less vulnerable than sawn edges, so the latter should be protected by a joint overlap wherever possible.

Vitreous enamelled steel is always coated after forming. Although it offers an extremely durable surface, the panels are prone to chipping around fixings, which therefore need to be carefully designed and installed.

Composites
Sheets or boards are available in which the thermal insulation is integrally bonded to the inside of the outer protective layer. These are sometimes known as sandwich panels, although they may not always be a true sandwich with the soft core skinned on both sides. Such composite panels were often used in the spandrels of otherwise glazed façades. They can of course be subjected to high temperatures because of the proximity of the insulation, and behaviour very much depends on the particular materials and design (Figure 3.55).

Some aluminium or steel sheets are supplied already bonded to thermal insulation (eg expanded polystyrene, polyurethane and polyisocyanurate) and are designed

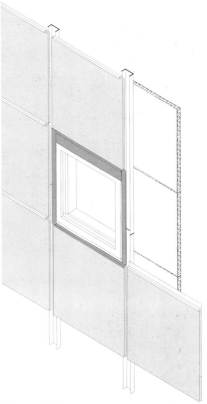

Figure 3.54
Here boards are fixed over shaped metal channels to catch and redirect rainwater penetrating the ventilated rainscreen

Figure 3.55
These composite spandrel infill panels have remained relatively flat, though they are discolouring

Figure 3.56
Secret fixing for shaped metal rain-screen panels

to be fixed with a cavity. The cladding is relatively easy to install, but careful detailing is required to avoid thermal bridges at returns. The metals should be protected to the same standards as unbonded sheets.

Fixings
Thick adhesives such as polymer emulsion modified cementitious mixes can be used to stick insulation boards such as expanded polystyrene to concrete panels. These adhesives are sometimes supplemented with large-headed plastics or metal pins locked in place by hammering into drilled holes in the concrete. Thin adhesives, which depend on surface contact, will not be suitable for use on exposed aggregate panels or on those with ribbed surfaces.

Adhesives, however advanced the formulation, must not be relied on to fix the suspension system for the outer skin of any overcladding.

Metal or plastics pins hammered into predrilled holes at appropriate centres probably form the main method of fixing for lightweight expanded plastics boards finished with thick render on lath. Provided the original structure is sound, pull-out strengths well in excess of dead and live load requirements are available. Pin material should be chosen to be compatible with the materials fixed, and in this respect it may be useful to specify a closer spacing of a lower strength pin if it gives better thermal insulation

performance (ie it does not form a serious thermal bridge through the insulation). Stainless steel, nylon and polypropylene fixing pins are obtainable, so durability should not be a problem, except in fire. Nylon and polypropylene fixings melt in fire, and it is therefore recommended that at least one fixing per square metre of any kind of overcladding should be of metal.

Suspension systems for rails carrying sheeted overcladding systems may well warrant heavier expansion bolt-type fixings, provided the concrete is good enough to accept them. Even in good concrete, care must be taken that the concrete is not fractured on tightening. Indeed, the condition of the concrete may preclude that sort of solution.

Shot firing hardened steel pins into precast reinforced concrete panels should be avoided. The fixings may be insufficiently durable because the firing destroys the thin protective plating on the pin, though hammered pins may find a place for fixing battens in low-rise sheltered situations where the consequences of possible early failure may be more acceptable.

Self-tapping screws into metal should be also avoided, but corrosion resistant wood screws are an acceptable fixing into timber battens. On shaped metal trays, secret pin and cam fixings (Figure 3.56) giving easy removal for inspection and repair or replacement have much to be said for them, although, arguably, they should be supplemented with at least one bolted fixing per panel. Certain of the thicker board systems can be part drilled from the back to receive secret fixings concealed within the board thicknesses. Total security, as with all kinds of fixings in the last resort, depends on frequency and the tensile characteristics of the board materials.

Large pop rivets are used in some systems, usually for fixing thin flat panels to metal bearers, sometimes through a gasket material. With care in the choice of metals for rivet and bearers (and panel if of metal),

satisfactory performance ought to be expected, although consideration should be given to the effect on durability of any dissimilar metals which may remain in the rivet. The rivet heads can be concealed by plastics caps matching the panel colour.

Corners etc
It is usual for fibre cement sheeting products to be include sheets of various shapes. Specially shaped metal sheets may be available too.

Openings and joints
In all cases, externally applied insulation under light cladding should cover all vertical surfaces as consistently as possible to avoid thermal bridges. The suspension system for the board or quilt materials in the external cladding in most cases needs to be taken back to the panels, at least at intervals; there will therefore be some loss of insulation where the uninsulated cleats act as fins. Insulation should be cut and fitted as tightly as possible round these points.

If new windows are to be fitted, it may be necessary to consider making them smaller than the originals in order to accommodate thermal insulation at reveals.

Two-stage joints
For rainscreen overcladdings on existing LPS construction, it is important that the original construction does not allow air to pass through. Therefore any newly applied insulation should either be sufficient to provide an air seal over previously open drained joints, or a separate air seal should be added to the outermost part of the original concrete panel over each old joint. If no catchment trays are provided behind the overcladding joints, for satisfactory performance it is essential that the new overcladding joint widths are closely controlled. Experimental evidence has shown that, with open vertical joints of 2.5 mm (±1 mm) and horizontal joints of 25 mm (±4 mm), very little water will cross a 25 mm cavity. Accuracy of this order, however,

over the whole building is not very practical, and it is therefore arguably better to provide catchment channels. Figure 3.57 gives the necessary sizes of catchment trays and their position in relation to the cavity, both of which depend on the joint widths chosen.

It is of course possible, in principle, to form channels by overlaps on the edge profiles of adjacent panels, both vertically and horizontally; the panels will therefore not be symmetrical. Lapped panels are more difficult to install than unlapped panels and are also more difficult to disengage when replacement is needed. With horizontal channels it is crucial that the joint does not fill with water.

One-stage joints

Sealed joints should be designed, as far as possible, to lap rather than butt. This gives protection to the gasket or sealant from solar radiation, and some protection from driving rain should the seal fail prematurely through, say, ageing. Against this, replacement of the seal will not be so easy if the joint is lapped. In any event joints should be designed to accommodate movements and deviations so that stresses on the jointing products are kept within acceptable limits.

Joints of boards against continuous bearers are usually backed by flexible purpose-made neoprene strips, and horizontal joints without continuous bearers are usually flashed with aluminium chair-section flashings.

Main performance requirements and defects

Choice of materials for structure

Cladding normally consists either of a thin sheet material in subframes or a sheet material which has sufficient strength to span between fixings or studs by itself. Typical examples are glass, fibre reinforced plastics, GRC, metal sheeting, cement based composites, thin stone panels, and sheathing spanning between timber or metal studs.

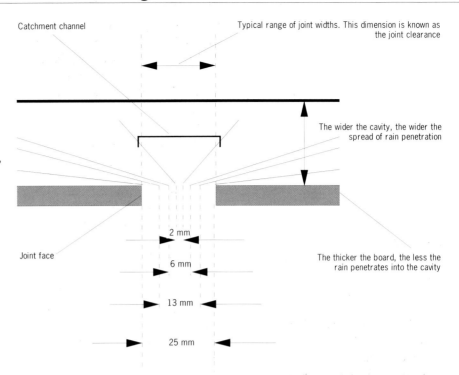

Figure 3.57
Sizing catchment trays for the vertical joint in drained and ventilated rain-screen cladding using 5 mm thick boards

Figure 3.58
The sheet cladding has here been used above pedestrian reach height, while a more robust solution is used at ground level

Sophisticated cladding systems are available using factory made metal cladding panels of outer weathering sheet, foamed insulation and an inner lining panel bonded together. They have the advantage of providing the complete cladding in one operation and most types eliminate the need for through the sheet fixings.

Strength and stability

Generally each panel of the type covered in this chapter will only be required to support its own dead weight. Dead loads of the cladding system are normally taken to the structure via support fixings at the base of each sheet or frame or ties acting in shear. Cladding, however, will also be required to bear wind load by spanning between support and restraint fixings or to subframes.

Impacts, however, are a significant risk to thin sheet claddings. It is customary therefore to put them out of reach of this risk (Figure 3.58).

In Cruden system dwellings, some bowing of panel clad walls has occurred following corrosion or loosening of fixings. There is no alternative to examining the panels in detail, and replacing defective fixings. Some bolts were reported missing from the frames of Hawthorne Leslies.

Reconsideration of wind loads when overcladding is contemplated
The ability of the building as a whole to withstand wind loads will have been accounted for in the original structural design. The addition of overcladding, provided it does not substantially alter the external shape of the building, will not significantly alter the design wind loads, but these wind loads will now be applied in part to the external skin of the

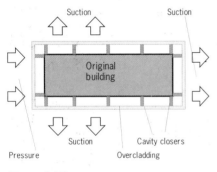

Figure 3.59
Diagram of the effect of wind forces on the end of an overclad building. The provision of cavity closers will limit the transfer of external wind pressures within the cavity

Table 3.1			
Testing for fatigue resistance			
No of cycles	**Percentage of design load**		
1	90	)	
960	40	)	
60	60	) applied	
240	50	) 5 times	
5	80	)	
14	70	)	
1	100		

overcladding. Depending on the porosity of the overcladding, part of the external pressure may leak through to act directly on the building surface while the remainder will be transmitted through the fixings to the building (Figure 3.59) The distribution of the fixing loads will also depend on the volume of any void between the overcladding and the building, and on the position of cavity barriers in this void.

In assessing the wind loads, the overcladding can be considered to fall into one of two categories:
● installations with a void or cavity between the overcladding and the building – panels usually are fixed to the battens or a grid of cladding rails attached to the building
● installations with no void between the overcladding and the building – insulating panels usually are bonded or mechanically fixed directly to the building surface, forming an impermeable outer skin

A more detailed description of the effects of winds on each of these categories is given in *Overcladding external walls of large panel system dwellings*[148].

In many cases the degree of permeability of the building surface is indeterminate and it is safest to assume that the overcladding must transmit the full wind loads through any adhesive bond and mechanical fixings. Bond and fixing strengths may be determined by testing small sections or by applying a proof suction load to the prototype panel using a test rig such as that described in BRE Information Paper IP 19/84[150].

A further factor to consider will be local deformation of the surface of the overcladding under wind loads, which may alter the geometry of the joint. Certain kinds of overcladding joints, eg unfilled joints in rainscreen systems, are more tolerant of changes in joint geometry than are face-sealed systems.

Dead loads
The dead weight of overcladding systems will vary according to the design and the materials used, but even apparently insubstantial systems can be as much as 50 kg/m².

Fatigue
Continual flexing of panels and fluctuating wind conditions can lead to fatigue and consequent cracking, particularly of sheet metals. There could also be loss of bond between the insulation and its composite sheets. This potential problem should be raised with suppliers and a satisfactory solution obtained. It is suggested that prototypes are subjected to a simulation test based on criteria set out in Table 3.1.

Fluctuations in external surface temperatures can also affect performance of materials at the interface in metal skinned sandwich or laminated panels used for cladding external walls. This has been known to cause local delamination which has adversely affected both appearance and durability. BRE tests have shown that it is possible to examine the risks of delamination[151].

Dimensional stability, deflections etc
Cladding is more responsive to changes in external conditions than the relatively more stable building fabric on which it is normally carried. It is often of low thermal capacity because of its thin section and is more directly exposed to external conditions. The temperature of lightweight claddings can change markedly over short time periods, with corresponding size changes also occurring over short time spans. The following points can be relevant to diagnosis of movements causing damage to the cladding:
● rapid size changes are unlikely to be accommodated by creep in the materials
● shading (eg by frame members such as mullions and transoms) can produce large local temperature differences with correspondingly high local thermal stress

- thin sheet claddings may relieve rapid stress development by bowing (oil-canning)

For coefficients of linear thermal and moisture expansion etc see Chapter 1.2.

Thermal movement may be one of the principal actions; the fixings and joints between panels must allow in-plane expansion or contraction to occur freely, otherwise excessive unintended forces can develop. In thicker materials, especially if highly insulating on the central or interior laminate, bowing movements can occur due to the differential movement between outer and inner layers (Figure 3.60) and very high and destructive shear stresses may develop in poorly designed systems.

It is important that movements are allowed for in the fixings of fibre cement sheets, and in no circumstances should tightly butted joints be permitted.

Weathertightness, dampness and condensation

All cladding should resist rain and snow penetration to the inside of the building, and the cladding itself not be damaged by rain or snow. In order to satisfy this requirement the sheet cladding may be either:
- a lap-jointed moisture resisting outer layer (eg tiles, see Chapter 9.1)
- backed by a moisture resisting inner layer
- backed by a clear cavity across which rain will not transfer

All joints, whether sealed or open, should be designed to prevent rain and snow reaching the inside of the building.

Winds tend to drive run-off sideways across the façades of buildings, and consequently the water load on vertical joints is not necessarily any lower than that on horizontal joints (Figure 3.61a). Vertical ribbing of the surface will to some extent divert sideways flow (Figure 3.61b), but this cannot be quantified into simple design rules.

Case study

Rainwater leakage in infilling spandrel panels

The BRE Advisory Service was asked to carry out an investigation of damp penetration problems in a concrete framed seven storey building located near the coast. Exposed concrete beams and columns were clad with mosaics. The seaward-facing southern elevation had been infilled from first to sixth floor with PVC-U framed windows and insulated spandrel panel units. During the conversion to an hotel the original infill framing between floors and columns was replaced with new glazing and spandrel units, but original metal sub-frames were left in place. Head and jamb sub-frames had been covered with PVC-U trims, but the original aluminium sills were retained as functional components at the base of the new panels.

Problems with water penetration were evident soon after refurbishment, showing as damage to wall plaster and paint, mainly concentrated at column positions, and at outside edges and corners of the partition walls.

Externally, the infill panels appeared to be well sealed to the surrounding structure, albeit with cellular PVC-U trims and profiles covering the existing metal sub-framing. There were only a very few places where deficiencies seen in the silicone sealant provided possible leakage points and some of these had been caused by previous investigation work by others.

Removal and dismantling of a complete infill frame confirmed the following points:

- crudely drilled drain holes had been provided in the spandrel panel transoms which drained the front part of the transoms ineffectively. The rear sections of the transoms could hold water which would be pumped through past the inner gasket seals. The inner gasket seals of the spandrels were not very carefully installed at the lower corners and a screw fixing from the windowboard penetrated right through into the top gasket, disturbing its seal
- rainwater penetrating round the opening windows collected on the windowboards and ran through the unsealed joints between windowboards and plastered reveals. It then entered the cavity behind the old metal subframes and wetted plaster, partition walls etc and also built up at the floor edges and ran into the rooms below
- routes for rain penetration directly through the spandrel panel gaskets into the cavity and blockwork backup wall were identified
- sealing of the panels to the concrete structure appeared to have been executed quite well and rain penetration through edge seals was only thought to make a minor contribution to leakage

To achieve a weathertight wall, replacement of the spandrel infill panels was likely to be a preferable option to modification and repair of the existing construction.

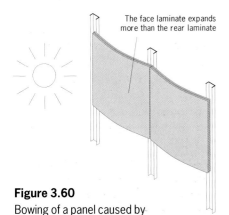

Figure 3.60
Bowing of a panel caused by differential thermal movements in layers bonded together

The face laminate expands more than the rear laminate

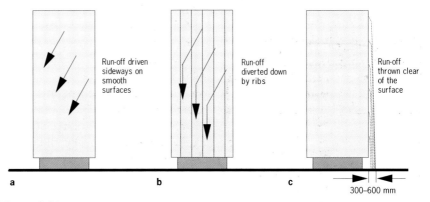

a   Run-off driven sideways on smooth surfaces

b   Run-off diverted down by ribs

c   Run-off thrown clear of the surface

300–600 mm

Figure 3.61
Runoff of rainwater on vertical surfaces

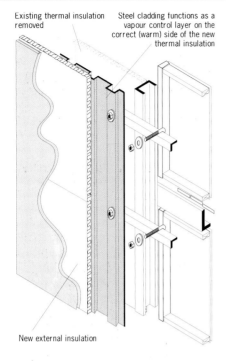

Existing thermal insulation removed

Steel cladding functions as a vapour control layer on the correct (warm) side of the new thermal insulation

New external insulation

Figure 3.62
External insulation of steel sheeted walls of a BISF house. The cavity should not be filled on any account, and the existing thermal insulation should be removed

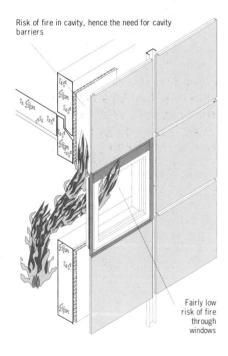

Risk of fire in cavity, hence the need for cavity barriers

Fairly low risk of fire through windows

Figure 3.63
Fire risk from combustible thermal insulation in overcladding systems

However, water flows are bound to be concentrated at vertical ribs, where the run-off rate can be many times the average. Because of this, where there is a choice, it is preferable not to specify vertical joints in re-entrant corners unless there are nearby vertical ribs to deflect sideways flow downwards.

Horizontal projections at regular intervals over the height of the building will also help to throw water clear of the façade, and therefore will reduce water load on horizontal joints (Figure 3.61c). Much of this run-off falls as a curtain of water some 300–600 mm clear of the façades, depending of course on wind conditions.

Thermal properties
In the case of systems employing steel sheet cladding, the risk in using cavity insulation lies in the fact that the steel forms a very effective vapour control layer on the incorrect (ie cold) side of the thermal insulation, thus risking severe condensation on the inside of the cladding. In theory the risk can be reduced by providing a vapour control layer on the inner (warm) side of the insulation, but these layers need to function as complete barriers and this is very difficult to achieve in practice. Several installations have been inspected by BRE where the cavities of steel system dwellings have been filled with blown fibre, but the long term prognosis is not good.

Internal insulation is not recommended for use where the wall is clad with sheet steel unless there is a ventilated cavity behind the impermeable cladding.

Many Atholls have been insulated, some internally and some externally, with apparent success and no further corrosion of the steel sheeting being observed. The internally insulated examples had a polythene sheet vapour control layer on the warm side of a 60 mm insulation quilt, replacing the original inner lining.

Perhaps the lowest risk of condensation is experienced using exterior insulation. In this case the existing steel cladding functioning as a vapour control layer is on the correct side of the thermal insulation (Figure 3.62). Factors influencing the choice of external insulation include the degree of overhangs of verges, eaves and sills which can master the additional thickness.

Hawthorne Leslies are prone to thermal bridges in gable ends, with ensuing condensation. In 5Ms, sagging of the mineral fibre thermal insulation blanket has occurred, leaving areas uninsulated.

Most of the systems designed in the interwar period will be deficient in thermal insulation. Filling the cavity is definitely not advisable, since the insulation could easily form a damp poultice round the steel, accelerating the rate of corrosion.

Fire
Dwellings built with industrialised systems of construction have been particularly identified as needing careful consideration of likely performance in fire situations. For example steelwork fire protection has been found missing in a number of Arcal system buildings.

Large areas of cladding on the side walls of buildings which are close to a boundary may represent a risk of fire spread to adjacent buildings, and are subject to control under building regulations. However, there is no current control on timber cladding if it is at least 1 m from the boundary. Where timber cladding has been used in buildings closer than 1 m, they probably were built under the old byelaws.

The performance of light cladding in fire is difficult to predict. There is generally little risk when non-combustible materials are used for additional insulation or overcladding except where the existing structure of the building being overclad presents a risk.

A risk of increased vertical (and also horizontal) fire spread has been identified during laboratory testing of cladding systems incorporating combustible insulants (Figure 3.63). Sheeted systems usually have designed or fortuitous cavities behind the cladding. Where the cladding is sheet aluminium, the laboratory tests have shown that a fire within the cavity can melt the aluminium and burn through to the surface several storeys above the fire. These emergent flames could re-enter the block via windows. Sheet materials may need to have an acceptable surface spread of flame characteristic on the side facing the cavity as well as on the outside.

The risks depend mainly on combustibility of materials used in the thermal insulation, and hence spread of fire in any overcladding. There is an obvious risk of spread on the outer surface, which is already limited in high-rise dwellings under the Approved Document, but a small risk also exists for fire propagation within any cavity containing combustible insulation; for example with polyisocyanurate foam or polystyrene (Figure 3.64).

See also Chapter 1.8.

Combustible insulation and no cavity

Most installations falling within this category will be renders over thermal insulation applied direct to the existing external wall. They are therefore dealt with in Chapter 9.2.

Cavity barriers

Where combustible insulation is exposed within a ventilated cavity of a building above three storeys in height, horizontal cavity barriers are recommended, at least at every other storey, to fill or cut off the cavity completely. In rain-screen designs, cavity closure will be needed in any case to limit the size of cavity for weathertightness reasons.

Lightning protection

New conductors over replacement cladding are needed, preferably of aluminium rather than copper as the latter tends to stain the surface and accelerate the corrosion of other metal components. Large areas of metal in systems will cause problems if not electrically bonded to the conductor. The most important areas to connect are at the top and bottom of the building, and cladding and conductors should be bonded electrically at these positions. Render stop-beads should be wired to metal lathing, and metal flashings fixed with metal fastenings – all these items should be bonded to the conductors. Electrical contact with metals and meshes can be improved by increasing the area of contact with clamp plates or welded connections.

Noise and sound insulation

Unwanted noise from thin cladding can occur, and some attempt should be made to assess this risk before undertaking any remedial work or replacement whenever possible. While noise is often heard, it is rarely complained about.

Drumming and whistling

Hail and heavy rain will drum on relatively thin sheet materials, especially of metal, and some damping may be necessary. This phenomenon happens also with sheet metal roofs but, in the case of overcladding, the existing wall construction will provide some sound insulation protection to occupants. Roofs have also been known to hum for long periods at low frequencies, but the risk of this in cladding is unknown. Provided some damping is included, or alternatively deep profiling of the panels which gives sufficient stiffness, the risk of disturbance to occupants is probably low (Figure 3.65).

Whistling or moaning of wind in ventilation slots is also a possibility. It is almost impossible to eradicate in existing construction without destroying the effectiveness of the ventilation.

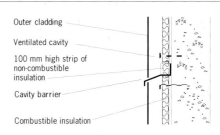

Figure 3.64
Cavity barriers are needed at compartment walls and floors, and at every floor level in domestic buildings

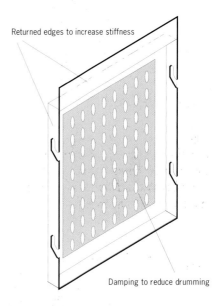

Figure 3.65
Metal sheeting can be formed with returned edges to increase stiffness and reduce vibrations in wind, but oil-canning may occur in consequence (see page 158)

Figure 3.66
Deterioration of surface coating at cut edges and sharp radius bends

'Stick-slip'

When relatively large, thin panels are subjected to solar heating they will expand rapidly. If these panels are fixed by means of cleats so that movements will take place between metal surfaces, the expansion can result in the type of intermittent movement known as stick-slip. Stresses at a fixing build up until there is a sudden slippage, accompanied by a loud report. Where this risk is identified, the fixing should incorporate a coated surface (eg of plastics). Where noise does occur it is known to be a great nuisance, and it is virtually impossible to trace the areas that give trouble.

'Oil-canning'

Another phenomenon which might possibly occur with relatively thin metal sheets, especially those having returned edges, is known as oil-canning. This is where the centre of a panel expands or contracts more rapidly than the perimeter and a sound like a drum beat results. Damping the back of the panel will reduce this possibility, as also will stiffening (eg by ribbing or by profiling the surface). There is no easy way of assessing the risk; assumptions must be made based on careful examination of prototypes and personal experience.

Durability

Light cladding systems of low mass are usually placed on the outside of thermal insulation, and so will respond comparatively rapidly to solar radiation. The maximum and minimum temperatures reached will also be slightly more extreme than those reached with high-mass claddings.

Provided metal sheets are protected to a satisfactory standard – for example, not less than that given in the appropriate Agrément certificate, and there are no abnormal pollutants – a basic life of about 30 years ought to be expected, though not necessarily without maintenance. Steel is protected by galvanising and powder coated paint systems, and by stoved and vitreous enamels. Aluminium protected by powder coated paint systems is repaintable if deterioration ensues.

Metal skinned composite panels used in cladding in the past have not always proved to be wholly satisfactory because there has been local delamination affecting both appearance and performance (Figure 3.66).

Two main points need to be remembered during the specification process for light cladding: surface temperatures can be substantially higher (or lower) than air temperatures, and surface temperatures (for any given orientation) will depend on colour. The darker the colour, the higher the temperature. The rule therefore for maximum durability, other things being equal, is the lighter the colour the longer the life. In some circumstances, specular (mirror-like) surfaces can pose a hazard to traffic by reflecting the sun. A suitable compromise would be reflectances in the range 40–65% (approximate Munsell values 7–8.5). Surface temperatures of up to 72 °C have been measured by BRE on sandwich or steel panels, and they could go up to 80 °C. However, BRE Digest 228[33] indicates the values to be taken for calculation of movements as typically –25 °C to +60 °C (a range of 85 °C) for lighter coloured low-mass materials tight to thermal insulation, and –25 °C to +50 °C (range 75 °C) for similar freestanding panels.

Other agents of degradation

Whether any particular agent of degradation will affect the life of light cladding depends for the most part on the materials which are used, in what combination, and where the building is situated. For example, while stainless steel fixings would be satisfactory for aluminium cladding for most circumstances in most rural and urban locations, their use together in a marine climate demands care. As a brief guide, it is unwise to use mill finish in a marine atmosphere, and anodised finishes must be regularly washed. Type 316

Figure 3.67
A pair of BISF dwellings: stanchions may be corroding

stainless steel is very durable, but with Types 302 and 304 stainless steels in marine conditions corrosion is a risk. And whatever the location, local industrial pollution may also be a problem. Because the lee side of a building is the most susceptible, there should be as few ledges as possible to harbour run-off ponds.

In system built dwellings are, no problems with loss of integrity of the structural frame have been reported on Atholls and Telfords, but some BISFs have had serious corrosion of stanchion bases (Figure 3.67). Also in BISFs, corrosion of the sheeting rails from condensation within the cavities has led to protrusion of hook bolts.

Extensive corrosion of sheeting on Atholls built post-1945 has occurred to a much greater extent than on interwar examples. Some Atholls also had timber framing in the external wall which is at risk of decay if rainwater penetrates. The structural frames of Open System dwellings were galvanised – a comparative rarity in dwelling systems, although less of a rarity in other building types. In Livett-Cartwrights jointing material has deteriorated, allowing rain penetration.

All systems in this category had some breakdown of protective coatings to the structural steel components, though in no case had this progressed sufficiently to allow more than superficial corrosion of the steel. However, the situation was rather different with steel sheet claddings, where some severe corrosion was found (Figure 3.68).

Rot in ply sheathing following rain ingress has been reported in Arcals and 5Ms, and in the latter case, also the plywood box beams at eaves level.

Hawthorne Leslies normally had galvanised steel framed external cladding panels. On some examples the galvanising was missing, leading to corrosion. In this system too there has been deterioration of the surface finish on the cladding boards. Repair has not proved practicable, and removal and replacement has been the preferred alternative. This system originally used felt pitched roof coverings which have not proved durable. In any replacement programme, care must be exercised not to increase the loads on the roof structure. On some 5M dwellings, an excessive number of unprotected steel shims was used to pack column bases to level. These were rusting and required attention.

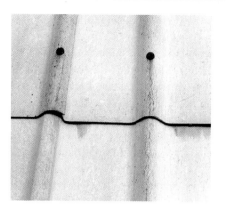

Figure 3.69
Deteriorating asbestos cement sheet cladding to an industrial building

Durability of boards

Asbestos cement sheets used to be judged to have lives of between 30 and 45 years, depending on the exposure and pollution to which they were subjected, before becoming unacceptably brittle and subject to impact and other damage – for example from movements under wind loads or corrosion or loosening of fixings (Figure 3.69).

Undecorated replacement sheets of other fibre cements will have similar lives, though applied finishes will have shorter lives; perhaps as little as 10 years.

Fibre reinforced corrugated and flat sheets are relatively fragile; some of their properties are described in BRE Information Paper IP 1/91[152]. The characteristics of these sheets are different from the asbestos cement sheets they replace. Cracking may occur, even in sheets as little as two or three years old, though its cause may be as much due to fixing deficiencies or inappropriate specification as to material deficiencies. Adequate allowance for movements is necessary to avoid cracking.

Durability of fixings

Durability of fixings into concrete must be assessed carefully. The risk of corrosion is greatest with chloride bearing concretes: in these concretes, in the presence of moisture, even 18-8 stainless steels are subject to crevice corrosion. Type 316 stainless steels should be specified.

Figure 3.68
Steel sheet cladding to housing in Glasgow. Significant corrosion has occurred at low levels, and rather less under the protection of the verge

Particular care should be given to the specification of self-tapping screws. Self-tapping screws are generally manufactured from carbon steel in order to achieve the desired mechanical properties. Corrosion protection has to be applied without dulling the cutting edge. This is normally provided by a thin zinc or cadmium coating. Unfortunately the life of such a coating can normally be measured in months rather than years when exposed externally. In dry conditions these fixings may perform adequately, but in damp conditions their durability is suspect. The outer (or head) part can be protected, given an adequate standard of workmanship, with an O-ring and plastics cup, but the back is vulnerable, and so long term durability could be threatened. Repeated loading under wind action may cause partial unscrewing. Self-tapping screws should not be considered as a permanent fixing, therefore, where regular inspection would be difficult and the consequences of failure might be serious.

Sealants

No sealant can be expected to match the full 30 year life of the cladding without replacements. Choice of sealant together with its renewal dates will need therefore to be made against a knowledge of the remaining design life of the building.

Provided protection can be afforded from ultraviolet or direct sunlight (eg by cover mould or shape of the profile), lives of up to 20 years can be expected from one or two-part polysulfides or polyurethanes and one part silicones. The elastic sealants such as the polyurethane and silicones will be found to perform best with the relatively lightweight sheet materials of high thermal conductivity, though compatibility should be checked before specifying.

The range of materials and profiles for gaskets is very wide, and it is not practicable to give guidance on their durability in this book. Information should be sought from manufacturers.

Maintenance

Few surfaces are truly self-cleaning under the action of rainwater since streams of run-off will preferentially follow certain routes rather than others. Early in the specification process, therefore, it will be necessary to consider whether uneven dirt adherence can be tolerated, whether periodic cleaning will be needed, and whether or not it is likely to be carried out – some systems require washing down several times a year.

Trapezoidal profiled or sinusoidal corrugated sheet will self-clean much better if the corrugations run vertically than if they run horizontally.

Anodised or mill finish aluminium should be washed periodically in any case to preserve its integrity. In practical terms this will often exclude it from the range of cladding options, though it has performed well in particular circumstances.

The potential side effects of overcladding an existing structure should be examined. Vapour barriers in the wrong place, leading possibly to condensation, have been mentioned already, but there will be other side effects. Damage may also be done to the panels while determining fixing points, and the overcladding may spall or lose exposed aggregate or surface finish after completion of the fixing.

It is sometimes thought that overcladding can provide some protection to pedestrians from falling concrete or aggregates, and some systems have been installed especially to do this, with a mesh secured over the original cladding prior to overcladding. While this practice may help in some circumstances, it must be remembered that any case of dislodgement becomes a potential source of failure in the overcladding, leading to detachment of insulation and surface material, and then to leakage of the envelope. The original structure must therefore be repaired to a satisfactory standard before any overcladding is undertaken, irrespective of the design to be used.

In spite of taking all reasonable precautions to ensure longevity in the base structure, it may still be prudent to choose a replacement sheet cladding or overcladding system which allows access for inspection and monitoring. However, certain sheet cladding techniques do not facilitate inspection at all, so that this requirement will limit the options available. The frequency of inspections will need to be based on engineering judgement. (BRE suggests intervals not exceeding 3 years.)

HSE Guidance Note EH 36[149], dealing with the cleaning of old asbestos cement roofs, also applies in some degree to walls. One of the most important considerations is that asbestos cement must not be brushed when it is in a dry state. It needs to be thoroughly wetted to minimise fibres being released from the material, and dust masks of the appropriate rating should be worn.

Redecoration of board finishes could be required. Rain alone will not wash surfaces clean and periodic washing with a mild detergent may be necessary. Stains can be treated with high pressure hot water.

Work on site

Ease of repair is likely to be critical near to the ground or at access levels at any height in the building, and it should be possible to take out and replace individual panels, for example, without removing and replacing a whole run. Special one-off replacement panels with unique surface profiles can be very costly to supply.

Some means of access by operatives to the cladding will be needed for the replacement of damaged panels; indeed the action of access in itself could increase the risk of damage. There may therefore be a case for installing a permanently available access system as an integral part of any new cladding, preferably including some means of anchoring suspended cradles against sideways movements.

Joints

It is most important that satisfactory designs are prepared in advance for crossover and tee-joints, and especially where vertical and horizontal joints are not in the same plane, to check for continuity. It is also advisable to have standby designs available for the extremes of variability which might be found in assembly.

Graphical aids for tolerances and fits [153] gives comprehensive guidance on the relationship between the sizes of joints and components in external walls.

Storage and handling of materials

All materials must be stored flat on pallets under cover and protected from the weather and the activities of other trades. Any moisture penetrating between stored boards will cause permanent surface staining. Protective paper or plastics sheet stuck to decorative faces should not be removed until after fixing.

Troughed sheeting in light gauge steel relies upon a combination of PVC based and galvanized coatings to give corrosion protection. It is therefore essential that the material is carefully handled and stored. One of the main dangers results from dragging one sheet over another which can mean the protective coating on the lower sheet being scratched by sharp edges.

Restrictions due to weather conditions

The extent to which cladding work is likely to be delayed or interfered with by bad weather will depend on the method of access and type of construction, but the risks of hold-ups should be given some consideration.

Work on low and medium-rise buildings

For work on low and medium-rise buildings using prefabricated cladding panels of no more than one storey in height, it will normally be most efficient to work from access scaffolding, with the materials lifted onto the scaffolding platforms in advance. There will be safety risks associated with handling the panels in strong, gusty winds, and due to slipperiness when there is ice or snow on ladders and scaffold boards. Assuming work stops under these conditions, an idea of likely lost time is given in Table 3.2. The data shown in this Table are based on wind speed at the standard height of 10 m. For higher buildings, even more interference may be expected.

Work on high-rise buildings

For work on high-rise buildings, or where prefabricated overcladding panels larger than one storey in height are used, it will normally be more practicable to lift the overcladding into position by crane at the time of fixing. For this, a more onerous wind-speed limitation will apply, considering the difficulties of safely handling and positioning large, lightweight panels. Restricted visibility due to fog will also prevent safe operation of cranes, and slipperiness criteria will also apply. In these conditions, a rule of thumb for lost time is to double the hours lost for low and medium-rise.

There is also the possibility that certain types of operation, such as joint sealing, may not be practicable when the weather is wet or cold, presenting a further possible cause of delay.

The likely time of year when installations will take place, and hence the moisture and temperature conditions of both original structure and overcladding, should be taken into account in the specification. To illustrate the point, the sizes of overcladding panels will generally be at a maximum in high summer, and clearances will need to be set accordingly so that tolerances on joints, for example, are not exceeded when the panels are at their smallest in winter.

The original building will have been constructed with very high dimensional variation; deviations are often considerably greater than expected. Before any overcladding is specified the building should be measured, as special provisions may be required.

Table 3.2

Work time lost on low and medium-rise buildings owing to bad weather [*]

	Hours lost per day	
	January	July
Heathrow		
worst in 5 years	4.5	1.5
average	3.4	0.9
best in 5 years	2.1	0.2
Glasgow		
worst in 5 years	6.2	1.8
average	5.0	1.2
best in 5 years	3.6	0.6

[*] Predicted average lost time in hours per day between 07.00 and 17.00 h GMT, assuming work stops when mean hourly wind speed exceeds 12 m/s, and there is snow, sleet or hail falling.

Workmanship

No matter which system of cladding is used, it must be capable of correct installation if its performance is to be guaranteed. Many of the metal sheet systems are highly sophisticated designs using combinations of different materials and skills which are highly sensitive to error during fabrication or installation. The missing gasket separating different metals, or the wrong type of fixing, may easily negate months of careful investigation and design work. These systems may need specialist skills not normally associated with construction, those of general engineering and metal working being more appropriate.

Fixing devices should have sufficient adjustability built in to cope with all but the grossest of assembly deviations of the existing façades. Deviations are often considerably greater than expected. The buildings should be measured if at all possible, but if not, then a total longitudinal adjustment range of 75 mm will be needed to take account of deviations within one storey, and probably as much as 50 mm in addition will be needed to take account of deviations over the height of the building.

Sealants must only be applied in appropriate weather conditions. Silicones should not be used against bare metal because they emit acetic acid in the curing process.

Chapter 3.5

Glass fibre reinforced polyester and glass fibre reinforced cement

This chapter deals with cladding made from glass fibre reinforced polyesters (GRP) and cements (GRC). These are undoubtedly products of the second half of the twentieth century. Other reinforcing materials may be occasionally encountered, such as steel or carbon fibres; but rarely, so they are not dealt with in this book.

Although infrequently used for housing, the best known GRP system was Resiform (Figure 3.75) of which some 1,800 dwellings were built, the material was more often used as cladding for other building types, including some prestige buildings. GRC is now more widely used in France and Spain than in the UK.

Characteristic details

Basic structure
Glass fibre reinforced polyester (GRP)

GRP is a composite material consisting of a resin reinforced with glass fibre, sometimes with fillers and pigments of various kinds. On its own the resin is brittle, but the inclusion of a glass fibre produces a tough, high tensile strength material of low weight.

Panels of GRP are produced by building up layers of liquid resin and glass fibre mats or fabric, and then compacting the whole with a roller. Alternatively, the material can be built up by simultaneously spraying the resin and the fibres. The resin reacts with added hardener, causing it to set. Glass fibre reinforced polyester panels can be produced to a wide variety of sizes, shapes and surface finishes. The material particularly lends itself to panels with returned edge profiles and concave or convex surfaces, giving strongly modelled façades (Figure 3.76).

The manufacturing process produces panels with a wide variety of properties. A major difficulty in assessing this material, as used in cladding, is that seemingly similar panels can have widely varying properties, depending on the resin type, glass content, compaction efficiency and curing conditions.

The Resiform system referred to above consisted of a resin laminate approximately 15 mm thick into which timber studs were set, making storey height panels in a range of widths up to around 7 m.

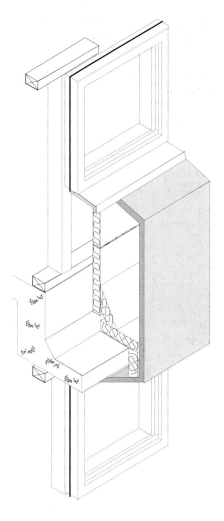

Figure 3.76
GRP panels can be formed into a wide variety of sizes, shapes and surface finishes

Figure 3.75
Three storey Resiform GRP clad housing

Figure 3.77
Precision moulded panels form the walls
and roof of this cabin

Glass fibre reinforced cement (GRC)

GRC is a cement mortar reinforced with around 5–9% alkali resistant (AR) glass fibre, with the alkali resistance given by the addition of zirconium. The standard E-glass used in GRP cannot be used in cement mortars. AR glass was developed by BRE in the late 1960s, and the process of manufacture has been available commercially since the mid 1970s. Although AR glass is not immune from alkali attack, the process takes place at a reduced rate.

Thin panels of GRC, sometimes as thin as 12 mm, can be moulded to a wide variety of shapes. The mix can incorporate various fillers, such as fine sands and pigments, up to around 50% of the total mix. Profiles of fabricated panels, usually formed by spraying the mix, may include single or double skins; also they may be flanged or ribbed and contain thermal insulation. Although there are comparatively few applications in the UK there have been many abroad.

External and re-entrant corners

It is usual for both GRP and GRC panels to be available in special profiles to turn corners without the use of cover moulds (Figure 3.77).

Main performance requirements and defects

Choice of materials for structure

Typical mixes for GRP are 2.5:1 resin:chopped strand fibre or 1:1 resin:woven mat fibre. The outer surface is normally protected with a gel coat of special polyester resin which may also contain additives to enhance weathering characteristics.

GRC typically consists of hydraulic cement and fine aggregate mortar with a fibre content of around 4% by weight.

Strength and stability

Panels in both materials are normally required only to support their own dead weight and wind loads, and are fastened back to the supporting structure normally at each floor level.

Some GRP panels which incorporated timber studs could be used up to three storeys for loadbearing purposes. Because of the gradual reduction in strength of GRC over time, the very high initial strengths cannot be exploited, and suitable factors of safety need to be applied in design.

Dimensional stability, deflections etc

For coefficients of linear thermal and moisture expansion etc, see Chapter 1.2.

In addition to linear movements, slight bowing may occur with both materials, especially in double skin panels. It should have been allowed for in the original design, either by providing adequate restraint, or by using appropriate joints which can accommodate the movement without exceeding their movement accommodation factor (MAF).

Weathertightness, dampness and condensation

All GRP and GRC cladding should resist the penetration of rain and snow to the inside of the building, and the cladding itself should not be damaged by rain or snow. As with other forms of light cladding, in order to satisfy the former requirement the cladding may either:
- be in itself, with its joints, resistant to rain
- be backed by a moisture resisting inner layer
- be backed by a clear cavity across which rain will not transfer

All joints, whether sealed or open, should be designed to prevent rain and snow reaching the inside of the building. The Resiform system already noted incorporated PVC baffles set into rebates in the moulded edges of the GRP panels, with a second line of defence using mastic within the depth of the joint.

Walling elements of both materials need vapour control layers on the warm side of the thermal insulation (Figure 3.86).

Thermal properties

Panels of both materials needed to have adequate thermal insulation to meet past (and current) requirements.

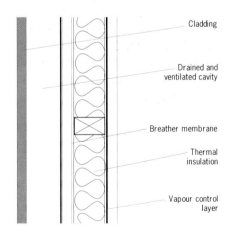

Cladding

Drained and ventilated cavity

Breather membrane

Thermal insulation

Vapour control layer

Figure 3.78
GRP and GRC cladding both need adequate back-up walling, protected against rain penetration, and providing thermal insulation and other functions

Fire

GRC is non-combustible and is considered to have Class 0 spread of flame.

GRP is another matter entirely, and GRP cladding would not be permitted on the side walls of buildings which are close to a boundary and which may represent a risk of fire spread to and from adjacent buildings. Attempts to increase the fire resistance of GRP mixes may prejudice durability.

Noise and sound insulation

Both materials are normally available in comparatively thin panel form. By themselves, thin panels usually have less than adequate performance and need appropriate back-up walling. Some double skin GRC panels incorporating special laminates and quilts are claimed to have up to 55 dB(A).

Durability

Provided the surface gel coat of GRP panels is well cured and well bonded, there is no reason why a 30 year life should not be expected, although there could be colour changes and slight loss of gloss. Dark colours may show slight blooming. Panels of GRP are only recommended for rain-screen designs if care is taken to obtain a satisfactory surface finish on the inside of the panel which will withstand intermittent wetting.

GRC will suffer from progressive embrittlement over a period of a few decades, and this should have been taken into account in the design stage.

During the late 1970s, some cracking of fibre cement (GRC) panels on a few sites was recorded. Most of the affected material was on south and west facing panels, and some bowing was also observed. The cracking represented a serious problem at some sites, and remedial action was needed to prevent detachment. It was concluded that most of the cracking could be attributed to design not taking account of all the relevant factors, and that in some cases manufacturing deficiencies contributed[154].

Most of the early failures of GRC occurred with double skin panels, and there seem to be fewer durability problems with single skinned panels incorporating metal studs.

Maintenance

Intervals for cleaning the external surfaces of GRC panels may be increased if they are treated with silicones. Toxic washes may be necessary to remove organic growths.

Painting the surfaces of both GRC and GRP is common, though great care is needed over specification. For GRC, paint suitable for painting other cement surfaces should be satisfactory; for example, alkali resistant primers, though they must be carefully applied to avoid trapping moisture within the panel which can cause bubbling of the painted finish.

Joints should be checked on a regular basis, and displaced gaskets renewed.

Work on site

Chapter 4   **Glazing and curtain walling**

Figure 4.1
'The view afforded of the outside world': buildings fronting the River Thames

A brief historical review of the development of windows as building components was given in the Introduction, Chapter 0. Glazing is not specified exclusively for the controlled admission of light – the admission of sunshine, heat and fresh air are important too, as well as the view afforded of the outside world (Figure 4.1); also, incidentally, it may provide interesting features in the townscape (Figure 0.39).

However, there are negative factors at play as well as positive. Glazing can also admit too much light, heat or air; it needs cleaning regularly; many kinds of frames need periodic maintenance; glazing may be a weak point in the façade, admitting rain, snow or intruders; it may be the focal point for condensation; and it may be vulnerable to accidental damage from the forces of nature and of man.

This chapter deals with the performance of all kinds of vertical glazing in all building types, mainly in the more traditional form of windows, but also in the form of that invention of the twentieth century with a rather chequered history, curtain walling. When a window becomes a piece of curtain walling is a matter not only of size, but also of structure and method of support.

Chapter 4.1

Domestic windows

Figure 4.2
Edwardian timber double hung sash window

This chapter deals with domestic scale windows in all materials, and incorporates also comments on the more common glazing materials.

The design of domestic windows has become much more sophisticated since the 1970s. Choices of types and materials have proliferated. The performance of modern windows may now be measured accurately under laboratory conditions, and weathertightness and other criteria closely defined.

Around 1 in 10 of the total dwelling stock in England have wood sash windows as the predominant window type, with 4 out of 5 of these dwellings dating from before 1918[2] (Figure 4.2). Just under half the total have wood casements. Between one third and one half of all dwellings in each age category have wood casements, so the type is predominant in the total stock. Over 2.4 million dwellings have single glazed metal windows as the predominant type, and there are now nearly 1 million dwellings with double glazed metal windows. There are nearly 4 million dwellings now with PVC-U double glazed windows. The number with wood double glazed windows is 1.3 million, or around 1 in 6 of the total stock.

In Scotland, around half of all 2.2 million dwellings have wood windows, with 1 in 10 metal and 1 in 3 PVC-U[3].

The figures for the other two countries in the UK will have been recorded in their house condition surveys, but have not been abstracted for this book.

Characteristic details

Basic structure

Wood as a material for windows reigned supreme for domestic scale work until almost the twenty first century, with virtually no competitor until after the turn of the eighteenth century. With the Industrial Revolution, iron frames were introduced, extending to other building types. Since then there was first a massive growth in the use of steel, and later still an even more phenomenal growth in plastics.

Stone frames

Stone, both natural and reconstructed, has been used widely in domestic windows, both in earlier centuries and on a massive scale in late Victorian and Edwardian times (Figure 4.3). It is used not so much to house the glass directly, but to provide structural support over wide windows. BRE investigators on site have seen central stone mullions removed when replacement windows have been installed, with disastrous effects on lintels.

Wood frames

Whether it is a simple window in the plane of the wall or a projecting oriel carried on brackets or a bay founded at ground level, construction techniques for wood windows have been very similar over the years. Shaped mullions were tenoned into sill and head, some to accommodate opening casements and some to accommodate fixed lights. Both fixed lights and casements were in earlier times subdivided into smaller apertures by wood glazing bars or

Figure 4.3
Victorian double hung sash window frames with a stone central mullion carrying a decorative lintel in the same material

astragals, either broad (2 inches or more) or narrow (down to half an inch), depending on date and fashion, or by wrought iron standards and stay bars or transoms to which lead cames were wired to take the small pieces of glass (confusingly also known as quarries). In later times, when glass became more easily available in larger sizes, fewer glazing bars were needed.

The double hung sash window began to replace the side hung casement in popularity in late Stuart times; from Georgian times to Edwardian, and even later, it took over as the preferred design of the domestic window. Most were counterbalanced with cords and weights accommodated in the sash boxes on the jambs; many continue to give satisfactory service, though cords need renewing from time to time. Some, however, in the poorest dwellings, have no such mechanisms, and need to be opened by brute force and held by wedge or pin.

After the 1939–45 war, the manufacturers took the lead in producing standard designs. The English Joinery Manufacturers' Association (EJMA) sections used for casements, and vertical and horizontal pivots, can still be seen surviving in common use.

It has been estimated that UK joinery manufacturers annually produce around 1.5 million wood windows[155]. Increasingly, hardwoods such as European oak are being specified rather than softwoods, though European redwood, which is comparatively easy to treat with preservatives, is still popular. The current standard for wood windows is BS 644-1[156].

Steel frames
Early cast iron windows were sometimes used in domestic work, but mostly in other building types. However, one occasionally comes across early rolled section windows.

Window frames in rolled steel have been manufactured in very large numbers since the interwar period, and were popular in speculatively built housing of this period, having cockspurs and stays in bronze. (Figure 4.4). They were standardised in BS 990[157]. Projecting hinges were introduced in the 1930s, the so-called easy-clean hinges, making the windows more useful in multi-storey developments.

So far as replacement windows are concerned, hot-rolled galvanised steel sections do not lend themselves well to the production of made-to-measure windows. Also, they have not been suitable (in domestic sizes) for double glazing units, although steel windows which can take wider airspace double glazing units are now available. Polyester powder factory-applied coatings which overcome the problems of site painting of galvanised steel, and the regular repainting that was needed, are now the normal finish[158].

Steel windows can be the appropriate choice for replacement of old steel frames in pre and postwar houses and flats where the distinctive slim frame appearance of the windows must be kept. They are available in matching sizes and patterns to the old ones. Thermal break frames are not widely available. The current standard for steel windows is BS 6510[159].

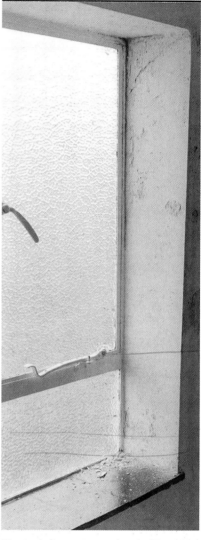

Figure 4.4
Galvanised steel window set directly into the masonry with the widow-to-wall joint having little protection from driving rain

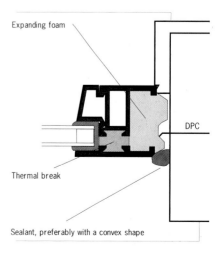

Figure 4.5
A thermal break window

Aluminium frames

Aluminium is a good conductor of heat (ie a poor insulator) and thus aluminium frames need to incorporate a 'thermal break' plastics insert to reduce the amount of heat which is transmitted to the outside of the building (Figure 4.5). This helps prevent condensation on the frames in cold weather.

Since the late 1980s, aluminium frames fitted directly into openings have largely superseded the types which required timber subframes. The normal finish for aluminium frames is a white polyester powder coating which should be durable and need no maintenance other than washing. Other colours may be available.

The current standard for aluminium windows is BS 4873[160] and for finishes BS 6496[161].

PVC-U Frames

Since the 1970s the majority of the replacement domestic window units installed are made of PVC-U. The reasons for this popularity include good overall performance, with hardly any maintenance required. PVC-U compound is extruded into lengths of 'profile', which are then cut to length and assembled into window frames. Corner joints are heat welded; other joints may also be welded or may use screw fixings.

Most types of window can be obtained in PVC-U: inward or outward opening, sliding sashes, bay windows, door frames and conservatories. They do not usually require subframes, but suitable loadbearing elements will be needed for bay windows.

Basic formulations of PVC-U are rather brittle, particularly when cold. PVC-U for windows is especially formulated for improved toughness. Other additives are used to limit the deterioration caused by ultraviolet sunlight. Different manufacturers use different formulations, and so the quality of the PVC-U is important. PVC-U is not as stiff as wood or metal, so profiles are made with a fairly large cross section. Metal reinforcement may be placed within the profile to stiffen it. The corner welds are a potential weakness.

Many materials can have their colour changed by exposure to sunlight. It is a potential problem for white PVC-U and imitation wood grain, just as for paints or polymeric finishes on window frames of other materials. In all cases, manufacturers need to test their materials for how well they hold colours.

Ultraviolet light from the sun will tend to break down the surface layer of white PVC-U. At first it will become rather more brittle. Then a

very thin layer will 'chalk' away. The window will lose its shine and fresh white pigment will be revealed. If suitable stabilisers have been added to the PVC-U, the loss of thickness will be too slow to matter.

For wood-grain finishes, the coloured film on the surface must last the whole life of the window. It is important that the dyes that give the wood-grain appearance do not fade. Ultraviolet light can make the film more brittle, further weakening the whole profile.

The dark colour of some windows causes them to become quite hot in summer sunshine. PVC-U expands substantially when heated, and so dark windows must be designed so that this expansion will not cause their opening lights to jam. Temperatures can approach levels at which PVC-U begins to soften and distort. Profiles must be resistant to this, and reinforcement is suggested for all dark PVC-U windows.

Windows should have profiles made to BS 7413[162], to BS 7414[163] if white, or to BS 7722[164] for a wood-grain finish. White windows should also be made to BS 7412[165].

For further information, see BRE Digest 404[166].

Glazing materials

A brief mention of the historical development of glass was given in Chapter 0. Annealed glass has been, and still is, the commonest glazing medium, in spite of its vulnerability to breakage. However, safety glasses are increasingly being specified, particularly after the introduction of legislation on safety in the workplace.

Ordinary laminated glass is, thickness for thickness, no stronger than annealed glass; its virtue is that broken shards tend to be held by the interlayer and, if recommended glazing techniques have been followed, the glass should be retained until safe removal is possible. Wired glass is weaker than unwired annealed glass of corresponding thickness but, as with laminated, the shards tend to be held. Thermally toughened glass is

Figure 4.6
Sealed double glazed units mounted in hardwood frames

stronger than annealed, both in terms of impact resistance and of thermal stress. When broken the glass will not hold together though the pieces are relatively small. The best of both worlds can be obtained with sheets of laminated toughened glass. Laminated glass for domestic work is commonly made with two sheets of glass laminated together under heat and pressure with a Polyvinylbutyral (PVB) interlayer. Laminates with acrylic interlayers may also be used.

There is further comment on risk of breakage of glass, particularly in large sheets of security glasses, which may be laminated or toughened, in Chapters 4.2 and 4.3.

Sheet glass is available in many varieties which it is not practical to list in detail here, though they range from methods of strengthening – toughening, laminating and wiring – to coatings to alter the light and heat transmission characteristics, and surface patterning and modelling. Fashions change, and matching old patterns for replacement purposes may be a problem. It has been considered important to try to retain fire finished glass in historic buildings wherever possible.

Plastics glazing materials are less prone to breakage than glass, but most are softer and more liable to surface damage such as scratching. They also may have quite large thermal movements. Those most commonly used are polycarbonate and acrylic.

Polycarbonate

Polycarbonate is not prone to discolouration so much as surface erosion due to weathering. This tends to lead to loss of surface gloss and reduction of clarity of transmitted images. Polycarbonate has relatively good retention of impact resistance and dimensional stability on weathering. Surface coated varieties are available with significantly enhanced weatherability.

Acrylic

Acrylic sheet (polymethyl methacrylate) has very good retention of optical clarity on exposure to weathering but can be prone to crazing and embrittlement in the long term, especially when subjected to undue stress.

Double glazing

Sealed double glazing units became generally available in the United Kingdom in the early 1960s. Their use in new housing has been boosted since the late 1980s by changes in the building regulations which allowed trade-offs between the opaque and glazed areas of a building.

Double glazing has a lower heat transmittance (U value) than single glazing, and can improve the comfort of building occupants in areas of rooms near to windows. Little difference is made to noise insulation.

A double glazing unit normally comprises two panes of glass held at a fixed distance apart by a continuous spacer bar located around the perimeter of the glass which is then sealed (Figure 4.6). Most types of glass can be used. The spacer is normally manufactured from mill finished aluminium, although other materials such as galvanised steel and plastics are used. If the unit is to perform satisfactorily over a long period, the sealant or combination of sealants must be appropriate and the unit must be properly glazed in a suitable frame. If only one sealant used, then the unit is termed a single seal double glazing unit, but if there is an additional seal compressed between the spacer bar and the glass, the unit is termed a dual seal double glazing unit.

In the UK and Europe, polysulfide is currently the most commonly used secondary edge-seal material. Other types of sealant that are also used in considerable quantities include polyurethane, silicone, hot melt butyl and polymercaptan.

In all double glazing units a desiccant is held within the hollow spacer bar. This desiccant absorbs water vapour sealed in the double glazing unit at the time of

manufacture and also absorbs moisture that permeates through the edge seal. The desiccant is intended to prevent misting within the air space during the service life of the units. Typical desiccants used in the United Kingdom are molecular sieves, silica gel or blends of both types[167].

Glazing techniques

The two principal glazing systems for both single and double glazing are known as drained-and-ventilated and fully bedded systems. The drained-and-ventilated is the preferred method, particularly for installing double glazing units of all types into frames and uses either a drained or a drained-and-ventilated system. In these systems weather proofing is provided by gaskets or by foam strips with a sealant capping in intimate contact with the glass. Any moisture which breaches these seals is not allowed to pond in the rebate but is drained away via a sloping rebate and drainage holes. Holes may also be provided in the upper section of the frames to encourage air to flow around the perimeter of the unit and to equalise air pressures in the glazing rebate. Particular attention should be paid to the type and position of setting blocks, location blocks and distance pieces to ensure that they do not inhibit either drainage or ventilation (Figure 4.7).

Drained-and-ventilated glazing systems can be designed for all types of window frame, including those

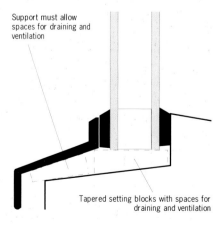

Support must allow spaces for draining and ventilation

Tapered setting blocks with spaces for draining and ventilation

Figure 4.7

The drained-and-ventilated glazing method

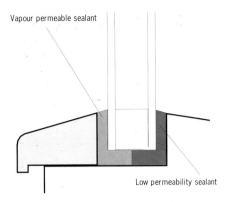

Vapour permeable sealant

Low permeability sealant

Figure 4.8
The fully bedded glazing method

made of aluminium, steel, PVC-U and timber.

The alternative method, the fully bedded method, means just that. In order to function correctly, and keep water from lying in contact with the edge seal, the bedding must be perfect (Figure 4.8). Fully bedded glazing systems should only be used where drained-and-ventilated systems are impractical or are unavailable.

Controls on opening lights

Hinged casements are normally fitted with pegstays, friction stays and cockspurs, and should be within easy reach for the average person. Horizontal sliders are normally fitted with pull-and-push finger plates integrated with the stiles, and fastening by hooked latches into keeps on the outer frames. Vertical sliders rarely have handles as such, and they fasten by a cam operating against a pin or shaped housing, the action of which is to wedge the leaves firmly into the aperture. Top hung hinged windows normally have only a pegstay which also functions as a handle. Pivot windows may have a variety of handles or cockspurs, and are stayed in the open position by means of friction pivots. Where used above ground level they should have opening limiters.

For those windows out of manual reach there is a variety of manual controls, mostly cord or rod operated worm drives, and these may even be motorised. Information for specifiers of automatic controls is available[168].

Window-to-wall joints

The joint between window and reveal is particularly important from the point of view of weathertightness. Also of importance is where the jamb is positioned in relation to the outer face of the wall. Although it is doubtful whether it was done for weathertightness reasons, it is interesting to note that by Act of Parliament, after 1 June 1709, in London, no door or window frame could be set nearer than four inches to the outer face of the wall[14]. A pity the law was repealed!

The traditional sash window was invariably set in a rebate in the surrounding masonry, and this gave some protection to the joint, even in the absence of mortar. The first galvanised steel windows were often set in timber surrounds, and they too could be set in rebates, but when these windows began to be used without subframes, it became common practice, at least in England, to set the windows as far forward as possible in the wall, so that single sills could give the required projection without recourse to subsills. It was also said that occupants liked this practice because it gave them a wider window board internally. Throughout practically the whole of Scotland, and in some other areas too, the practice of setting the windows back from the outer face still continues, and there is no doubt that the window-to-wall joint benefits from this extra protection.

Attempts to overcome the increased risk of air and water leakage in non-rebated joints led to the practice of sealing the joint with mastic, sometimes with doubtful results. BRE investigators on site have rarely seen this done neatly.

Figure 4.9
Remarkably, much of the glass is intact. Few wall ties can be seen in the remaining inner leaf (Photograph by permission of Portsmouth and Sunderland Newspapers Ltd)

Main performance requirements and defects

Strength and stability

Provided annealed glass of the correct thickness for the exposure and size of window has been used, breakage from gusting wind loads is very rare, even though on occasions building occupants may be surprised by the 15 mm or so designed for maximum deflections which occur in strong winds. Where the panes are small, it may be the whole window which is at risk (Figure 4.9).

However, impact damage is another question entirely. BS 6262[169] gives requirements (Sections 4.7 and 5.7) which govern minimum standards for the design of glazing and choice of glass in certain areas for safety because of the risk of breakage and injury as a result of human impact; these have now been embodied in building regulations (Figure 4.10). The risk areas include glazing in doors and side panels above 1.5 m and glazing wholly or partially below 800 mm from floor level. These risk areas may need greater thickness or different types of glass than would be required simply for resistance to wind loading. Where glazing immediately abuts a door, it must be safety glazing for 300 mm each side of the door up to a

height of 1.5 m. Alternatively, annealed glass is permitted where the smaller dimension of any pane is less than 250 mm.

Windows at low level may also be protected by balustrades at least 800 mm high, which do not allow a 75 mm diameter ball to reach the glass surface, and which satisfactorily resist a lateral load of 1100 N.

For double glazed units, where glazing is below 800 mm from finished floor level, BS 6262 requirements for glass in the individual panes of double glazed units are:
● Class C of BS 6206[72] in areas where many people are likely to be gathering, circulating or passing
● BS 6262, Clause 5.7.2.3, Table 11 in other situations. This table specifies allowable area limits for glass thicknesses in the situations where the risks to people are not considered sufficient to necessitate the use of a safety glass. For 4+4 mm units the maximum area allowed is 0.5 m², for 5+5 mm units this is 1.2 m² and 6+6 mm units are permitted up to 2.5 m².

Other health and safety requirements for windows and glazing relate to the safety and ease of cleaning the glass, either from the inside or the outside of the building, but preferably from the inside. These aspects were originally dealt with in BS CP 153-1[170], now replaced by BS 8213-1[171].

With such a vulnerable material as annealed glass commonly used in buildings, the question of adequacy of guarding will inevitably arise. The Building Regulations Approved Document Part K2 covers pedestrian guarding. An amendment to Part K2 relates to the guarding of openings of windows, giving minimum height for sill or guarding height. BS 6180[172] is also relevant. Where windows are used as means of escape there may be a conflict with other safety considerations.

Glazing breakages in a leisure complex
Breakage of glass in timber framed windows of some five and seven year old buildings had indicated that safety glazing had not been installed everywhere it was needed. Potentially, all 700 accommodation units were likely to be affected.

A typical arrangement consisted of four windows in timber frames (with pane size approximately 1200 x 1700 mm) over a low sill, only 500 mm above floor level, and alongside a patio door with full height glazing. Patio door sealed units were clearly marked with manufacturer's BS 6206 corner marks indicating that they were toughened. There were no similar markings on the four window unit panes. It was noted that the double glazing unit spacers bore a manufacturer's code.

The internal sill height varied between 740 mm and 790 mm above floor level, and the height of the bottom of the glass varied between fixed and opening lights.

The daylighting conditions at the time of inspection allowed investigators to use polarising filters which showed stress patterns from toughening clearly. Stress patterns did not show, though, on the dining area window glazing.

Although the glass in full height glazing was in toughened safety glass and its compliance with BS 6262 was therefore not in question, other windows with glass below 800 mm from floor level did not comply with BS 6262. These were apparently neither BS 6206 safety glazing nor, on the evidence of the broken units inspected, did they meet the basic requirements, being made of 4 mm annealed glass for which the maximum permitted area is 0.5 m². For the sizes of panes seen, 6 mm glass in a 6+6 mm sealed unit would have been necessary.

The main recommendation from this assessment was that the glazing of the windows with sill heights below 800 mm from floor level would need altering to comply with BS 6262 requirements.

Another aspect of safety relating to openable elements in the fabric is the avoidance of creating hazards to people, some of whom might have impaired vision, using paths immediately adjacent to the building. This concerns windows and doors which, when open, project out onto circulation routes (Figure 1.85).

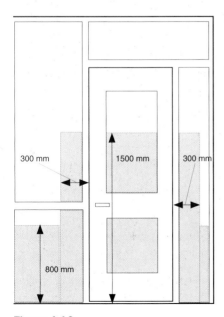

Figure 4.10
Safety glazing requirements for new domestic construction

300 mm 1500 mm 300 mm 800 mm

Figure 4.11
The 'standard burglar' at work, breaking and entering

Illegal entry

Windows and glazed doors are common points for gaining illegal entry. Unsecured windows are particularly vulnerable to attack by intruders. The intruder can easily enter by breaking the glass and opening unlocked windows, and even the smallest opening light to reach other fasteners. Despite the ease with which normal glass can be broken, illegal entry by breaking the glass is not one of the main methods of entry by the opportunist thief because of the noise created. It is often easier to lever or pry open a window (Figure 4.11).

Windows therefore should:
● be glazed with appropriate materials, firmly fixed in position
● be constructed from materials of suitable strength and sizes
● be securely fixed in position in the surrounding wall
● have a minimum of opening lights, except where they form a means of escape
● have as large a pane size as possible

and, where opening lights are required:
● be fitted with locks and concealed or tamper proof fixings and hinges

The main standard is BS 7950[173].

The following points assist in preventing unauthorised entry.
● Glazed areas should be as large as possible, consistent with other requirements. The noise created by breaking glass is a deterrent
● Small opening lights which provide access to locks or catches on adjacent windows and doors should be avoided
● Louvred windows are vulnerable
● Trickle ventilators should be used to provide ventilation
● Laminated glass should be installed in all vulnerable windows or glazed doors
● Push button locks, or locks which need a key to open them, should be installed on all opening windows. Since this may interfere with means of escape, the fire officer should be consulted
● Where an opening light forms part of a fire escape route, a key should be placed in a suitable receptacle next to the window where it cannot be reached by an intruder through a broken window
● Glazing should be firmly bedded in its frames. External glazing beads should be firmly fixed and not able to be levered free by a knife or screwdriver. It should be impossible to knock out internal beads by kicking, hammering or pushing on the glazing
● The framing sections of PVC-U or aluminium double or triple glazed windows should be reinforced next to the locking gear to prevent the frame from being deflected into the glazing cavity, allowing the lock to disengage. Otherwise a hook bolt or long-throw bolt should be used
● Frames should be adequately fixed to the surroundings and fixings concealed and made tamper-proof
● Hinge pins should be used which cannot be removed or driven out of hinges which are accessible from the outside

---

Case study

Typical test on a wood window to measure operational characteristics

The objective was to measure and assess the operation and strength characteristics of the window against the methods and performance requirements specified in BS 6375-2[174] for a side hung window.

The window was softwood framed with two outward opening, side hung casements, a top hung vent and a fixed light. There were two lockable operating handles on the stiles of the two side hung casements, and a peg stay on the bottom rail of the top hung vent; the lights closed against wiping seals. Hinges were variable geometry on the side hung and two butt hinges on the top hung. Also fitted central to the hinge side on the side hung casements were interlocking blocks. The window had a trickle vent mounted above the top hung vent.

For the operation and strength tests the window was secured against a rigid steel frame. Operation and strength tests were in accordance with BS 6375-2 Appendix A, tests A2 to A8, comprising:
● ease of fastener operation to A2
● ease of movement of sash to A3
● resistance to accidental loading to A8
(Tests to A5, A6 and A7 were not performed on this window.)

The window satisfied the following criteria:
● ease of fastener operation: fastener engaged in the designed position
● maximum permitted torque to release sash 10 Nm: measured torque 2.75 Nm
● ease of movement of sash, maximum permitted force to initiate movement 80 N: measured force opening 62 Nm, closing 42 Nm
● maximum permitted force to sustain movement 65 N: measured force opening 42 Nm, closing 102 Nm (not passed)
● resistance to accidental loading: fastener engaged in the designed position
● resisted 1000 N for 1 minute

The operation and strength characteristics did not meet all the BS 6375-2 performance requirements for a side hung window. The force to maintain movement of both the side hung sashes in the closing direction exceeded those given in the standard. The forces peaked when the interlocking plastics blocks on the hinge side started to meet. It is possible that these plastics blocks expand marginally when they are not interlocked, particularly when the sash is left open in warm weather. How much the normal closing action is hindered depends on how the interlocking blocks were set originally.

Where necessary, windows can also be fitted with electronic sensors to detect illegal opening, or attempts to open, or be monitored with closed circuit television.

Dimensional stability, deflections etc

For coefficients of linear thermal and moisture expansion etc, see Chapter 1.2.

Frames holding glass need to have limits on deflection of $^1/_{175}$ of the span or 15 mm, whichever is the smaller.

Weathertightness, ventilation and condensation
Weathertightness

If a window is not weathertight, the surrounding materials and finishes may become damp and deteriorate. Many window-to-wall joints have the potential to leak. Site inspections during 1991–94 indicated that nearly half the vertical window-to-wall joints inspected were deficient in weathertightness, and also lacked provision for stop ends to lintels in around 2 in 5 cases (Figure 4.12).

Windows may leak air and water around the backs of frames if they are positioned close to the building face and if not provided with a properly housed joint. Damaged, cracked or inadequately projecting

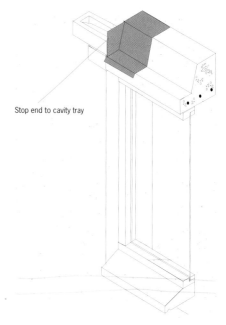

Stop end to cavity tray

Figure 4.12
Lintels in cavity wall construction need stop ends to the cavity tray to prevent rainwater penetrating the outer leaf from saturating insulation in the cavity

sills can result in penetration under a window. Cavity wall construction without a vertical DPC or cavity tray can be at risk of penetration at the window head or sides. Failed glazing compound or beads may allow water penetration around glass and into the fabric of a door or window.

Some areas of the UK experience more severe wind and rain conditions than others, and it is important to choose windows with an adequate degree of weathertightness for their location. Windows can be classified on their weathertightness performance by testing them in accordance with the appropriate British Standards (eg BS 5368[175]). Windows are classified in an 'exposure category' which is defined in BS 6375. The 'exposure category' for any particular location in the UK is related to the wind speeds during the most severe storms in the area. It is expressed as a design wind pressure in Pascals (Pa) and there are five exposure categories – less than 1200 Pa, 1200 Pa, 1600 Pa, 2000 Pa, and over 2000 Pa. The two lowest categories differ only slightly in the degree of resistance to rainwater penetration

Typical weathertightness test on a softwood framed window
The objective of the tests was to assess the weathertightness of the window against the performance requirements specified in BS 6375-1 for a design wind pressure (exposure category) of 2000 Pa. The air permeability, watertightness and wind resistance of the window were measured in accordance with BS 5368-1 and BS 5368-2 respectively.

The window was softwood framed with two outward opening, side hung casements, a top hung vent and a fixed light, glazed with 14 mm (4 + 6 + 4 mm) glass units, in Butyl 66 non-setting compound with external softwood beads. There were two lockable operating handles on the stiles of the two side hung casements, and a peg stay on the bottom rail of the top hung vent, and the lights closed against wiping seals. Hinges were variable geometry on the side hung and two butt hinges on the top hung. Also fitted central to the hinge side on the side hung casements were interlocking blocks. The window had a trickle vent mounted above the top hung vent.

For the tests the window was mounted in the BRE test rig, to form one wall of a pressure box, with the outside face facing into the box. A spray grid was mounted in the pressure box to apply water to the outside face at the rate of 2 l/m² per min in accordance with BS 5368-2 (EN 86) preferred spraying method no 2.

Weathertightness tests were in accordance with BS 6375-1, and BS 5368-1 to 3 and comprised:
● air permeability to BS 5368-1 (EN 42) with a maximum positive pressure differential of 300 Pa in pressure steps of 50, 100, 150, 200 and 300 Pa

● watertightness to BS 5368-2 (EN 86) using spraying method no 2 with a maximum positive pressure differential of 200 Pa in pressure (Pa)/time (min) steps of 0/15, 50/5, 100/5, 150/5, and 200/5 (+ 300/5 where over 2000 Pa)
● wind resistance to BS 5368-3 (EN 77) for a design wind pressure of 2000 Pa
● repeat air permeability tests at 300 Pa positive pressure only
● repeat watertightness test

Results of tests:
● air permeability at 300 Pa, max air leakage rate 16 m³/h per metre run: 2.32 m³/h per metre run measured
● watertightness at 200 Pa, no leakage permitted: leak measured at 150 Pa
● wind resistance at 2000 Pa, no damage permitted: no damage measured
● deflections not to exceed $^1/_{175}$ of length or height of glass: measured within criterion*
● air permeability (repeat test) at 300 Pa, no significant increase: 2.31 m³/h per metre run measured
● watertightness (repeat test) at 200 Pa, no leakage permitted: leak measured at 150 Pa

According to the values measured, the window met the air permeability requirement, but failed to meet the watertightness and possibly failed to meet the wind resistance requirement.

* BS 6375-1 specifies that during the wind resistance test the deflection of members retaining insulating glass units must not exceed 1/175 of the length or height of the glass

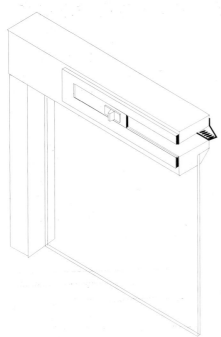

Figure 4.13
A trickle ventilator set in a timber window head

and cover most windows in buildings up to three storeys high in built-up areas and sheltered countryside. The 1600 Pa category applies to windows in low-rise buildings on exposed hillsides, high ground and in open country. The two highest categories only apply to low-rise buildings in extremely exposed locations which experience the most severe storms and driving rain conditions in the UK (eg north west Scotland and some south west facing coastal areas in England).

The concept of 'driving rain index' for a site is not very relevant to the process of specification testing and performance of windows.

Testing of windows is concerned with gusting of wind over **shorter** periods. When water is in contact with joints during rain, such pressure fluctuations may make windows leak by overcoming seals or pushing water over upstands.

The driving rain index concept is for absorbent masonry walls where **longer** period 'spells' of wind and rain combined may saturate brick and blockwork, causing penetration.

To ensure windows are adequately weathertight, purchasers of replacement windows should check that samples have been tested in accordance with the British Standards and given an exposure category which matches the area where the windows are to be used.

Ventilation
Old draughty windows may have ensured a high rate of natural 'trickle' ventilation but a change to multi-locking weatherstripped units, while improving comfort and reducing energy requirements, may inadvertently promote internal condensation and higher levels of pollutants in internal air.

Windows can make a contribution to the balance of heating and ventilation and therefore control of condensation in buildings only if the building occupants find the controls easy to reach, convenient to use, and precisely adjustable so that air flows can be regulated to best advantage in windy conditions. The arrangement of fixed and opening windows should aim to minimise draughts by providing ventilation to the upper levels of a room, preferably above 1.75 m above finished floor level. The small top hung opening light has the advantage in allowing small adjustments which would be difficult to match with, for example, large pivots. Fine control of ventilation is, however, also possible by incorporating unobtrusive trickle ventilators in the heads of frames (Figure 4.13).

When following the provisions of the Building Regulations 1991 Approved Document F[176], trickle ventilators installed in window frames and glazed areas will need to be of 8000 mm^2 or 4000 mm^2 depending on whether the room in which they are situated is habitable or not. If they are slotted the minimum open dimension should be 5 mm, and if they are holed 8 mm. These dimensions will not exclude insects. Any mesh size introduced as a remedial measure should not reduce the ventilation opening size below that given by building regulations.

Alternative provisions for ventilation will be found in BS 5925[177] and BS 5250[41].

Condensation
Condensation occurs to a greater or lesser extent during cold weather on single glazing and metal parts of windows without thermal breaks (Figure 4.14). The extent to which it occurs relates to several factors including heating, ventilation, and the amount of moisture generated by those living in the dwelling. Condensation on glass, providing it is drained effectively to the outside, can be beneficial; the glazing acts as a dehumidifier, tending to reduce the moisture content of the internal atmosphere and the risk of condensation on other parts of the building fabric.

There are no agreed standards by which building designers can specify a particular level of condensation resistance in the same way that they can with other aspects of window performance[178].

Condensate on glazing or frames that accumulates inside the building or saturates adjacent construction (further lowering its insulation value) can lead to problems, particularly with metal frames which should preferably be fitted in a timber (or plastics) subframe with suitable drainage.

Rehabilitation works may increase the risk of condensation if ventilation rates are reduced by removing chimneys, substituting solid for suspended floors, or providing draughtstripped windows and doors, unless substantially

Figure 4.14
Condensation occurring on a single glazed timber window. There is no drainage to outside provided, and the pool of water extends over the whole of the window board

improved levels and distribution of heating and alternative controllable means of ventilation are provided.

Thermal properties

Frame members of metal without thermal breaks are poor insulators (a thermal break window was illustrated earlier), but the thermal performance of windows also depends very much upon the glazing.

Double glazing units are installed for a number of reasons of which improved energy efficiency is the main one. How important this is depends upon the individual property and the priorities of the individual building owner, but all installations of replacement double glazing for single glazing will lead to energy savings. The cost savings from double glazing are difficult to estimate, being dependent on the type and level of space heating in a dwelling, and the types and quality of old and new glazing systems. The greatest savings can be made in properties that have expensive or inefficient heating systems.

Sealed units with 16 mm or 20 mm airspace (the optimum gap for thermal insulation is 20 mm, but 16 mm is only marginally worse) have become the industry standard, fitted to aluminium and PVC-U frames. Wood and steel frames will not normally take these wide gap sealed units. Secondary glazing in double windows, with wide air gaps and good draught sealing, may be the answer for historic buildings where the original frames cannot be changed.

Typical double glazing units have a thermal transmittance value (U value) of 3.0 or 4.2 W/m^2K, depending on frame materials and air gap sizes, compared to a typical single glazed figure of 5.45 W/m^2K. To achieve even lower transmittance figures low emissivity glazing can be used, and the space between the panes filled by inert gases such as argon or krypton. The low-emissivity coatings allow light to pass through the glazing in one direction, but reflect the majority of long wave radiation or heat, re-emitted from

inside the building. This increases the 'greenhouse effect' of glazing, allowing solar radiation into the building but trapping heat inside. Filling the cavity between panes with special gases reduces the heat loss through the cavity. Coatings and inert gases reduce the overall U value of the glazing part of the window to 2.0 W/m^2K, or even as low as 1.4 W/m^2K.

Heat losses through windows are significantly affected by radiator siting (either under the window or on a side wall), sill depth, curtains, double glazing and low-emissivity glazing. However, trickle ventilators have little effect on heat losses[180].

For spaces with high heat gains in

summer, it may be worth considering 'heat mirror' low-emissivity glazings, which have a lower total solar transmittance than conventional low-emissivity glazing.

Light transmission

Some mention of the early history of domestic windows was given in Chapter 1. There is no doubt that early windows did not admit a great deal of light, being small and glazed with, at best, translucent materials. The window taxes introduced in 1695 tended to reduce their numbers too, and it was not until these taxes were abolished altogether in 1851 that designers could specify freely.

The sizes of frames and glazing

Case study

Draught and condensation problems with single glazed pivot hung windows

The windows had given rise to continuing complaints from occupants, and inspection by others had not shown any clear reasons for these problems. During the BRE site visit, a number of different problems with the windows were seen:

- considerable gaps in the weatherstripping around the closed pivot windows, visible as daylight at the top corners of the frames. They could be responsible for draughts in windy weather. This defect would seem to be relatively easy to correct by replacement of weatherstrips, which in this case were standard 'off-the-roll' materials (eg cellular EPDM and PVC and nylon pile)
- looseness of framing and bowing of long sections
- difficulty in window opening caused by deep pile seals. The BBA Certificate showed smaller dimension pile seals than those seen on site
- heavy condensation on glazing. Some occupants were aware of neither slot ventilators, nor the top hung vents sited above the pivots. These were rather high for easy operation and insufficiently secure for ground floor use
- it could be seen from the BBA Certificates that the test pressure classes for side and top hung windows were considerably higher than those for the pivot hung windows. It is not so simple to weatherproof horizontally pivoted windows with their half inward opening, half outward opening, configuration as the outward opening side or top hinged types of frames

Nevertheless, the test pressure classes stated for these pivot windows complied with

considerably more than the minimum levels in the latest version of BS 6375-1 when applied to these buildings. The grading in BS 6375 covers wind resistance, water tightness and air permeability tests, but not the fixing or sealing of individual windows into a building.

Remedial measures seemed reasonably clear and could be summarised as follows:

- inspection and careful sealing where necessary with a low-modulus BS 5889-type[179] silicone sealant of all gaps seen at the top and bottom of window coupling mullions
- adjustment to weatherstrips at the head of windows to close the gaps and ease the problems with opening. Draughtstrips must be bonded on to the timber sections of these windows; mechanical fixings penetrating the PVC sheathing would not be acceptable
- persuading occupants to ventilate moisture generated at source better (kitchens, bathrooms and tumble driers) and to use more background ventilation through the vents provided. The window design was quite thorough in this respect, providing occupants with a number of options for ventilation. There should be no need to open the larger pivot windows for satisfactory ventilation in wintertime. However, with single glazing, some condensation nuisance would always be present
- the defects seen were not very severe and the only certain benefit from replacement would seem to be the possibility it offered of upgrading to double glazing

Figure 4.15
This Victorian double hung sash window has very slender glazing bar and frame sections which take around 15% of the window area

Figure 4.16
A modern window (of similar size but smaller opening area to that in Figure 4.15) in which the framing takes over 35% of the available area

bars, and even their shapes, have a significant effect on light transmission. In this respect, windows in older buildings often had very slender glazing bars with splayed sections (Figure 4.15), whereas more modern designs are crude by comparison (Figure 4.16). The standard EJMA frames were around 30% of total areas, depending on size.

Clear glass, when clean, transmits over 90% of the light falling on it perpendicular to the face, but the proportion transmitted reduces as the angle of incidence reduces. Tinted and solar control glazing reduce heat gain but also have low levels of light transmission too. Studies in offices in the USA have indicated that transmittances below around 35% may make the view out look gloomy. With absorbing glasses, reflections may not necessarily increase.

Fire

This is not dealt with in detail, since each set of circumstances will differ. Windows on the side walls of buildings which are close to a boundary may represent a risk of fire spread to adjacent buildings. Windows, whether fixed or opening, are classified as unprotected areas set into the external wall. The major limitation on use of unprotected areas is within one metre of the site boundary, though unprotected areas of less than 1 m^2 may be disregarded[52].

Windows are, however, used as emergency exits in ordinary houses and in loft conversions. Given suitable hard standing, turntable ladders will normally permit escape from up to the sixth floor, and some hydraulic platforms can go higher. In such cases, means of safe egress from the building may need to be provided – for example through windows with opening lights of minimum clear space 850 × 500 mm – subject to safety devices which prevent accidental opening.

There have been concerns that, being plastics, PVC-U windows would be particularly dangerous in the event of fire. In fact, tests carried out by the Fire Research Station for the British Plastics Federation have shown that PVC-U windows do not create new hazards in fires.

Noise and sound insulation

Typical sound insulation values for closed windows against traffic noise are:
- single windows, 28 dB(A)
- sealed double glazed units, 33 dB(A)
- secondary windows, 37–40 dB(A)

Against aircraft noise, double windows give about 3 dB(A) improvement over sealed units.

Although replacement window companies sometimes make claims about improved noise insulation for their replacement windows it is only properly designed secondary glazing, with wide air gaps (at least 100 mm) and well sealed, that gives optimum sound insulation for external traffic and aircraft noise. Conventional replacement frames with double glazing units instead of single glass, even well draughtstripped, will make only a marginal improvement in noise insulation. The noise insulation performance of an air filled double glazing unit improves only slightly with increasing air space width. Worthwhile improvements in noise insulation will only come from the use of a secondary window.

Durability
Timber windows

Until the 1960s, softwoods such as European redwood, although classified as non-durable, had proved remarkably durable in practice, especially if their paint finish was regularly maintained. The increasing demand for housing in the 1960s led to an increased proportion of sapwood in the timber used for joinery, together with changes in design and workmanship which placed greater reliance on regular maintenance. This situation resulted in a significant increase in rot in window joinery, essentially because of paint failure and the subsequent water penetration into exposed joints. This mainly affected the lower horizontal sections and their adjacent stiles in both sash and casement windows (Figure 4.17). To overcome this, wood preservation was recommended and now softwood windows are normally treated with preservative.

The hardwood commonly used in sills, European oak, is classified as durable, and would not normally require preservative treatment unless sapwood is present. Some imported hardwoods are also naturally durable, but others are supplied as mixed species (eg commercial supplies of meranti) with widely differing durabilities, and in these cases treatment is normally specified[155].

New European standards classify wood preservatives according to effectiveness in different environments. The level of treatment to be achieved is classified by penetration into the wood and the retention of loading in a specified zone, irrespective of treatment method or wood species[181].

Steel windows

When steel windows were first introduced, many were ungalvanised. Provided maintenance was meticulous, these windows could have a reasonable life, but many rusted, causing fractured glass and broken window furniture as increased force came to be used to close them. Hot dip galvanising largely cured this problem.

PVC-U windows

The material from which these windows are fabricated has been much improved since first introduction. Quality of manufacture is important, particularly in relation to the quality of corner joint welds, but discolouration can also occur. Dark coloured units can suffer from distortion under solar heating.

Glass

Glass can become etched and aluminium frames, and zinc and lead flashings, can suffer corrosion staining by lime leaching from fresh mortars onto glass or frames.

Figure 4.17
Neglect over many years has led to progressive deterioration of this timber window – not only paint, but also putty has failed, leading to opening of the joints

Double glazing units

Apart from rot of non-durable timbers, one of the main problems that BRE comes across with wood windows is failure of double glazing units. This is because most existing wood frames do not allow for drained-and-ventilated glazing methods and the fully bedded glazing used can reduce the durability of sealed units by allowing moisture to be trapped between the frame and the unit. Drained-and-ventilated glazing for wooden doubleglazed windows avoids this problem. If this is not possible, glazing that complies with BS 8000-7[182], and the recommendations of the Glass and Glazing Federation and British Woodworking Federation, should be considered. If windows have to be glazed on site they should be glazed in accordance with BS 8000-7.

BRE is frequently asked about the lifetime of double glazing units. A number of variables affect lifetime, but given the design, eventual failure is inevitable. There have been problems with double glazing units where they have been installed in frames that were only suitable for single glazing. Such frames should not be used. They cannot take the

Case study

Deterioration of PVC-U replacement windows
Brown coloured PVC-U windows in a house distorted to the extent that some of them had to be replaced. Fifteen years of south facing exposure had produced unsightly whitish deposits on the windows, but appeared not to have eroded the surface substantially. However, the high susceptibility of the profile to heat reversion, combined with solar heating of the dark surface, had led to distortion of the frame to the extent that the windows were no longer operable. The use of reinforcement in the profile and greater numbers of fixings to the wall, might have limited the distortion to an acceptable level.

weight and cannot be properly glazed to protect the double glazing unit from rainwater and sunlight.

BS 7543[74], on the durability of building components, states that a design life of a sealed unit is intended to be 30 years, but notes that their service life can vary between 5 and 35 years. In a survey carried out for BRE, some 30% of units had failed prematurely.

The most common failure is misting between the glass panes. Some misting failures have been found to be the result of volatile emissions from components of the double glazing unit itself or from glazing compounds, but most are a result of the condensation of moisture within the unit. Failures can also include glass breakages.

Any weakness in the edge seal which allows water or water vapour to penetrate into the unit will cause the dew point of the air within the double glazing unit to rise, and failure can ensue. Silicone based seals are prone to failure when used alone in single sealed systems, as also are linseed oil putty fully bedded systems which allow water to lie in contact with the units.

Water may condense on the inner face of the outer pane when the outside temperature is low. This misting may not be present during the day when the building is occupied and the temperature is higher. Nevertheless, regular intermittent condensation can lead to changes in the inner surface of the outer pane known as scumming. Condensation between the panes of a double glazing unit at any time is unacceptable[167].

Manufacturers' warranties generally cover either 5 or 15 years. However, the conditions of the warranties should be fully checked as failure to maintain windows could be a cause of default.

If a double-glazing unit fails, resulting in condensation between the panes, it cannot be repaired and needs to be replaced. However, the cause of failure must first be determined so that the same mistake is not repeated. Once the cause has been established, replacement can involve either a better quality unit or glazing in such a way that premature failure will not recur.

It is important that consumers demand BSI kitemarked[183], or similarly quality-assured, double glazing units. If they are kitemarked they will bear BS 5713, the manufacturer's name and the year of manufacture all stamped on the spacer bar. This can easily be checked as they arrive on site. To gain a kitemark the manufacturer must have a quality control system in place in the factory (ISO 9000) and have units audit tested to the requirements of BS 5713. In future, European Standards will replace BS 5713 and the indications are that these standards will provide more stringent tests for double glazing units.

Maintenance

Careful maintenance is needed to preserve the appearance and integrity of timber windows, particularly those installed in heritage buildings. If poorly maintained, timber windows are more disposed to movement, shrinkage, cracking and distortion; they are also more likely to be at risk from wet rot, particularly in the case of untreated softwood windows.

Steel windows corrode as a result of the failure of protective zinc or paint systems. Aluminium or PVC-U windows may have deteriorated in use. Painting of installed PVC-U windows is not normally recommended; in particular, white windows should not be painted in dark colours. All windows will have associated hardware which may be worn after many years of use or may have corroded or seized.

A number of window designs rely on gaskets of various kinds to ensure adequate weathertightness, and these will need to be inspected and replaced if they lose resiliency or are damaged in use (Figure 4.18).

Many traditional windows retain the glass panes with linseed oil and whiting putty which if inadequately protected dries, cracks and eventually falls out. Replacement of

---

Case study

Dewpoint test on sealed double glazed units

It was clear that all the small double glazing units in the timber windows on a housing estate were showing condensation within the units in a comparatively short period of time, and further failures were daily being observed. In particular, some small double glazing units exposed to large amounts of driving rain were observed to have a high rate of failure. Since moisture was present between the panes, the desiccant in these units was obviously exhausted and further failures could be anticipated.

The units were externally glazed with beads and butyl glazing compound in a fully bedded system. Gaps between the ends of the beads and the frame of 2–4 mm were found on a number of window panes and gaps had also been left where there should have been a full bed of butyl, in turn allowing water to permeate through to the rebate area in contact with the edge seal of the double glazing units. Protrusion of the spacer bar above the sightline of the glazing bar frame was observed for many units.

A dewpoint test was carried out on sample units, recording information on the room orientation, window, double glazing unit, time to dew formation, and dewpoint temperature.

Moisture was present within all the double glazing units. The range of dewpoint temperatures found for the large double glazing units was from –28 °C to –45 °C, much lower than the dewpoint temperatures for the smaller units. The average time to failure of these larger units was estimated to be between 2 and 5 years. The remaining small double glazing units all had high dewpoints, indicating that they had a high probability of early failure.

It was concluded from the site inspections and the tests that the earlier deterioration of the small units as compared with the larger units was a function of a number of factors.
- The west facing small units were subject to more driving rain than the east facing larger units
- The early failure and high dewpoints of the small double glazing units was a function of poor glazing technique and inappropriate materials and methods
- The protrusion of the units into the sight line had allowed UV light and heat to affect the edge seal

the 'back putty' will generally require removal of the glass. Beaded glazing or gasket glazing may be found in more modern windows. Both systems may deteriorate with time and may require maintenance.

Limited deterioration of timber windows is repairable; however, skill, care and specialist preservative treatment and repair adhesive/fillers are required. Rotted areas should be cut out down to sound wood, the cavity flooded with a suitable preservative (not creosote) and allowed to thoroughly dry before priming and filling with a proprietary filler. If the solvent of the preservative is not allowed to dry it may affect the durability of subsequent painting. It may be beneficial to use an unpigmented primer designed to achieve deep penetration of 'de-natured' wood.

Although many paint systems perform poorly when applied to even carefully prepared weathered surfaces, some of the newer exterior paints available can provide a longlasting maintenance coat (see also Chapter 9.4).

For replacement purposes, wood should be specified from well managed resources. The rebates should be suitable for accommodating 16 mm wide double glazing units. When replacing windows, the joint between new frames and the existing wall should be carefully considered, remembering the following points:
- providing two lines of defence against wind or rain penetration
- frames should preferably be set back from the face of the building, and protected with rebated reveals

All gaskets, glazing tapes, sealants and glazing compounds should be checked, internally and externally, for watertightness, and all drainage and ventilation holes checked and cleaned out on a regular basis. The main aim should be to ensure that water is prevented from lodging next to the edge seal.

Work on site

Storage and handling of materials
Particular care must be taken while transporting, handling and storing the double glazing units on site to ensure that they are not damaged; the edge seals of the units and the edges of the glass panes are especially vulnerable.

Workmanship
When replacing windows in rehab work, it should go without saying that the integrity of the surrounding wall, and the window-to-wall joint, should be maintained. BRE site inspections show that this often does not happen.

Cutting and fixing of glass should always be done in accordance with manufacturers' recommendations. In particular, cutting laminated glass incorrectly, where the interlayer is damaged, may lead to premature deterioration.

Waterborne paints should not be used on linseed oil putties for at least four weeks, or failure may occur.

Double glazing units should be installed in accordance with the provisions set down in British Standards BS 6262 and BS 8000-7[181], and in the Glass and Glazing Federation's *Glazing Manual*[185]. Particular attention should be paid to the type and position of setting blocks, location blocks and distance pieces, especially in the case of drained glazing to ensure that they do not inhibit either drainage or ventilation. Similar conditions apply also to the installation of laminated glazing, and linseed oil putties should on no account be used.

Wherever possible wood windows should be repaired rather than replaced, particularly in heritage buildings. Joinery skills are still available, and specialist firms exist.

Overcladding
Windows are normally replaced at the same time as overcladding is installed, and should be considered integrally with the overcladding. It may be possible to fix the windows into a sheet cladding system, or at least to the suspension system, instead of to the original structure, and consideration will need to be given to compatibility of materials. The new windows may also be fixed before the old ones are removed,

Figure 4.18
There are twin gaskets on this vertically pivoting window. The crossovers (not shown in the photograph) where the window profiles change from open-in to open-out, are particularly crucial to performance

reducing inconvenience to occupants.

There are some theoretical advantages in fixing windows to inner leaves which are insulated on their outside. Since these leaves will be in a more stable environment (and warmer), the stresses on the windows will be reduced compared with windows fixed into the overcladding. There is less need to worry about weatherproofing and thermally insulating the joint edges, though a thermal break in the window frame itself, where appropriate to the material, would still be advantageous. On the other hand, if the windows are fixed to the original structure, there may be sizeable differential movements between structure and overcladding which will tend to fracture joints. Robust weathering details are essential. Fixing to the cladding will need some means of masking the gap at reveals, leaf or sills with a flexible material.

Window-to-wall joints

It is always better in principle to seal windows at the back of the window-to-wall joints, and to have some kind of mechanical joint, overlap, or bead at the front, in order to create, effectively, a two-stage joint. This applies whether or not the joints in the surrounding cladding are one or two-stage, and whichever leaf contains the window.

Opening windows should be checked for of warping and twisting which can lead to poor airtightness or inadequate wind resistance. Diagnosis of a weathertightness problem is often complicated by condensation. Site testing of windows may be justified on large schemes.

Most defects are readily observed, but rotted timber may need testing with a small pointed instrument (Figure 4.19). Wood surfaces that have been exposed to the weather may be soft and noticeably darker or lighter than the surrounding wood, and this is indicative of wet rot. A grey surface, with no softening, indicates surface degradation by sunlight.

Figure 4.19
Here the softwood sill , though adequately painted on its upper surface, has been saturated from its unpainted underside by the moisture laden exhaust from the boiler. Rot in the sill has resulted

Double glazed units should be inspected for defects in the edge seals and the glass panes of units. Where there is some doubt about the long term behaviour of the seals, dewpoint testing may be justified. It should be noted, however, that predicting the failure of double glazing units is not an exact science.

Dewpoint testing

Dewpoint is the temperature at which condensation will occur. The dewpoint test measures the dewpoint of the air in the space between the glass panes of double-glazing units. Clear glass is easiest to test, but it is possible to test obscured glass.

After manufacture a double glazing unit should normally have a dewpoint of less than –45 °C; in time the dewpoint will rise as moisture permeates through the edge seal and the desiccant in the spacer bar becomes saturated. The likelihood of failure increases as the dewpoint rises to around –10 °C.

The method involves a methyl alcohol filled insulated metal container with a polished metal plate. The alcohol is cooled to

a temperature of –75 °C using liquid nitrogen. The surface of the polished metal plate is then sprayed with alcohol, together with the glass surface, and placed against the room side surface of the glass. The time period to dew formation inside the unit at the polished plate is then measured. The shorter this time then the higher the dewpoint temperature and thus the more moisture there is in the airspace. Graphs are used to convert the recorded time to a dewpoint temperature. If no dew forms within two minutes the dewpoint is assumed to be below –45 °C and it is probable that the unit will have good potential durability.

The problems to look for are:
◊ inadequacy of daylighting
◊ inadequacy of ventilation
◊ rot in timber frames
◊ corrosion in metal frames
◊ no thermal breaks in metal windows
◊ condensation on either frames or glass
◊ malfunctioning opening limiters
◊ opening lights out of reach
◊ missing or misplaced vertical jamb DPCs
◊ lintels missing or defective
◊ failure of putty
◊ flush sills
◊ sills having no drips
◊ glazing set in inadequate reveal depth
◊ no condensation drainage to exterior
◊ non-safety glazing at low levels
◊ casements or pivots open over pedestrian routes
◊ failure of double glazing units
◊ rain penetration (performance inadequate for exposure)
◊ draughts – no weatherstripping
◊ trickle vents inadequate
◊ sound insulation inadequate where required
◊ no child-proof fastenings to casement and pivot windows above ground level which limit the initial opening to 100 mm
◊ stair windows not within normal reach for cleaning and opening.
◊ glass not in accordance with BS 6262 or with Building Regulations Approved Document N
◊ side hung casements not having easy-clean hinges with a clearance of 100 mm minimum
◊ stair windows which are cleaned under contract having opening lights lower than 1.12 m from the floor and which are not fitted with budget locks with a visual indication whether the lock is open or shut
◊ windows which are reversible for cleaning have no locking bolts to fix them in the fully reversed position
◊ a height of 2 m is exceeded for hand operated controls
◊ finger traps in window hinges or controls

Chapter 4.2   Non-domestic windows

Figure 4.20
A very large double hung timber sash
window used as a shop front (Photograph
by permission of B T Harrison)

This chapter deals with windows
used, for example, in shopfronts,
public buildings, offices, churches
and factories.

Many of the performance
requirements and solutions for non-
domestic windows are similar to
those for domestic windows except
for those relating to size (Figure
4.20), and this can be seen in the case
studies in this chapter. The chapter
therefore does not repeat relevant
items from Chapter 4.1.

Characteristic details

Basic structure

Stone frames
Windows created by rebated stone
mullions, transomes and other
tracery was a technique well known
to medieval craftsmen, and
employed widely in ecclesiastical
architecture and occasionally, from
the late fifteenth century, also in
large scale domestic architecture.
Special care is needed in repair; for
example in respecting the bedding
planes of replacement stone (Figure
4.21). Conservation methods
cannot be dealt with simply, and
therefore are largely outside the
scope of this book.

Cast iron frames
Cast iron frames were introduced
quite early in the Industrial
Revolution although they did not
rival wood windows, either in cost or
popularity. They were widely used in
factories and small industrial
buildings, but also to some extent in
other building types where the
additional costs of non-ferrous
window frames (eg bronze) were not
justified. They reached their zenith
in 1851 in the Crystal Palace.

Frames of other materials
Large windows, as in shops, are
sometimes to be found framed in
wood but more often than not in
metals such as aluminium or bronze.

Louvred windows, although they
are occasionally found in domestic
construction, are more likely to be
found in non-domestic and, for that
reason, are mentioned here. Frames
are usually in extruded aluminium
alloy, with blade holders in
polypropylene. Louvre blades are
normally in 6 mm thick glass, under
200 mm wide with a limited range of
patterns, and sealed to the adjacent
blades with weatherstripping when
in the closed position.

Glazing
In general, larger panes mean
thicker glass, and a broken window
should always call for a
re-assessment of whether the glass
was correctly chosen and installed.
Windows in building types other
than domestic sometimes offer
larger areas suitable for decorative
treatments such as stained glass or
etching (Figure 4.22). See the same
section in Chapter 4.1 for a basic
description of glazing materials.

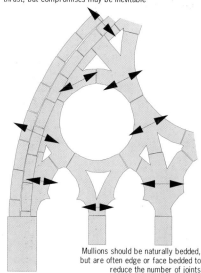

Bedding in tracery generally should be at right angles to
thrust, but compromises may be inevitable

Mullions should be naturally bedded,
but are often edge or face bedded to
reduce the number of joints

Figure 4.21
Bedding planes should, wherever possible,
be laid in accordance with the direction of
maximum thrust

Window-to-wall joints

As with domestic scale windows, it is the window-to-wall joint which is crucial to performance. Since the windows are normally larger than domestic, and the area exposed to sideways driven rain is correspondingly larger, the extra shelter to the window-to-wall joint to be gained by setting the windows in deep reveals is only marginal (Figure 4.23). Cornices, string courses, hood and label moulds were the traditional methods of providing more shelter.

Figure 4.22
West window, Coventry Cathedral

Figure 4.23
Horizontally pivotted windows, the Orangery, Margam Abbey

Main performance requirements and defects

Strength and stability

The larger the window, the more likely it is to suffer deflection in strong winds, though this ought to have been taken into account at the design stage.

The Building Regulations 1991, Approved Document N, Glazing – materials and protection[186] – now requires BS 6206[72] 'safe breakage' or screen protection in critical locations, which include glazing between finished floor level and 800 mm above that level. Under these regulations thick annealed glass not complying with BS 6206 is restricted to a minimum of 8 mm thickness, typically used in shop fronts, showrooms etc. There are limits on total areas relating to thickness, with the larger windows requiring 12 or 15 mm thickness.

The primary requirement for windows is that glazing, if it breaks, shall break in a safe manner and not cause injury to building users (inside or outside). This is governed by BS 6262[169] for glazed panels beside doors, glazed doors themselves, low-level glazing (below 800 mm from the floor), staircases, circulation areas and buildings with special uses (eg schools, pools and sports halls). Glazing materials (whether glass or plastics) are assessed according to internationally agreed impact loading tests which are set out in BS 6206, and the main glazing Code of practice, BS 6262.

The Workplace (Health, Safety and Welfare) Regulations 1992[187] are retrospective, and now require all glazing in relevant non-domestic buildings which is at a vulnerable height to either be protected or to be in a safety glazing material. Annealed glass may be suitable provided it is thick enough.

Protection against unauthorised entry

Shop windows and entrance doors are increasingly prone to attack, and special glazing should be considered for replacements.

Special safety glazing includes:

- laminated glass has considerable impact resistance, depending on thickness and number of laminates, given that rebate depths are also adequate. There may, however, be implications for means of escape in cases of fire
- bandit resistant glass, usually laminated with a thick interlayer
- bullet resistant glass, usually heavy gauge laminated glass with several glass sheets and interlayers of a total thickness of 25 mm or more
- plastics (eg polycarbonate sheet) presents a formidable obstacle to impact but is vulnerable to heat and is marred by scratching in service. Plastics is also much more flexible, and needs much deeper rebates than even laminated glass for safe retention, more especially so in large windows

Toughened glass can be easily shattered with a centre punch, and is not recommended for high risk areas. If risks require it, and sometimes at some cost in reduced light transmission, additional protection can be afforded by:

- grilles or shutters (see Chapter 5.3)
- intruder detector systems

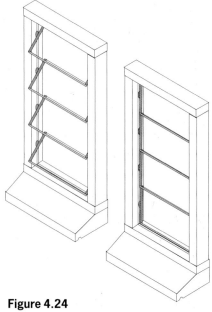

Figure 4.24
Early versions of louvred windows sometimes leaked

Dimensional stability, deflections etc
For coefficients of linear thermal and moisture expansion etc, see Chapter 1.2.

Weathertightness, ventilation and condensation
Weathertightness
Some of the first louvred windows on the market did not perform very well when subjected to driving rain, and had high air leakage characteristics, even in the closed position. Later versions with improved blade seals performed better in situations up to severe exposure (Figure 4.24).

See the same section in Chapter 4.1.

Ventilation
When following the provisions of the Building Regulations 1991 Approved Document Part F[176], trickle ventilators installed in window frames and glazed areas will need to be of 4000 mm^2 or 400 mm^2 per m^2 of floor area, depending on whether the room in which they are situated is occupiable or not. If they are slotted, the minimum open dimension should be 5 mm, and if they are holed 8 mm. These dimensions will not exclude insects.

Figure 4.25
Use of both internal and external light shelves for enhancing daylighting levels in a Midlands office building

The majority of the more troublesome larger varieties of insects can be excluded by provision of a 3–4 mm mesh over ventilation slot, but any smaller mesh size introduced as a remedial measure must not reduce the ventilation opening size below that given by building regulations.

Alternative provisions for ventilation are specified in BS 5925[177] or in CIBSE Guides A and B[43].

Thermal properties
Similar deficiencies exist with inappropriate bedding of larger double glazed panes and non-thermal break window frames, as with the smaller windows covered in Chapter 4.1.

Daylighting
It is important to achieve adequate levels of daylighting for what can sometimes be fairly deep buildings (although having the benefit of cooler conditions in summer). Fashion in design at one time dictated the use of very large windows. After a spate of complaints in the late 1950s and early 1960s, sometimes counteracted by the installation of air conditioning, it became common practice to cut down the size of windows in office and educational accommodation in particular, and to rely on the use of permanent supplementary artificial lighting to restore the daylighting deficiency. Energy costs now drive a further swing towards higher efficiencies in daylighting designs, a reduction of fuel for lighting and air conditioning, and an increasing interest in novel daylighting solutions (Figure 4.25). See Chapter 1.6 for a list of factors to be taken into account.

In art galleries and museums, and also, for that matter, in heritage buildings of various kinds, it is necessary for conservation reasons to cut down the amount of natural light reaching items which might deteriorate under UV action. Specialist advice should be sought in these circumstances.

EPDM gaskets and edge seals in double glazed aluminium windows
The BRE Advisory Service was asked to inspect and report on the condition of double glazed window gaskets in brown coated, aluminium framed windows with alternate fixed and opening lights infilled between concrete panels.

A number of locations in the building showed signs of rain penetration. These were over the heads of windows:
- against plastered internal concrete columns
- abutting brickwork flank walls

These positions of leakage did not correspond to any of the sealant adhesion failures seen during the inspection.

Window frames had been sealed throughout the building with brown sealants, subsequently identified as polyurethanes, which were not in very good condition. They had extensively weathered surfaces with much fine crazing and seemed to have age-hardened considerably. There were a number of areas with adhesion loss to window framing, although sealants were generally well bonded still to brickwork or concrete. This was most noticeable in vertical joints and at ends of sills and window modules where greatest movements were likely to be concentrated. These sealants had mostly been applied as triangular section fillets rather than butt joint seals. Although fillet seals are inevitable with this type of aluminium frame, in places they did not carry far enough on to the edges of the frames to provide a reliable bond area.

The internal glazing seals were square edged, ribbed gaskets which were in good condition, but there were some gaps visible at gasket corners which were not obviously mitred or sealed.

Life expectancies of sealants are quoted in BS 6093, Code of practice for design of joints and jointing in building construction – ie polysulfide and silicones up to 25 years, polyurethanes up to 20 years, neoprene and EPDM gaskets up to 20 years.

The conclusions from this investigation were:
- that the double glazed window gaskets appeared to be in good condition and likely to meet their normal life expectancy
- the polyurethane sealants used for window perimeter sealing were showing signs of surface degradation, age hardening and adhesion failure, and required some remedial sealing

Figure 4.26
A PVC-U assembly showing signs of corrosion of the steel reinforcement caused by inadequate welding of the joints

Shop windows form a category where reduction of reflection might be thought desirable, and indeed some inclined windows with black soffits over, or even made of glass curved or circular in section may be encountered, albeit that they are space consuming. There is discussion of alternative designs in *Windows*[188], pages 22 to 26. Glass with low reflectance coating is now available.

Fire

Glass as such is not classified as fire resisting – it is the glazed element as a whole which can be fire resisting or not when tested. Some glazed elements containing laminated glasses have been tested and found to provide a degree of fire resistance. There is some discussion of the parameters which influence the performance of glass in fire in BS PD 6512-3[189].

Windows in proximity to escape routes do need to be fire resisting.

Fixed (ie non-opening) windows which have passed the 30 minutes integrity and 15 minutes insulation fire resisting tests may possibly be classified in future as not being unprotected areas.

Noise and sound insulation

Laminated glasses, especially those with several interlayers as used, for example, in shop fronts, will provide a marginal improvement over single annealed glazing of the same thickness.

See the same section in Chapter 4.1.

Durability

Unfortunately the same range of problems exists with large windows as with small, though the consequences could be more serious (Figure 4.26).

See the same section in Chapter 4.1.

Maintenance

See the same section in Chapter 4.1.

Work on site

See the same section in Chapter 4.1.

Inspection
In additional to those listed in Chapter 4.1, the problems to look for are: ◊ bowing of glazing on large windows due to wind pressure ◊ loosening of wrought iron transoms in stone windows ◊ erosion and cracking of stone tracery

Chapter 4.3

Curtain walling and glass blocks

Curtain walling, and its derivative, the so-called but confusingly named 'structural glazing', is undoubtedly a development of the twentieth century. The true curtain wall has no loadbearing function other than to carry its own loads and to transmit the wind loads on the face of the wall to the primary framed structure behind. The industry is imprecise about the definition of curtain walling. However, for the purposes of this book, it is taken to be a form of cladding with the majority of its surface area consisting of glazing. The main impetus behind its development came with the need to reduce the dead loads on high-rise office buildings while at the same time providing an external skin which met all the necessary functional requirements (Figure 4.27).

It is convenient also to mention glass blocks in this chapter. Although they are not quite in the same category as curtain walling, being essentially masonry in character, they have some affinity – they are not structural, are commonly contained within a frame of some kind, and have the property of light transmission, though with the addition of increased security and fire resistance. Their use became popular in the period between the wars, especially by pioneer architects of the Modern Movement. Largely falling into disuse after the 1939–45 war, there is some indication that they are currently returning to popularity (Figure 4.28).

Figure 4.28
Glass block walling built in 1997

Figure 4.27
A familiar high rise landmark: London's International Finance Centre

Characteristic details

Basic structure
Curtain walling in essence is the cladding of a building with glazing contained within or superimposed on a grid of framing members. It may be provided on all faces of a building or on selected façades, with the other façades usually of heavier construction such as masonry or

Figure 4.29
A complete skin of tinted glazing

concrete. The curtain walling may be prefabricated off site into assembled frames complete with infilling, or it may be built up on site from separate framing members and infill panels. Site assembly, sometimes called stick building, provides more opportunity for adjustment of dimensions to fit the structure, but at the same time more opportunity for deficiencies in workmanship leading to deficiencies in performance.

Patent glazing may be used vertically. (Patent glazing is described in Chapter 5.2 of *Roofs and roofing*[24].) The greatest difference affecting performance of patent glazing used vertically instead of on the slope is that most of the vertical component of loads comes on the bottom edge support instead of being shared with the bar. On the other hand there are no snow and access loads to consider, and the wings to the bar play a lesser role in preventing rain penetration – there is no second line of defence. This places emphasis on the integrity of the seals.

Most pressed or cast glass insulating blocks produced up to the 1950s were square on face about 200×200 mm or smaller. They were available in a variety of patterns including ventilating and radiused corner blocks. Similar units, up to 300 mm square, are available currently in a wide variety of patterns, a few basic colour tints, and shaped and curved units.

Main performance requirements and defects

Choice of materials for structure

Curtain walling can employ many different kinds of materials for both frames and infilling. Perhaps the most popular installations have consisted of steel or aluminium frames, with infilling primarily glazing, but backed where appropriate; for example at floor levels to cover the structure, with coloured sheets of fibre reinforced board. Since the 1970s there has been a growth in the use of toughened glass supported from framing members by edge or corner cleats or by threaded and collared pins through holes in the glass; the joints between the glass panels are face sealed with a durable mastic or gasket (Figure 4.29). 'Structural glazing' has been used to describe this form of curtain walling – a misnomer if ever there was one. No frame as such is exposed on the outside; only perhaps the occasional cleat at the corners.

For those systems based on face sealing, the performance of mastics and gaskets would have been a critical decision on the part of the designers, and in the early days there was much wishful thinking on the part of designers on the capabilities of available materials.

Glass blocks used to be laid in stack bond in a 'fatty' lime mortar. Since they are non-loadbearing, the panels of blockwork were often limited in height to around 6 m. Large panels need reinforcement, which can be found usually laid in the horizontal joints to resist wind and minor impact loads. This reinforcement would consist of normal open mesh brick reinforcement, preferably in stainless steel. Advice used to be that the reinforcement would need to be continuous through the movement joints which were provided at the perimeters of panels, which were in turn to be bedded in or fixed to the main structure. This could be a source of disruption of the block panels; in replacement or new work a sliding edge connection in a channel fixed to the jamb would be a better proposition, especially where the panels exceed around 3 m in either width or height. Pointing is usually in a white mortar or mastic.

A comparatively new development are blocks laid with proprietary plastics jointing sleeves which, since their fire resisting properties are less than mortar, should be specified with care.

Strength and stability

Curtain walling is non-loadbearing; that is to say it should not receive loads from other parts of the construction. Loads from curtain walling itself and from wind loads are usually taken back to the structure at floor and roof levels, although for very large expanses with no intermediate support, some form of stiffening is needed (Figure 4.30).

The fixings used will normally be of the kind which provide some adjustability at the time of assembly

Figure 4.30
Here at the Sainsbury Centre at the University of East Anglia, fins of glass stiffen the very large glazed area

to accommodate dimensional deviations in the main structure. These fixings will need to provide positive fixity after adjustment.

Dimensional stability, deflections etc

For coefficients of linear thermal and moisture expansion etc, see Chapter 1.2.

Distortions in the mirrored images of reflective claddings often indicate where local stresses in the claddings occur (Figure 4.31).

Weathertightness, ventilation and condensation

Weathertightness

For conventional curtain walling, that is to say not designed to act as a permeable rain-screen, where reliance is normally placed on the provision of an impermeable skin to resist rain penetration, the joints between the framing and the infilling assume considerable importance. Even a pinhole in a mastic seal can admit copious amounts of water when under pressure high up on a façade. It is much better if some further protection can be given.

When curtain walling first became popular in the late 1950s and early 1960s, the Building Research Station, as it then was, carried out a series of case studies on curtain walling and light cladding, which showed that

many examples were deficient in respect of rain penetration. Many examples placed considerable reliance on oil based mastics to provide seals between frames and infilling. These seals could, at best, tolerate only limited movements in service, and typically broke down after about three or four years. When the investigation team revisited some of the sites after an interval of several years, it was found that some buildings had been re-sealed several times with the identical materials, which, of course, had again broken down. No attempt had been made to rectify the defect causing the problem. No wonder that curtain walling got a bad name in the early days!

Things may not have improved much in the intervening years. In April 1997 a study by Taywood Engineering revealed that two thirds of curtain walls failed to keep out water when first installed[190]. Unfortunately, there is still much wishful thinking on the ability of inexpensive sealants to accommodate large amounts of movement.

Airtightness of existing installations may need to be examined. It is recommended that air infiltration should not exceed $1 \, m^3/m^2$ per hour when tested to BS 5368-1[174].

Case study

Sealants used in a curtain wall
The BRE Advisory Service was requested to inspect and report on the condition of mastic and gasket joint sealant materials used in curtain walling in a four storey building situated near the coast. There was considerable evidence of leakage inside the curtain wall framing.

Some external corner vertical joints were sealed with insert gasket strips which were in good condition. Elsewhere, glass was sealed to its framing by front fillet seals of a black polyurethane sealant. In some areas, this sealant was very badly degraded, with the crazing and weathered surface having eroded the shallow edges of the fillet seal completely away. In addition, adhesion to the glass and frame edges was very poor such that it could easily be pulled away by hand. This severe degradation was worst on the lower parts of the exposed south side of the curtain wall, but showed to a lesser extent on the west side of the building. Some black glazing sealant was in good condition, but this may have been different material or a remedial application.

In some joints in parapets, plant room walls and louvres, an oil-based sealant had been poorly applied over the top of polyurethane and silicone sealants. The oil-based material was completely brittle and did not appear to be performing any useful function.

The conclusions from the investigation were that the polyurethane sealants used for curtain wall sealing were showing some sign of surface degradation, age hardening and adhesion failure, and re-sealing would be justified.

Thermal properties

Some of the early examples of curtain walling had minimal thermal insulation provision, and by now many installations will have been refurbished to current standards. Thermal bridging was identified in the Building Research Station light cladding surveys of the 1960s as a significant problem, since many early systems had no thermal breaks in the framing members (Figure 4.32 on page 190).

Figure 4.31
Distortions in the reflected image given by cladding

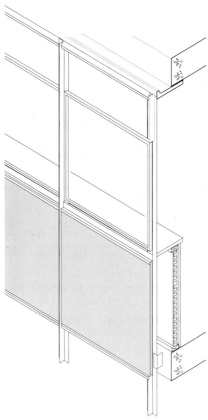

Figure 4.32
A typical early curtain
walling installation with no thermal breaks in
the exposed framing and little thermal
insulation in the backup wall

Fire
Cavity barriers need to be provided
to limit the sizes of cavities in certain
doubleskin curtain walling
installations, and in refurbishment it
is important to review their provision
in relation to the design of the
weatherproofing – in particular
whether the system relies on cavity
ventilation for drying off rainwater
penetrating the outer skin.

   Non-ventilating glass blocks should
give around 1 hour's fire resistance in
relation to integrity, but probably not
in relation to insulation. The
manufacturer should be consulted.
   See also Chapter 1.8.

Noise and sound insulation
A single skin of lightweight curtain
walling will only give sound
insulation approximately the same as
that for a single window.

   Typical sound insulation values
for sealed curtain walling against
traffic noise are:
● single glazing, 28 dB(A)
● sealed double glazed units,
  33 dB(A)
● secondary glazing, 37–40 dB(A)

Against aircraft noise, double
construction gives about 3 dB(A)
improvement over sealed units.
Where there is the opportunity,
breaking up the internal cavities of
double construction and lining with
absorbent material may help.

Daylighting, solar control and glare
Many installations of curtain walling
consist of solar control or tinted glass
which has reduced light
transmission, and which is highly
reflecting. These installations can
produce significant glare from the
sun, sometimes annoying or
dangerous for motorists (Figure
4.33). A discussion of the geometry
involved is to be found in Chapter
1.6.

   Curtain walling installations are
increasingly being specified with
some form of solar protection
integrated into the framework (see
also Chapter 1.7).

   Light transmission of glass blocks
varies from around 50% to around
75% according to pattern and tint.

Figure 4.33
Glare from the sun reflected from this sloping curtain wall

Durability

Specification of glass (particularly the heat absorbing and reflecting types) needs careful consideration to reduce the risk of cracking from temperature difference across panes at the edge covers and from shadowing. It is recommended that thermal safety checks are carried out by the glass manufacturers or suppliers for all situations where their products are being used.

Many types of glass are available which are suitable for use in curtain walling, and reference should be made to the basic descriptions in Chapter 4.1. However, where there is a perceived risk of injury to people, special care should be taken at the original design stage and with specifying replacements.

With thermally toughened glass, there have been some instances of spontaneous breakage which may constitute a safety risk. The risk is greater in overhead roof glazing, although even these cases have been comparatively rare[191]. The problem occurs because of contamination of the glass by nickel sulfide inclusions. There is nothing that the surveyor can do to guard against the risk in existing construction since it is one which occurs in manufacture. When specifying replacements, however, only glass which has been heat soaked subsequent to toughening will allow the phase change which causes fracture to take place before installation.

Other causes of glass breakage include inadequate attention to clearances on installation (at least 3 mm for single glazing, rising with sheet size, and more for double glazing and plastics glazing), and difficulties of access for routine maintenance and cleaning. There should be no direct contact between the glass and any hard temper metal-rebated supporting sections unless the supports are sprung. Large sheets of toughened glass are sometimes fixed with clips or cleats instead of rebates, and it is very important that any replacement is carried out competently.

Materials other than glass will sometimes be encountered in glazing of short life industrial or agricultural buildings, and will be liable to deterioration. The durability of various forms of plastics glazing is discussed in *Roofs and roofing*, Chapters 1.7, 4.3 and 5.3.

Maintenance

Curtain walling installations on tall buildings should have some in-built means of access for cleaning. Many materials need to be washed regularly to remove particulates causing corrosion.

The silicone sealed all-glass wall is frequently used on prestige buildings. These buildings should be regularly examined for disfiguring dirt streaking; this is inevitable, both at the eaves and verges where installations commonly have no projection to throw rainwater run-off clear of the vertical façade, and at the support points.

Work on site

Inspection
In addition to those listed in Chapter 4.1, the problems to look for in curtain walling are: ◊ deterioration of sealants and gaskets ◊ debonding of bonded glazing ◊ cracking of glass at points of high stress (eg at fixing cleats and lower edges of patent glazing) and for glass blocks: ◊ lack of soft joints at perimeters of panels ◊ corrosion of bed joint reinforcement

Chapter 5　**External doors, thresholds and shutters**

Although external doors, as part of the external envelope of buildings, in theory should conform as nearly as possible to the performance requirements of the remainder of the envelope, the extra requirements of access and the wear and tear that this involves have inevitably tended to make them more of a compromise than that of the wall in which they are situated (Figure 5.1). Many have been compromises between lightness in weight, for ease in opening, and robustness to resist unauthorised entry.

This chapter deals with all kinds of external doors used for both pedestrian and vehicle access, and with shutters and gates which provide additional protection to both doors and windows against unauthorised access.

Figure 5.1
Glazed entrance (and emergency exit) doors at a school in the days before security became a real issue

Chapter 5.1

Hinged, pivoting and sliding pedestrian doors

Figure 5.2
'Carefully designed to be in harmony with the remainder of the detailing'

Doorways, particularly main entrances, have, over the years, invariably been a major element in the architectural design of building exteriors, and have usually been very carefully designed to be in harmony with the remainder of the detailing. Frequently the doorway was meant to impress, or even overawe, visitors, even when the need for defence was past. Door heads were superimposed by ornamental or segmental pediments signifying the status of the owner (Figure 5.2). In Georgian times an elegantly glazed tympanum might fill the space between the door head and the relieving arch over.

Door leaves too could be elaborate. Fifteenth century framed door leaves, for example, might be decorated with arched or linenfold carved panels. Where impressions of strength were required, a clapboarded door leaf might be studded with metal. By the late eighteenth century some of the finest entrance doors were framed up to double square proportions, in painted wood. Wood held sway until the Industrial Revolution, when metal frames became practicable. Glazed doors are for the most part a feature of the twentieth century.

Characteristic details

Basic structure
Leaves of timber hinged to frames of timber have were the basic construction for the majority of domestic doors, and for many non-domestic doors, until the second half of the twentieth century; indeed, a considerable number of timber doors have survived from the sixteenth century, particularly in ecclesiastical buildings.

Framed and panelled main entrance doors, depending on their height and width, might be of six, eight, ten or twelve panels. Stiles, rails and muntins might be ogee moulded, or just chamfered, with panels sometimes flush and sometimes raised. In any event, the panels would normally be made up with the door, and not beaded in, glazing fashion. The fashion in rail, muntin and moulding sections changed over the years, sometimes narrow, and sometimes broad.

It is only during the twentieth century that alternative materials came into common use for both leaves and frames, firstly steel, and then aluminium, and now the practically all-glass door (Figure 5.3) Hinging was not always appropriate for these designs, and pivots were introduced, sometimes incorporating hydraulically operated closers.

Attempts to reduce the infiltration of wind and rain through open doors led to the revolving doorset, smaller ones hand operated (Figure 5.4), but larger versions power operated.

Power operation devices for hinged or sliding doors are of two kinds, either fully automatic, where an electronic sensor detects the presence of a person, and operates the door, or where the pedestrian's initial contact with the door is sensed before the mechanism assists.

Figure 5.3
Glass doors, Sainsbury Centre, University of East Anglia

Mechanisms should not operate too quickly, for example a maximum speed of 0.5 m/s at the leading edge.

Further information for specifiers of automatic controls is available[82].

Main performance requirements and defects

Choice of materials for structure

External doors of European oak have been in common use in the UK for many years for all building types. The first doors were ledged and boarded, with the bracing function for the most part carried out by long strap hinges of wrought iron. Tudor and Jacobean doors were commonly of broad oak planks. In many cases the door was left to weather naturally to shades of grey. Boarded doors without effective bracing can lozenge.

Softwood framed doors, mostly of European redwood which needed to be painted, became popular in Georgian times. Since the 1970s, domestic door leaves framed up in the red hardwoods such as meranti and lauan have become popular, though some designs are of rather doubtful pedigree. Where extra strength is required, doors may be of

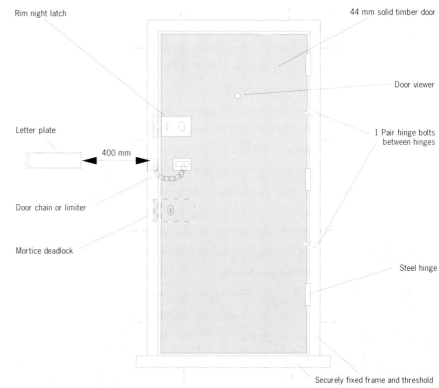

Figure 5.5
Hardware provision on final exit doors

solid flush timber or of steel.

Non-domestic main entrance doors are more likely to take different forms with, usually, a high proportion of glazing set in metal frames.

Standards for doors and doorsets

include:
- timber to BS 4787-1[192]
- steel to BS 6510[159]
- aluminium to BS 5286[193]

Installation of replacements is covered in BS 8213-4[194], and there is also a UEAtc Method of Assessment and Test[195].

Strength and stability

All doors providing access to a building should be robust and capable of withstanding rough treatment. Doors in concealed positions should be able to resist attack by tools (eg hammers and tyre levers). In domestic situations, 44 mm thick solid timber doors offer better protection than hollow cored doors, and steel doors even better.

Final exit doors

Front doors are the main entry and exit routes from a house and are normally the main escape route in an emergency. Such doors should be provided with:
- a mortice deadlock operated from either side only by a key
- a rim automatic deadlock

Figure 5.4
Revolving doors such as this assist in conservation of heat in well-used entrances

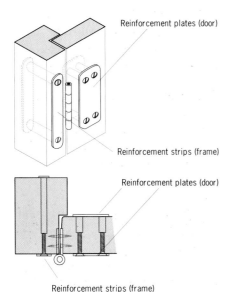

Reinforcement plates (door)

Reinforcement strips (frame)

Reinforcement plates (door)

Reinforcement strips (frame)

Section through door and door frame

Figure 5.6
Reinforcement strips and plates for hinges

● a door chain or limiter
● a door viewer
● a letter plate, if a suitable position is not available near the main entrance door. It should be positioned at least 400 mm from any locks to prevent manipulation of the locks. Additional bolts should not be fitted to the main entrance door since they may delay escape of the occupants in an emergency (Figure 5.5 on page 195).

Other external doors
Other external doors do not normally need to perform as escape doors and can generally be secured from the inside before occupants leave the building.

Sliding patio doors are a common point of entry for intruders. Unless precautions are taken during fitting, it is comparatively easy to lever them out of their tracks.

Side, rear or interconnecting doors should be fitted with:
● mortice sashlock
● two or more surface mounted or mortised bolts, preferably key operated versions

Twin leaf sliding doors (patio doors) should be fitted with:
● mortice sash lock, preferably with a hook bolt

● one pair of surface-mounted or mortice-type key operated bolts to each door
● anti-lifting devices to prevent them from being levered out of their tracks
● a mortice hook bolt, with either a fixed guide pin or small dead bolt to prevent the door from being lifted
● either a pair of push-to-lock key-operated bolts or a multi-point locking system

In addition, entrance doors should if possible meet the following requirements.
● Wooden doors should be solid and a minimum of 44 mm thick
● Wooden doors should be securely fixed to the frame by one and half pairs of steel hinges. They should incorporate one pair of hinge bolts, set between the hinges and be fitted with appropriate locks
● The frames and thresholds should be securely fixed to the surroundings, using a minimum number of fixing points of two per metre run of frame or threshold with a depth of penetration of the fixing screws or bolts into the subframes of 100 mm or more
● A glass panel fitted to a door or adjacent to a door should be of a size, position or glass type to prevent their being smashed to allow an intruder direct entry and to prevent manipulation of the locks or bolts fitted to the door
● All ironmongery must be fixed with screws of adequate size and length. 25 mm no 10 screws should be considered as the minimum
● Additional hinge bolts should be fitted where external hinges are fitted and the hinge pins are accessible from the outside (eg on outward opening doors)
● Metal reinforcement plates must be provided at appropriate places to prevent jemmy attacks on locks
● Reinforcement plates should be fitted around locks rebated into door leaves, and on the frames and doors at the hinge points (Figure 5.6) to prevent splitting of the timber. This is particularly

necessary for doors which may be subjected to prolonged and violent attack (eg back doors or interconnecting doors)

Glazing
Glazing in all new doors in domestic accommodation, and in all doors in other building types, must be safety glazing to BS 6262 up to a height of 1.5 m above finished floor level (Figure 5.7). Annealed glass is permitted provided the smaller dimension of the pane is less than 250 mm.

Dimensional stability, deflections etc
For coefficients of linear thermal and moisture expansion etc see Chapter 1.2.

Weathertightness
Laboratory measurements of water penetration through doors, using the same test methods as for windows have shown doors to be generally much less resistant to leakage than windows. However, in many cases, unfortunately it does not take a laboratory test to demonstrate this defect. It is difficult to design an

inward opening door for an exposed situation which will meet a high standard of resistance to leakage without making it difficult to open, or without providing extra protection such as a canopy or even a porch (see BS DD171[(196)]). Outward opening doors are much easier to make weatherproof, but even these may need the extra protection of a porch in the most exposed situations (Figure 5.8).

Inward opening doors may lack weatherbars or weatherboards, weatherbars may be incorrectly positioned (Figure 5.9 on page 198) and door frames may lack adequate, correctly positioned drainage channels. Sills and sub-sills may lack drips or throating on the undersides of projections. If doors are not weathertight, surrounding materials and finishes may have become damp and deteriorated. Diagnosis of a weathertightness problem is often complicated by condensation. There are some simple principles that can be employed to improve weather resistance[(197)]:

- rain check grooves at least 6 mm deep
- water bars fitted inboard of the rain check grooves or rebate. This requires the door to be rebated too

- trough sills with adequate drainage holes and that do not allow water to track across their stop ends (Figure 5.10 on page 198)

External doors should be checked for warping and twisting which can give rise to poor airtightness or inadequate wind resistance.

Figure 5.8
Stable-type door opening outwards, Saxtead Windmill, Suffolk

Thermal properties
The average thermal insulation value (U value) of a typical single glazed wood framed door will be around 5.0 W/m²K, depending on frame materials and air gap sizes. A solid unglazed timber door should give around 3.0 W/m²K and half glazed doors will give somewhere in between these figures.

Leaking hardwood doors
The BRE Advisory Service was asked to carry out an assessment of external door weathertightness problems in a house. Leakage of doors had been complained of by the owners of this property since completion. Despite extensive remedial measures involving additional sealing and weatherings, little improvement had apparently been achieved.

External doors had been supplied in American white oak, and were a mixture of inward and outward opening, both single and double leaf types, without weatherstripping, raincheck grooves in frames or weatherboards at the foot.

Doors and windows had been taped by the building owners to limit leakage. Removal of some of these tapes from fixed light framing revealed poor bedding of external beads and units, gaps in glazing compound and gaps at the joints in the sill. A crack in one timber door stile was reported to leak.

Doors had an excessively wide gap at the top, up to 9 mm, tapering to 3 mm at the base. A single door was seen with additional rubber seals fitted behind the line of the sill waterbar. Although not water tested at the time of the visit, this feature would allow easy leakage of water draining down to the bottom corners.

Was the American white oak used for the joinery a suitable material? BS 1186[(198)] indicates that American white oak can be used for exterior joinery without preservative treatment. It should not be confused with American red oak, which is not durable and therefore unsuitable.

The simple principles of designing weatherproof doors did not seem to have been taken into account by the architect or makers of the doors. Nor did the installation take into account established principles of design to minimise rain penetration on an exposed, open country site. Remedial measures involved replacement with proven designs incorporating weather resistant features.

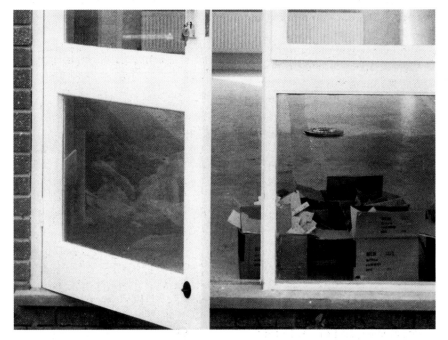

Figure 5.7
Not up to standard: a thin annealed glass in a door and low level glazing found on a BRE site investigation

Figure 5.9
This water bar is fitted in front of the jamb rebate so that water running down is directed to the interior

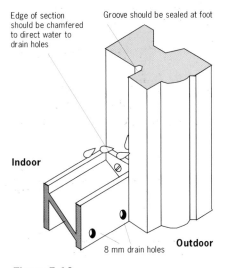

Edge of section should be chamfered to direct water to drain holes

Groove should be sealed at foot

Indoor

Outdoor

8 mm drain holes

Figure 5.10
Some proprietary trough sills may lead water inwards

Figure 5.11
This compression seal set in an aluminium section does not allow adjustment to ease opening difficulties

Fire

Doors on the side walls of buildings which are close to a boundary may represent a risk of fire spread to adjacent buildings. They are classified, like windows, as unprotected areas, and, in building regulations, design limits need to be placed on their use in these situations.

Outward opening external doors on escape routes should be openable by a simple device from the inside, and electronically controlled locks should fail safe (ie open) in the event of a power failure, or alternatively should be provided with a manual over-ride. Certain doors will need to be fitted with panic bolts.

Noise and sound insulation

It is very difficult to improve the performance of external doors to more than around 30 dB(A), and where this figure needs to be improved upon, consideration should be given to fitting a lobby. The use of two doors either side of a lobby should, depending on the absorbency of the surrounding wall surfaces and how well the walls are constructed, give around 45–50 dB(A).

Durability

Doors made in earlier centuries from softwoods such as European redwood, although normally classified as non-durable, have proved to be remarkably durable in practice, provided their paint finish was regularly maintained. Some whitewoods imported since the 1950s have not proved so durable: typically, painted but non-preservative treated doors have suffered extensive wet rot within around ten to fifteen years. Such doors are now normally treated with preservative.

The native hardwood, European oak, is classified as durable and resistant to treatment with preservative. On the other hand, while some imported hardwoods are naturally durable, they may contain a proportion of paler wood of less certain durability, and they are now normally specified to be treated[155].

Durability of doors in other materials is usually governed by ability to survive impact damage.

Maintenance

Doors which have very good weathersealing, especially those of timber with insufficient clearances, tend to be very difficult to open against their seals; close attention to maintenance is needed if they are not to become a nuisance (Figure 5.11). BRE Digest 319[199] provides further information.

For maintenance of finishes, see the relevant section in Chapter 9.4.

Work on site

Workmanship

The fitting of weatherstripping so as not to create difficulties in opening does need particular skill and care.

Inspection

The problems to look for are:
◊ rain penetration – doors not suitable for exposed situations
◊ no water bars or bars set in wrong places
◊ no weatherboards
◊ warping of leaves
◊ inadequate strength for security needs
◊ no drainage channels
◊ non-safety glazing at low levels
◊ deteriorating lintels
◊ no weatherstripping
◊ finger traps in pivoting leaves
◊ automatic doors malfunctioning
◊ force required to open is too great

Chapter 5.2

Thresholds

Figure 5.12
'Danger of tripping, rain penetration and thermal bridging'

The external door threshold has always been one of the problem items in the external envelope so far as danger of tripping, rain penetration and thermal bridging are concerned (Figure 5.12).

On the other hand, many doorways throughout the ages have been protected by porticoes and canopies, and, for the favoured few, by *portes cochères*, where rain penetration of the actual door of course does not occur.

Since standards for access by disabled people were introduced for public buildings, there have been many attempts to provide so-called 'flat' thresholds at entrances. Examples of flat thresholds in use which were examined by BRE investigators showed a wide range of solutions, nearly all of which could be criticised on construction grounds (eg thermal bridges,

inadequate DPCs etc). However, most cases of rain penetration have proved to be in the more exposed areas of the UK. Nevertheless, it is anticipated that, following proposals to make flat thresholds obligatory for new housing also, there will be more attempts to convert existing thresholds. It should be appreciated that there will inevitably need to be compromises, and what is best for access will not necessarily be best, or even adequate, for other functions.

Characteristic details

Basic design

The basic requirements for a flat threshold are summarised in Figure 5.13.

Four typical solutions in principle are shown for providing flat access for wheelchairs to these basic requirements and at the same time reducing the risk of rain penetration

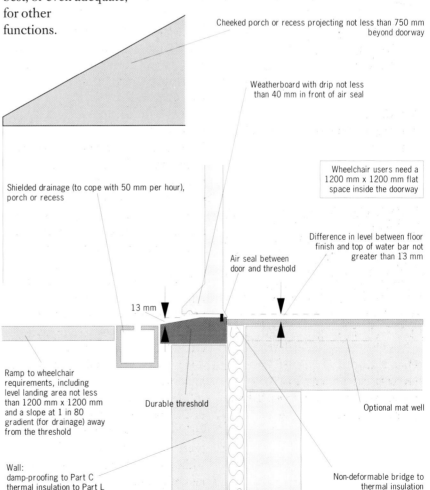

Cheeked porch or recess projecting not less than 750 mm beyond doorway

Weatherboard with drip not less than 40 mm in front of air seal

Wheelchair users need a 1200 mm x 1200 mm flat space inside the doorway

Shielded drainage (to cope with 50 mm per hour), porch or recess

Difference in level between floor finish and top of water bar not greater than 13 mm

Air seal between door and threshold

13 mm

Ramp to wheelchair requirements, including level landing area not less than 1200 mm x 1200 mm and a slope at 1 in 80 gradient (for drainage) away from the threshold

Durable threshold

Optional mat well

Wall:
damp-proofing to Part C
thermal insulation to Part L

Non-deformable bridge to thermal insulation

Figure 5.13
Basic requirements for a flat access threshold

and cold bridging:
● one 13 mm upstand, porch protection (Figure 5.14)
● two 13 mm upstands, one combined with slot drainage (Figure 5.15)
● one 13 mm upstand, bridge-over-moat drainage (Figure 5.16)
● one 13 mm upstand, drainage channel with grating (Figure 5.17)

Figures 5.14 and 5.15 show wiping blade seals on the foot of the leaf, and Figures 5.16 and 5.17 show compression cushion seals on the under surface. Other types of air seal are also possible.

Furthermore, the drawings show examples of thresholds adjacent to:
● a screeded concrete slab floor
● a screeded concrete raft
● a timber boarded concrete floor
● a timber boarded, timber joisted suspended floor

In principle all solutions are possible with all types of floor – not all permutations can be shown. All four Figures show a section through the entrance threshold; the construction need extend only just beyond the door jambs to either side of the door, say 100 mm.

Joint between threshold and door

Traditionally, upstands of around 15 mm (or just over) in height have been used. BRE surveys have shown, in practice, that many installations have exceeded this height, but these higher thresholds have always given problems of access for disabled people.

It is now assumed that wheelchairs can acceptably negotiate single 13 mm vertical upstands. From the point of view of resistance to rain penetration, the steeper the 13 mm upstand, the better – vertical is best. For access, the shallower the rise the better.

Where more than one vertical upstand of 13 mm is used, at least 125 mm should separate them horizontally.

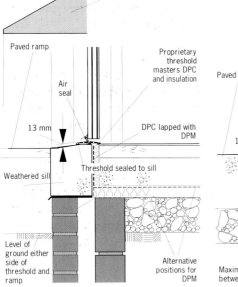

Figure 5.14
Paved ramp to a building: no drainage, but sheltered by a porch

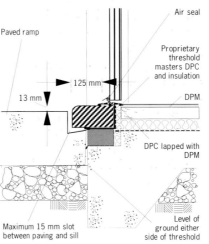

Figure 5.15
Purpose-built ramp to a building: with drainage slot

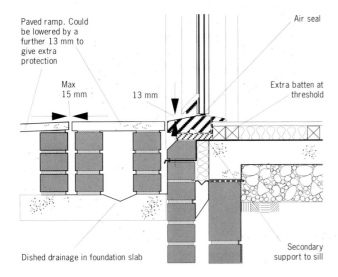

Figure 5.16
Paved ramp to a moat bridge

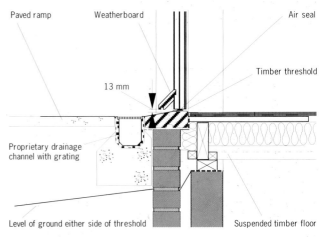

Figure 5.17
Paved ramp to a drainage channel: with grating

Main performance requirements and defects

Choice of materials for structure

If condensation is to be successfully avoided, adequate thermal insulation must be provided to link with that in the remainder of the external wall. The traditional stone or concrete threshold tightly butted to stone flagged floors, even when carpeted, provided a massive thermal bridge (Figure 5.18).

Dimensional stability, deflections etc

The most important consideration here is to provide adequate end clearance for water bars cast or inserted into stone or concrete thresholds. This is especially necessary on wide or double doors.

For coefficients of linear thermal and moisture expansion etc, see Chapter 1.2.

Weathertightness

Even some existing designs of conventional raised thresholds have proved to be inadequate in resisting driving rain in exposed areas of the UK, though most designs conforming with previously relevant Codes and Standards for weathertightness seem to have worked reasonably well.

It is undoubtedly the case that the higher the upstand, the better the detail will be in resisting water being driven over it by air leakage. In the absence of an air seal at the foot of the door, and therefore nothing to resist the passage of an air stream carrying water over the threshold,

even upstands of 25–50 mm will be vulnerable to rain penetration.

The water load on paved areas to be provided for by the design is 50 mm/hour (see BS 6367[200]).

Alternative means of reducing the water load on the area to be drained are met by providing:
● a shelter (eg a porch)
● a drainage channel or gutter in front of the threshold
● permeable paving

If any one of the above items is present, it is estimated that the chances of leakage for most of the UK will be reduced to an acceptable level. Porches need to be cheeked to prevent driving rain blowing in sideways. A projection of not less than 750 mm will prevent most driving rain from reaching the threshold, but it will obviously depend on the direction of the prevailing winds.

Drainage gutters need to have a compromise gap of around 15 mm for:
● safety – for example, to avoid trapping small section heels and umbrella ferrules
● reduction of risk of blocking by detritus

Paved areas outside entrance doors should be laid to a minimum fall to prevent water accumulating and then freezing. A minimum fall of 1 in 80 is conventionally accepted for this required slope. Because of site dimensional deviations, and the reduction of the possibility of backfalls and subsequent ponding, the nominal fall should be increased to 1 in 60.

The splash zone for driving rain is conventionally taken to be 150 mm measured both vertically and horizontally, although under certain weather conditions this will be exceeded. Ramps running parallel to the wall in which the threshold is situated therefore need to be kept at least 150 mm clear of any wall over the DPC (Figure 5.19).

Waterbars

There is no doubt that a waterbar is a satisfactory method of reducing, if not entirely preventing, rainwater from penetrating a threshold. Waterbars essentially are of two kinds; placed at the foot of the opening leaf, creating an upstand against which the leaf closes; the second prevents water being drawn through the horizontal joint between the underside of the

Figure 5.18
Thermal bridge at an external door threshold

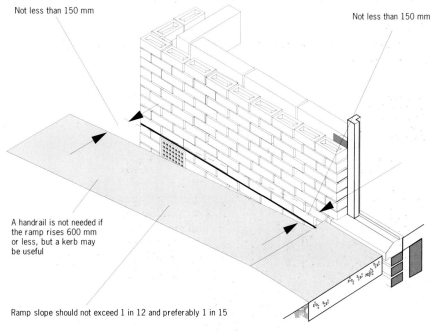

Not less than 150 mm

Not less than 150 mm

A handrail is not needed if the ramp rises 600 mm or less, but a kerb may be useful

Ramp slope should not exceed 1 in 12 and preferably 1 in 15

Figure 5.19
Ramps running parallel to the external wall need to be kept 150 mm away from it, and preferably more

threshold and the wall into which it is set. Clearly the first demands adequate resistance to impacts and wear, while the second, being protected, can be more flexible. The former does need to be in the right place, however, and BRE investigators on site have on several occasions seen it placed too far forward, so that leakage running down the rebates in the door frame falls inside and not outside the bar.

Some laboratory experiments were carried out by BRE in the 1970s involving a variety of sections of stainless steel, aluminium, neoprene and PVC waterbars cast into concrete beams.

The tests were severe, the first creating a 10 mm pond of water adjacent to the bar continuously for 24 hours, and the second a conventional pressure box test. Only seven out of 34 cases showed any degree of leakage in the ponding test; the pressure box tests however, showed, as expected, that the height of the waterbar to a large extent governed its performance. None were of sufficient height to entirely resist the flow of water carried in the air stream, emphasising the importance of an effective air barrier in resisting water penetration (Figure 5.20).

It was noted that some of the

Figure 5.21
An open cavity at a threshold in new construction inevitably attracts detritus

neoprene and PVC waterbars had developed a kink at one end after a period of hot weather, since they had been cast in tightly. The kink (caused by expansion in the bar) remained after cooling. In the case of the aluminium and steel sections, the expansion had taken place without apparent distortion. It is good practice, therefore, that the built-in ends of non-metallic waterbars are wrapped in a compressible material to provide some degree of protection against kinking, particularly when cast into concrete which shrinks on curing.

Thermal properties

Many thresholds seen in common use present problems with thermal bridging and do not conform to current thermal insulation requirements. One tendency noted on construction sites was that often a gap was left at the threshold during construction which would be filled by detritus. This compromises the function of the cavity in respect of both rain penetration and thermal insulation (Figure 5.21).

Maintenance

Regular checking is needed that the weatherproofing or seal at the foot of the door is working effectively and not jamming the door.

Work on site

Inspection

The problems to look for are:
◊ upstands greater than 13 mm where wheel chair access is provided
◊ no drainage channel on level access doorways
◊ condensation on thermal bridge at sills
◊ rain penetration
◊ frost attack on subsills

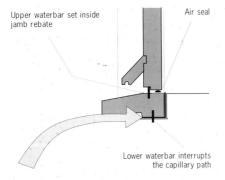

Upper waterbar set inside jamb rebate

Air seal

Lower waterbar interrupts the capillary path

Figure 5.20
Alternative positions and functions for waterbars incorporated into door thresholds

Chapter 5.3   Garage doors, shutters and gates

Figure 5.22
Fortified gatehouse to the Bishop's Palace, Wells

The shutter or gate in the form of the portcullis or the drawbridge over the moat have been used since Norman times to protect the main entrance doors of both military and domestic architecture, and to provide security against unauthorised access (Figure 5.22). Its modern derivative today is the metal collapsible or expanding gate. On the other hand, most doorways do not need this degree of protection, and some openings in external walls which are provided for the passage of pedestrians, or even vehicles, are deliberately not provided with doors at all. The archway fronting a passageway or ginnel through the front of a building to allow access to the rear is an example of this.

This chapter deals with the larger door, for example used for vehicular access, but also shutters used externally for security reasons. Shutters installed for solar protection seem to be more rare in the UK than in other European countries and these are not dealt with in this book.

This chapter does not deal with internal shutters found in heritage domestic buildings and which are frequently integrated with internal framed panelling. They are dealt with in Chapter 10.3.

Characteristic details

Basic structure
Small doors for vehicle access (eg for domestic garages) are of many different designs: hinged leaves of wood, up-and-over in wood, sheet steel or aluminium, to name a few.

Larger vehicular or aircraft access doors are normally of the rolling shutter or folding segment type (Figure 5.23). All require tracks or guides to be installed, usually fixed to supporting steelwork.

Main performance requirements and defects

Choice of materials for structure
Although framed wood is used for pairs of hinged doors of up to around 1.5 m width, the hinges may need to be of the strap kind. Protected steel or aluminium sheet with framing of the same material will be normal above this width. For these widths in timber or in other materials, the leaves will be narrow concertina-hinged, suspended from overhead and sill tracks. Most doors and all suspension systems save for hinges on domestic garage doors are proprietary.

Vision panels and wickets are sometimes incorporated.

Figure 5.23
Narrow slat rolling shutters on industrial accommodation

Figure 5.24
Motorised rolling security shutters in a high risk suburb

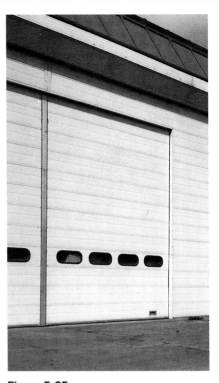

Figure 5.25
Large-slat rolling shutters with vision panels
on an industrial building

Strength and stability

Designs for doors vary so much that there is little useful information that can be provided. Specifiers will need to consult manufacturers.

Doors can be tested for impact damage risk in the same way as other building materials.

External shutters are sometimes specified for high risk domestic premises. Installations are proprietary, and usually consist of steel slats, which may be perforated to provide ventilation when in the closed position (Figure 5.24).

Dimensional stability, deflections etc

For coefficients of linear thermal and moisture expansion etc, see Chapter 1.2.

Weathertightness

Perhaps the main difference from smaller doors will be the larger areas being more vulnerable to buffeting wind pressure. Of course there may be special needs, too, for different kinds of traffic.

See also the same section in Chapter 5.1.

Thermal properties

For new buildings, the regulations suggest a U value of 0.7 W/m²K for vehicle doors, a figure which in practice demands separate provision for thermal insulation within the door structure itself. Since doors can form a large proportion of the external envelope of some industrial buildings (Figure 5.25), it may be worthwhile considering retrofitting thermal insulation.

For rolling shutters used for protection of domestic windows and doors, as already noted, slotted ventilation may be provided; there is therefore no additional thermal benefit from this kind of shutter.

See also the same section in Chapter 5.1.

Fire

Shutters of the rolling pattern may be required to provide resistance to fire. Testing is the only sure guide to behaviour. Collapsible gates of course have no fire resistance whatsoever (Figure 5.26).

Automatic shutters on escape routes should not be triggered by smoke, only by fusible link.

Noise and sound insulation

This property will depend on two factors, the weight of the materials covering the opening, and the amount of air leakage or gaps. Even the better installations cannot be expected to achieve much more than around 20 dB (A).

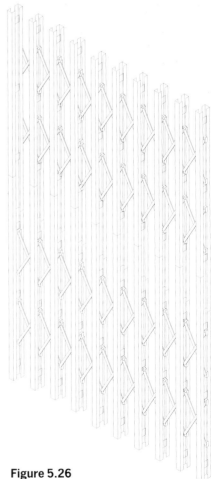

Figure 5.26
Collapsible gates have
no fire resistance. Nevertheless,
they do have a role in security situations

Durability

Wear on the edges of slats in rolling shutters, removing protective paints, is one of the main points to examine for corrosion in steel units.

So far as timber doors are concerned, movements are one of the main factors to take into account. These can be considerable on large doors, and even significant on small doors where specifications are ill thought out.

See also the same section in Chapter 5.1.

Maintenance

Since most systems depend on mechanical suspension systems, both for hand operation via gearing and for motorised drives, continued satisfactory performance will depend on regular maintenance (Figure 5.27).

Figure 5.27
The doors to this former airship hanger, now BRE laboratory at Cardington, are formidable structures in their own right. They need regular maintenance if they are to continue to function satisfactorily

Work on site

Inspection
In addition to those listed in Chapter 5.1, the problems to look for are:
◊ deficient thermal insulation
◊ risk of trapping fingers and clothing in expanding gates
◊ detritus blocking channels
◊ lack of maintenance of running gear
◊ smoke-triggered automatic shutters on escape routes
◊ wear on slats
◊ corrosion of slats

Chapter 6

Separating and compartment walls

This chapter deals with separating and compartment walls in all building types and in all forms of construction. The main performance criteria relate to sound insulation and fire resistance, though there are cases where separating walls become external walls where thermal insulation is also relevant (Figure 6.1). Deficencies do exist, however, in many cases: while it may be practicable to upgrade existing walls for increased performance, the most satisfactory solution is to have built them correctly in the first place.

The term separating wall has been used as the generic term. Where a separating wall is in common ownership it is termed a party wall, but the various quasi-legal responsibilities of surveyors acting for owners of party walls in pursuance of the Party Wall Act 1997 are not covered in this book.

Figure 6.1
Stepped-and-staggered separating walls under construction

Chapter 6.1

Brick and block, precast and in situ concrete

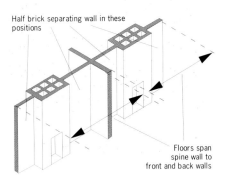

Figure 6.2
Typical situation where a half brick
separating wall might occur

One brick thick separating walls have for many years been a common way of providing a reasonable standard of sound and fire separation between adjoining buildings. However, many separating walls, particularly in blocks of flats built during the early years of the twentieth century where the floor spans are front to back rather than side to side, have separating walls only half a brick thick (Figure 6.2). Soon after the 1939–45 war there was a move to cavity separating walls in the hope that they would provide a significantly better standard of sound insulation, but in practice this was never realised.

Precast concrete separating walls were common in the systems of the 1960s. For example the separating walls in the Reema Conclad system were built of 7 inch thick solid concrete panels[116].

Characteristic details

Basic structure
Before the introduction of building regulations, most separating walls were of brick, with thickness depending on structural and fire requirements rather than on sound insulation requirements. As observed earlier in the book, though, one brick thick walls are common to most of the terrace and semi-detached dwellings in the country.

Under the influence of the London Constructional By-laws and certain local authority byelaws founded on the Model Byelaws, they were often continued above roof level (Figure 6.3).

In a surprisingly high number of cases, though, there was no continuation of the wall into the roof spaces of terraces.

It was reported that during the period 1991–94, there were approximately equal numbers of solid and cavity separating walls being built in new construction.

Abutments
The junction of the separating wall with the external wall is normally bonded or blockbonded. There should be no gaps into any cavity in the external wall to provide flanking sound transmission routes or routes for fire spread.

Compartment walls of course occur in many different building types. These walls may not even be rectilinear. There is no difficulty in principle in complying with building regulations in this respect, although the Building Regulations 1991

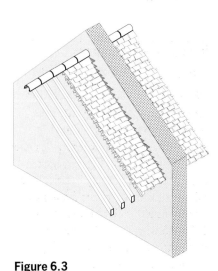

Figure 6.3
In Victorian and Edwardian times, and even later, separating walls were often continued above the roof slope

Approved Document B2/3/4 Appendix D2[(52)] states that *'a compartment wall which is used to form a separated part of a building should run the full height of the building in the same continuous vertical plane'.* However, there may be problems in installing cavity barriers where compartment floors abut a wall which is circular on plan.

Openings and joints

Normally there should be no openings as such, though occasionally it will be necessary to provide access; for example for means of escape.

Where large single dwellings are being converted to multiple occupation, separating walls are sometimes created from what were previously partitions. In a number of cases, BRE investigators on site have seen gross deficiencies in the performance of these walls.

Accommodation of services

In general, it is unwise to accommodate services within the thickness of masonry separating walls. Where chasing is necessary, it should be done according to prescribed rules. On several occasions when investigating new construction on site, BRE investigators have encountered chasing and socket outlets back to back in separating walls, to the detriment of sound insulation. (Figure 6.4).

Main performance requirements and defects

Choice of materials for structure

As already noted, brick was the universal solution for older buildings. Dense concrete block then became popular in the drive for increased productivity in bricklaying when skills were scarce during the years immediately following the 1939–45 war. These blocks needed two-handed lifts (approaching 30 kg), which have, in today's health conscious climate, been outlawed. Since the early 1990s, lightweight blocks with 75 mm cavities have

become popular, with considerable reliance being placed on the results of tests of prototypes for measuring sound insulation. Only rarely are tests carried out on completed construction, and there is no provision in building regulations for this to be done on a routine basis.

Strength and stability

Control of the thickness of separating walls (in conjunction with external walls) goes back to the rebuilding of London after the Great Fire. To give one example, in *'buildings fronting Streets and Lanes of note and the River Thames'* the separating walls of three storey houses, plus cellar and garret, were required to be two bricks thick in the cellar, one and a half bricks thick for all three main storeys, and a minimum of one brick thick for the garret (Figure 6.5). Other types of houses had other requirements[(201)].

Separating walls built under the old Model Byelaws were controlled by thickness according to height or horizontal distance between lateral supporting walls and piers. Thus can be found one brick thicknesses in walls not exceeding 30 feet height and length, one and a half brick walls in the first storey of walls exceeding 30 feet in length, two brick thick walls in the first storey of walls not exceeding 50 feet high, and so on up to three and a half brick thick walls in the first storey of walls not exceeding 120 feet high and 45 feet in length, with progressive reductions in thickness as the height increased.

Largely as a result of systematic testing, the strength and stability of loadbearing masonry walls can easily be calculated, and there are a number of examples of multi-storey construction of ten or more storeys.

Separating walls are often found to be inadequately bonded to front and rear walls. Wall- to-floor connections may be absent along spans which run parallel to the separating wall. Some old walls will incorporate timbers built into the masonry; these may be susceptible to decay if there has been dampness in the structure. Previous modifications to older properties

Figure 6.4
Electrical sockets set back-to-back in a separating wall

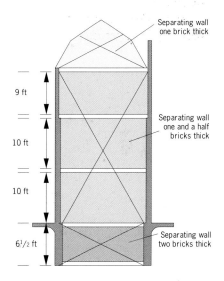

Figure 6.5
The cross-section of a house authorized by the Rebuilding Act of 1667

may have removed structural support to separating walls (for example chimney breasts or internal loadbearing walls). Excessively thin walls have been found where half brick (112 mm) brickwork rises two, three or even four storeys with little restraint.

Dimensional stability, deflections etc

Dimensional stability is not often considered since the separating wall is covered by the roof. However, see the case study on page 213.

For coefficients of linear thermal and moisture expansion etc, see Chapter 1.2.

Dampness and condensation

Problems may be found in large numbers of terraced houses where local byelaws required that separating walls were built to project through the roofs. Many houses with this feature can now be found 'flashed' with cement fillets or flaunching. There is no realistic substitute for a metal flashing over a secret gutter, or soakers, in this situation. If, however, a cement flaunching is preferred, it should include a gritty fine aggregate with some lime in the mix. In addition, the DPC may have become ineffective.

Where a pitched roof abuts a separating wall in a stepped-and-staggered situation, part of the roof becomes an external gable, and the outer leaf of masonry becomes an inner leaf below roof level. Particular care must be taken to ensure that

Figure 6.6
Lack of fire resistance at boxed eaves

rainwater is prevented from reaching the interior. There is further discussion of this problem in *Roofs and roofing*[(24)].

Rising damp in separating walls will generally prompt the same technical considerations as for other internal walls, but additional problems may be encountered because the wall usually separates dwellings in different ownership and because floor levels are often stepped either side of the wall. Where access is available to one side only of a wall, the provision of a fully effective DPC may be difficult to achieve. Where there is a stepped wall between dwellings, the wall on the lower side may need to be considered as if it is part basement.

Where rising damp has been diagnosed, the possibility should be considered that the external wall DPC is being bridged by a solidly bonded separating wall.

Older dwellings will often not have any DPC within separating walls. Where solid floors are used, the floor construction may cause bridging of any DPC which has been provided in the original construction. Where steps occur at ground level, a vertical dampproof membrane will often not have been provided. Injection type DPCs are not able to deal with step conditions and a vertical physical barrier must therefore be provided. Where rising damp exists, rot may be found in timber in contact with the wall – in skirtings and stair strings, for example.

BRE investigators on site have several times seen separating walls become external walls when the adjoining house has been demolished. It is not easy to provide for an acceptable degree of weathertightness without cladding the whole wall.

Fire

Soundly constructed masonry walls are usually able to provide fire resistance for periods longer than required by building regulations. This inherent resistance to fire is easily compromised, however, if any unsealed holes or gaps are left

through a wall. Walls may rely for their stability upon other parts of the structure, such as timber floors or roofs which are not so resistant to fire.

So far as solid walls of masonry are concerned, their fire resistance increases rather more than directly in proportion to thickness. Hence an unplastered half brick wall should give one hour, whereas a one brick wall should give well in excess of two hours – indeed, up to six hours. Sanded plasters give only marginal improvements, but lightweight plasters offer considerable improvement – for example increasing the resistance of half-brick walls to six hours. While such a separating wall might satisfy the fire criterion, however, it would not satisfy the sound insulation criterion.

Reinforced concrete separating walls at least 120 mm thick, with at least 25 mm of cover to any reinforcement, should give at least one hour's fire resistance.

Where a building contains several dwellings, each must be contained within fire resisting compartment walls and floors. When converting a large property into a number of smaller dwellings it is important to ensure that the compartment walls, together with any supporting structure, will be able to meet the period of fire resistance required. Only limited account should be taken of old lath and plaster in assessing the fire resistance of a wall.

A BRE survey of house design and construction revealed that fire-stopping at the top of separating walls is rarely effectively done. If the gap is not properly fire-stopped, fire can spread from one dwelling to the next (Figure 6.6). There is further discussion of this problem in *Roofs and roofing*. See also BRE Defect Action Sheet 8[(202)].

Internal walls used for compartmentation (or perhaps earlier conversion) may not provide the required fire resistance because the supporting structure will fail before the wall. Sealing of all cavities which bypass the wall (eg floor and roof voids and around pipes in ducts) is often overlooked.

Standard fire resistance tests on fire separating elements in buildings, such as walls, are undertaken on representative elements of construction of limited size. Modern buildings, however, can have elements which are many times larger and the real fire exposure can differ in timing, severity and extent. Careful engineering design and construction is needed to ensure that thermal movements and restraining forces do not impair the stability, integrity and insulation of large assemblies[203,204].

The resistance of no-fines separating walls to the passage of smoke (as indeed for sound too) self-evidently will depend on the integrity of the plastering.

Noise and sound insulation

Separating walls built under the old local byelaws created following the passing of the Public Health Act made no specific provision for sound insulation, since the stability and fire resistance requirements governed wall thicknesses in any case.

A half brick separating wall could be expected to give a single figure sound insulation value of around 44 dB when combined with the massive construction of fireplaces and flues on the separating wall[205]. This value does depend on frequency.

When complaints of sound transmission through a separating wall have been investigated, that part of the separating wall within the roof space has been found often to be a major contributing route. In some cases, BRE investigators have found there to be no separating wall whatsoever in the roof space. The material of a separating wall, or its thickness, may change above ceiling level, or there may be unfilled joints in masonry, or even large holes. Rough rendering the separating wall within the roof space may provide a partial solution.

A solid one-brick (225 mm) wall with plaster on both sides is usually regarded as adequate for sound insulation, but the actual performance achieved, measurable by standard test procedures, will depend also upon the surrounding construction. Single figure values of around 52 dB or slightly more can be expected. Cavity brick walls will give similar values to solid, although the variability will be somewhat greater and will be influenced, for example, by wall tie design and junctions with external walls[205].

Dense concrete separating walls of 150 mm thickness will give values ranging from 47 to 53 dB, and the normally thicker no-fines concrete walls will also give around these figures.

The sound insulation of separating walls is now covered in the various Building Regulations, Part E in England and Wales[206], Part H in Scotland[207], and Part G in Northern Ireland[208]. The Building Regulations require walls between dwellings to provide reasonable resistance to airborne sound. 'Reasonable' is made explicit in England and Wales in Approved Document El/2/3 by means of acceptable construction or performance criteria. While compliance with building regulations may not be mandatory for conversion or rehabilitation schemes, performance which does not reach the building regulation standard is a common cause of occupant dissatisfaction.

The following constructions should meet the requirement for sound insulation of masonry separating walls:
- solid dense concrete block of at least 415 kg/m² of wall area plastered both sides
- in situ or precast dense concrete of at least 415 kg/m² of wall area plastered both sides
- cavity brick or dense concrete block wall of at least 415 kg/m² of wall area plastered both sides
- cavity lightweight aggregate block (maximum density of 1600 kg/m³), 75 mm cavity and overall weight, including plaster, of 300 kg/m² of wall area
- single leaf brick or dense concrete core of at least 300 kg/m² of wall area lined on both sides with isolated lightweight panels of cored or glued plasterboard

It has been estimated that around one third of all separating walls in dwellings fail to meet the sound insulation standards of the Building Regulations Approved Document E. Some of the more common causes are:
- pressed bricks laid frog down
- walls of inadequate mass
- non-butterfly ties in cavity separating walls
- lightweight flanking external walls
- unfilled perpends in masonry, particularly within floor zones and loft spaces
- joists built into the separating walls with inadequate beam-filling between them
- deep chases for services cut back-to-back in the separating wall
- lack of dry-packing under walls in concrete panel systems
- lightweight linings touching the masonry core

Where sound insulation of walls consisting of units of adequate mass proves to be deficient, the cause may be found in unfilled perpends within the wall construction; perhaps where the bricklayer has simply buttered the arrises of the units before laying (Figure 6.7).

Holes or cracks through separating walls are often found and can seriously reduce the sound insulation objective. Walls of inadequate thickness or built of low-density blockwork may not meet performance standards. Even localised reductions in thickness can

Figure 6.7
Poor filling of perpends in separating walls can lead to deficiencies in sound insulation

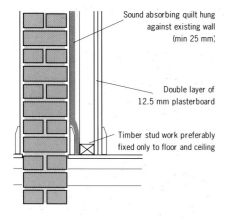

Figure 6.8
Separating walls: a new insulated stud wall constructed against the existing wall

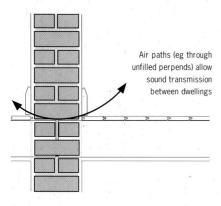

Figure 6.9
Separating walls: inadequate filling between joists provides sound paths

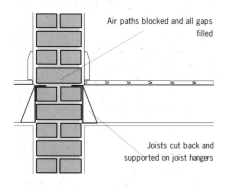

Figure 6.10
Separating walls: joists cut back and supported on joist hangers allows brickwork to be inserted to improve sound insulation

reduce overall performance significantly. Masonry walls that are dry-lined rather than plastered have a history of poor performance, largely owing to the lack of sealing of joints.

The actual performance achieved can be established only by field tests which are expensive and are not practicable if the building is in an incomplete state, or if access to both sides of the wall is not available for testing.

BRE tests demonstrate that the field performance of lightweight aggregate and aerated block cavity walls, when plastered, can meet the standards inherent in Approved Document E.

A comparison between the insulation ratings found in field measurements across stepped-and-staggered walls with those for similar in-line party walls confirms that displacement significantly enhances insulation[209].

Solid autoclaved aerated concrete block walls, plastered or dry-lined, perform very poorly, whereas cavity autoclaved aerated concrete block walls, plastered or dry-lined, are capable of good performance.

Information from site surveys indicates that the cavity closer between the external wall and the separating wall is often missing. It certainly needs to be present; it can offer up to 2 or 3 dB improvement in performance.

Strategies for upgrading
Where the lack of sound insulation is due to the design of the separating wall rather than to flanking paths, the following strategy may be effective.

A completely separate timber stud framework is erected on one side of the wall with a sound absorbing quilt between and two layers of 12.5 mm plasterboard fixed to the studs (Figure 6.8). Where space is at a premium, proprietary partition systems may be useful provided the mass is not reduced.

If the joists are built into the separating wall, inadequate beam-filling between the joists may cause air paths between dwellings (Figure 6.9). In the worst situations the joists should be cut back and supported on joist hangers built into the separating wall. Redundant holes in the brickwork should be filled (Figure 6.10).

Sometimes the separating wall in the loft space is only a half brick thick. This provides adequate fire resistance but may be inadequate for sound insulation if it is not well built. It is usually impracticable to render this wall in its entirety because of the proximity of the roof trusses, but the render that can be applied will make an improvement. Sound insulation can usually best be improved by reducing transmission through the ceilings by adding an additional layer of plasterboard to the ceiling on either side of the wall (Figure 6.11).

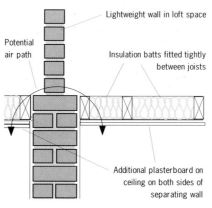

Figure 6.11
Separating walls: ceiling insulation helps reduce sound transmission between dwellings via loft spaces

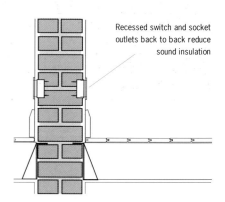

Figure 6.12
Separating walls: back-to-back electrical boxes may need relocating to reduce sound paths

Cracking in a brick separating wall between two semi-detached dwellings

The two four-bedroom dwellings were erected just before the end of the nineteenth century on London Clay. Large plane trees were situated 2 m from the external wall of one of the dwellings, and around 7 m from the separating wall. The plan of the dwellings was in the form usual for this period; that is to say two large brick boxes side by side each some 5 m wide in one brick thickness with further but narrower boxes, also in one brick thick walls, extending into the rear garden (Figure 6.13).

were made good cosmetically.

Some few years later, severe cracking developed in various parts of both houses, with two particularly serious vertical cracks, one in the separating wall (Figure 6.14), and one in the external wall, at the junction of a large box with a smaller box. Differential movement was of the order of 11 mm, and this was sufficient to disrupt drains too.

Geotechnical investigations revealed that the original foundations to the external walls

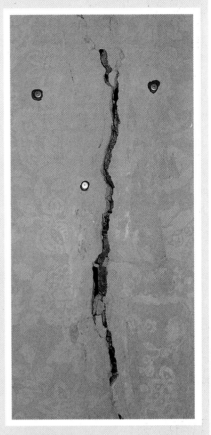

Figure 6.14
Cracking in the one-brick separating wall under investigation

Figure 6.13
Semi-detached houses under investigation. The masonry carcase only is shown, without the roof and internal non-loadbearing partitions

separating wall, was widening. It became so wide that fingers could be inserted into it, and the neighbours could do likewise from the other side. During later monitoring it closed slightly.

The loss adjuster, who was acting for both insurance companies, with technical advice from engineers, concluded that the cracking had resulted from the dry summers, exacerbated by the presence of the plane trees, in spite of the depth of the original foundations. After monitoring for one year, underpinning was prescribed, and both houses were underpinned in a single contract using a piled raft. The largest cracks were stitched, and the smaller ones grouted with epoxy.

The houses had survived relatively unscathed for 80 years, but a series of dry summers had then caused some movement to the external walls, showing as narrow horizontal cracks in the plaster at the head of the separating wall at roof level. The cracks

were set some 2 m below ground level and consisted, for the most part, of concrete trench fill. The foundation of the separating wall was shallower, estimated to be around 1 m deep. Monitoring every six weeks showed that the largest crack, in the

Sound insulation of separating walls which were dry-lined

Measurements were made in a row of traditionally built terraced houses, all with internal walls of studwork or paper cored-type partitions and separating wall surfaces dry-lined. The separating wall also contained back-to-back flue blocks for a gas heater in the living room area. Sound transmission tests made between two of the houses in the terraced block showed satisfactory results compared to the performance standard required by current building regulations.

Sound insulation of separating walls built of innovative bricks

Sound insulation between flats with separating walls of single leaf, vertically perforated, fired clay masonry units (Calculon 'C' brick) was investigated in three buildings. Measurements showed wide variations in performance. The poorest results, obtained in a particular set of maisonettes, were found to be partly brought about by the use of a resilient lining on the external walls. There was a resonance effect in the lining which led to inadequate performance in the adjoining part of the separating wall.

On rare occasions, problems may be caused by excessive chasing of the wall for electrical services (Figure 6.12 on page 212). In these cases, recessed switches and socket outlets may have to be changed for surface-mounted boxes and chases filled with mortar.

Separating walls built in precast concrete panels may sometimes be encountered. For instance, the Taylor Woodrow Anglian system of prefabricated two storey housing had 7 inch precast concrete separating walls (in four sections for easy crane handling) that was claimed to ensure a sound reduction of 50 dB[210].

BRE Digest 333[211] deals generally with the sound insulation of separating walls. BRE Digest 293[212] and BRE Defect Action Sheet 105[213] evaluate methods for improving the sound insulation of existing separating walls. General advice is also given in BS 8233[214] †.

Durability

Normally there should be no problems with masonry separating walls, unless they project through the roof. The durability of these projections was dealt with in *Roofs and roofing*.

However, one factor which can and does affect masonry separating walls is movement from the ground. These movements can be quite serious, as the case study on page 213 shows.

† Under revision.

Work on site

The stability of existing separating walls should be examined critically, even where the existing construction appears to be sound. The examination should consider, in particular, the condition of the masonry; the adequacy of connection to adjacent floors, roofs and other walls; and the thickness of masonry walls in relation to support from adjacent walls and floors. Differential movements should be investigated, and measures taken to ensure that the walls act as homogeneous panels and that structural interaction between elements is achieved.

Vertical cracking is often identifiable in unplastered areas of the wall such as the apex within the roof or in a cellar. Special attention should be given to the junctions with other structural walls where movement can show as cracking in corners. The direction of span of floor joists should be noted: where parallel to the separating wall it is possible that no structural connections have been provided to the wall. It will be necessary to identify whether parts of the original structure may have been removed (eg a chimney breast or internal wall).

Obvious signs of rising damp should be noted (presence of salts and tidemarks). Moisture levels on walls at low level can be measured, but the results obtained by electrical resistance-type meters should be interpreted with care.

The extent of subdivision in roof spaces and floor voids should be checked; also perforation and fire-stopping at junctions with other elements, if necessary with an optical probe.

The problems to look for are:
◊ missing walls within roof spaces
◊ unfilled mortar joints within floor thicknesses and roof spaces
◊ wall ties too stiff in cavity walls
◊ sound insulation deficient (walls too thin or density too low)
◊ fire performance inadequate (eg cavity barriers omitted)
◊ inadequate bonding to external walls
◊ services passing through (particularly in conversions)
◊ walls of inadequate mass (eg pressed bricks laid frog down)
◊ lightweight flanking external walls
◊ deep chases for services cut back-to-back in separating walls
◊ lack of dry-packing under walls in concrete panel systems
◊ lightweight linings touching the masonry core
◊ removal of buttressing effect of intersecting partitions during rehab

Chapter 6.2

Framed

Figure 6.15
This lath and plaster covered timber framed wall was successfully converted into a separating wall

Framed separating walls can be encountered in buildings of any age; even though they may not have started as separating walls, they can become so when conversions are carried out (Figure 6.15).

However, the vast majority of framed separating walls that will be found are those used in system built dwellings, and especially those constructed using timber frames. Such walls were used extensively in the 1960s; for example the Firmcrete system had a separating wall of storey height timber panels covered on each side with two layers of plasterboard totalling one and a half inches thick, and a glass fibre curtain hanging between the leaves[135].

Characteristic details

Basic structure
Two sets of steel or timber studs at 600 mm centres, with a cavity of at least 200 mm, clad on each face with plasterboard to a total thickness of 30 mm, and with an absorbent curtain or quilt incorporated in the cavity (Figure 6.16). The quilt needs to be unfaced, possibly be reinforced, and should have a density of not less than 12 kg/m^3. If it is hung centrally in the cavity it should be a minimum of 25 mm thick, and if hung from one face it should be a minimum of 50 mm thick. Within the roof space, it may be found that a single row of studs only has been used, clad on each face with double sheets of plasterboard.

Steel framed separating walls may also be encountered, though comparatively rarely (Figure 6.17 on page 216).

Abutments
Ideally the floor should not bear on the separating wall, and the cladding should be carried through the thickness of the floor.

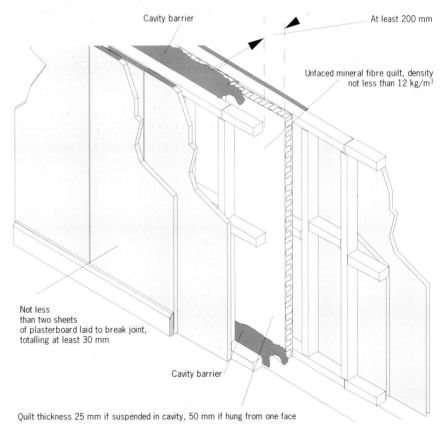

Cavity barrier

At least 200 mm

Unfaced mineral fibre quilt, density not less than 12 kg/m^3

Not less than two sheets of plasterboard laid to break joint, totalling at least 30 mm

Cavity barrier

Quilt thickness 25 mm if suspended in cavity, 50 mm if hung from one face

Figure 6.16
One type of framed separating wall in accordance with Approved Document E

Figure 6.17
Steel framed separating walls under construction in 1996 in the Isle of Skye

Figure 6.18
The diagonal bracing in the separating wall has prevented correct installation of the cavity barrier at the intermediate floor

Accommodation of services

Services should not be incorporated in the separating wall, but where it is essential that electrical sockets are incorporated, they should not be fitted back to back, and should be backed by plasterboard of the correct thickness.

Main performance requirements and defects

Strength and stability

Strength depends on the adequacy of the studs and how well they are nogged. Some separating wall leaves may not be adequately tied together, since ties might interfere with sound insulation.

Dimensional stability, deflections etc

There may be some shrinkage of the timber studs as they dry down to their in-service moisture content, which could lead to cracking at the junction with the ceilings and with the external walls. It is unlikely to be extensive. If the smaller section timbers are dried down to 20% moisture content at the time of grading, the risk of subsequent shrinkage is reduced.

Twin or triple sheets of plasterboard will give sufficient resistance to both deflections and to impact damage, particularly if they have been laid to break joint.

For coefficients of linear thermal and moisture expansion etc, see Chapter 1.2.

Rising damp and condensation

Rising damp from a slab is unlikely to occur provided the DPM is satisfactory. Condensation is a theoretical risk at the junction with the external wall if the thermal insulation is deficient, allowing a thermal bridge.

Thermal properties

The acoustic curtain goes some way to providing a reasonable standard of heat insulation on each side of the separating wall. There are no formal requirements.

Fire

Fire can spread between dwellings having framed separating walls where an effective fire separation has not been achieved. In some dwellings, the fire separation is impaired where the fill between the separating wall timber frames does not abut the roof. Where the roof space separating walls are lined with plasterboard, gaps have been noted at eaves and purlin positions. Boxed eaves are particularly likely to contain gaps which outflank the separating wall. (See also *Roofs and roofing*[24]).

Dwellings built in industrialised systems of construction have been identified as possibly having poor performance in fire situations. BISFs and Hawthorne Leslies need careful attention paying to fire-stopping at the junctions of separating walls and roofs. In the case of Steanes, large voids have been found in the separating walls. Steelwork fire protection has been found missing in a number of Arcals.

Cavity barriers will be needed at various places, for example at the junctions with other elements of construction, such as floors (Figure 6.18) and external walls, for both fire and noise reasons.

One hour's fire resistance should be obtainable from a framed separating wall built in accordance with manufacturer's instructions and faced with various combinations of wallboard and plank. Improved performance (one and a half or two hours) is also obtainable, but this may require the use of boards of enhanced performance.

Noise and sound insulation

A framed separating wall built with twin facings of plank and wallboard, and hung with a quilt of recommended density and thickness, should give a sound insulation of the order of 50–55 dB.

Work on site

Inspection

Inspections of all the following points after completion of the works will necessitate the use of optical probes.

In addition to those listed in Chapter 6.1, the problems to look for are:
◊ cavity barriers missing at junctions with other elements of construction
◊ ties used to connect stud leaves are too heavy
◊ air paths round back-to-back electrical sockets
◊ plasterboard joints not staggered
◊ quilt suspended incorrectly

Chapter 6.3

Stairway enclosures and protected shafts

Figure 6.19
A fire started deliberately has here breached the protected stairway in a hotel

Stairways may need to be enclosed and protected, since they are frequently designated as means of escape in case of fire. Indeed, all stairways in principle might be designated as means of escape, but not all might need to be enclosed[215]. Conditions might arise where communication is needed between compartments of a building; for example, in staircases in multi-storey dwellings, and also in other buildings such as shops, offices and hotels (Figure 6.19). Lift shafts and service ducts may also require protection where these connect compartments.

The concept for dealing with the protection for compartmentation is encompassed in the term 'protected shaft', and the structure which encloses the shaft used to be called 'protecting structure', though this term was removed in the 1992 edition of the Building Regulations Approved Document B[52].

Characteristic details

Basic structure
The enclosing walls of protected shafts may be constructed of any materials which meet the relevant fire resistance criterion. Walls forming the surrounding structure would have needed to be of limited combustibility if the fire resistance requirement was one hour or more, and the surface materials likewise, though there were exceptions. Reference should be made to construction described in other chapters of this book as appropriate to the situation.

Accommodation of services
There are additional limits on the uses to which protected shafts may be put. For example those containing stairways may not also be used for oil pipe runs or ventilating ducts; a dedicated protected shaft is frequently needed for these, and perhaps automatic fire shutters need to be fitted at inlets and outlets.

Main performance requirements and defects

Choice of materials for structure
See other chapters appropriate to the materials used in the construction of stairways and protected shafts.

Strength and stability, dimensional stability etc
See other chapters appropriate to the physical qualities required for stairways and protected shafts.

Fire fighting stairways are required to meet enhanced impact resistance criteria.

Ventilation
Protected shafts containing lifts or used for carrying gas pipes should be ventilated at their heads with a permanent ventilator.

Fire
The most important consideration is to keep fire away from stairways. However, there is no requirement for fire resistance in dwellings of one or two storeys. Fire safety for occupants of dwellings will depend primarily on stair layout, enclosure and final exit arrangements, and only

to a lesser extent on the fire resistance of the stairway itself.

Where a stairway is contained in a protected shaft, and that shaft has a requirement of not more than one hour fire resistance, glazed partitions separating a corridor or lobby from the protected shaft may be used. They should be fitted with glazing having a half hour's resistance, and can be used provided that the corridor or lobby is separated from the rest of the storey by construction also having a half hour's resistance; that is to say two half hour partitions between the protected shaft and the rest of the compartments (Figure 6.20).

Conversion work can result in a situation where stairs separating one flat from another, or a flat from a common area or escape route. In these situations the soffit of the stairs should be considered as a compartment floor and adequate fire protection provided. A stairway should be enclosed by separate fire resisting elements where it forms part of a formally designated escape route in flats or houses over two storeys high.

Permanent ventilation for smoke control purposes must be provided at the top of internal stairways serving more than one dwelling. Such stairways may lack protected lobbies at landings and may route stair users past ground floor flat entrance doors. Enclosed stairways may lack openable ventilators at each landing.

Lath and plaster, or plasterboard of inadequate thickness, which fails to provide sufficient fire protection if the stairs are part of a separating structure between dwellings, can be present. Although building regulations allow combustible materials to be added to the upper surfaces of stairs and landings (ie floorings), this does not apply to the vertical surfaces of walls.

Noise and sound insulation
See other chapters appropriate to the noise and sound insulation properties of stairways and protected shafts.

Durability and maintenance
See other chapters appropriate to these requirements.

Work on site

Reference to building regulations and the various codes on means of escape will indicate whether a stairway needs to be constructed of materials of limited combustibility and whether it needs to be enclosed by (or comprise part of) a fire resisting element of the construction. Visual inspection on site should reveal whether the appropriate standards are achieved.

Inspection
The problems to look for are:
◊ walls not having prescribed periods of fire resistance
◊ stair not of limited combustibility where required
◊ surface spread of flame classification not appropriate
◊ lack of permanent ventilation where appropriate
◊ conveyance of oil service pipes in stair enclosures
◊ glazing in corridors and lobbies not of half hour's resistance where required

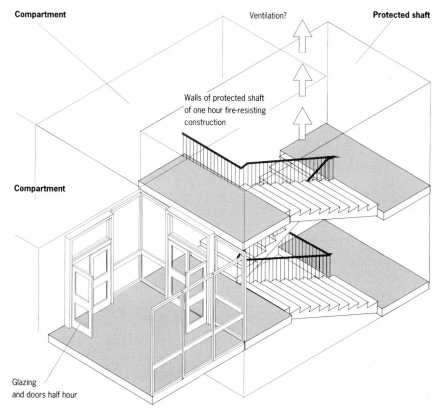

Figure 6.20
Glazed screens separating the lobby from the stairway in a protected shaft

Chapter 7 **Partitions**

This chapter deals with all kinds of fixed partitions, whether of heavy construction (such as of masonry) or light (for example stud and plasterboard); relocatable or moveable. In essence, therefore, the chapter deals with internal walls which have no separating or compartmenting function. They may be of similar construction to separating or compartment walls, and, insofar as that is the case, the information in Chapter 6.1 is not repeated. They may also, of course, be very different from those described in earlier chapters (Figure 7.1).

Figure 7.1
A freshly cleaned internal wall faced in buff coloured terracotta at the Natural History Museum

Chapter 7.1

Masonry partitions

Partitions constructed of brick and block may be found to be either loadbearing or non-loadbearing. Imposed loads may be imparted mainly by floors, but in some cases roof purlins may be strutted from partitions which run from foundations to roof. Effectively one of the main differences, so far as half brick or 100 mm block walls are concerned, will be whether the wall has been carried below the floor level to a separate foundation (Figure 7.2) or not. However, in the case of partitions built off rafts, even this distinction is difficult to draw. This chapter therefore deals with brick and block partitions, including those of cast plaster blocks, whether or not they are loadbearing.

Partitions have been built in a tremendous variety of materials over the years (eg hollow clay block and plaster slab reinforced with organic

materials) either erected using normal bricklaying techniques or cast in situ. Some partitions, in concrete cast in situ between shuttering, and consisting of coke breeze and Portland cement, may be found in older buildings.

Characteristic details

Basic structure

For many years loadbearing partitions have commonly been constructed in half brick thicknesses laid in stretcher bond. In the interwar years, the use of lightweight concrete blocks for internal partitions, both loadbearing and non-loadbearing, became an established practice, though timber stud partitions with plasterboard facings ran them a close second in popularity.

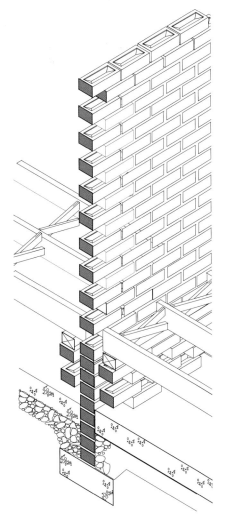

Figure 7.2
A loadbearing masonry partition

Figure 7.3
Loadbearing extruded blocks. Defects are in evidence (eg at joints and joist bearings)

Extruded clay partition blocks used to be made in the UK by a number of firms to a British Standard[216], with grooves cast on each of the larger faces to provide a key for plaster, but other pattern blocks may also be encountered (Figure 7.3). Hollow cored or solid gypsum blocks or panels for use in partitions are not now common in the UK but they are still made on a considerable scale abroad and it is possible some may be imported.

Abutments and corners
Partitions, especially non-loadbearing and less than 100 mm thick, cannot always be self-supporting. Some masonry partitions were provided with corner posts in timber to provide the necessary stability. In other cases, stability is provided by full room height doorsets pinned to ceilings, with the blockwork held against horizontal displacement essentially by the architraves. Otherwise, returns are necessary.

Main performance requirements and defects

Strength and stability
It may not be clear whether an internal partition exists only to subdivide space or to perform a loadbearing or buttressing role or both. An analysis of the structure should reveal the function of a partition. However, one should always be wary of any fortuitous contribution to overall strength, such as where an essentially non-loadbearing partition (ie having no foundation) offers support to a floor spanning over. If this sort of arrangement has performed well previously, and loads are not expected to change, then maintaining the status quo may be justified. Conversely, situations will be found where partitions impose excessive loads on a suspended floor which were not intended, perhaps resulting in floor deflection and settlement of the partition itself; this in turn could disengage any head restraint originally provided.

Masonry partitions may be overstressed by superimposed loads (eg water tanks in the roof). They may be too thin in relation to their height, not tied to other walls at ends and not fixed to the floor or roof structure above. It is bad practice to position a rigid masonry partition on a 'live' timber floor but such arrangements are common, and often result in cracking where partitions abut rigid enclosing walls (Figure 7.4).

The relevant provisions of BS 5628 lead generally to a conservative design for single leaf walls[218].

In tests, single leaf clay brickwork and concrete blockwork storey height partition walls have been subjected to static vertical loads representative of those in cross-wall housing construction up to three storeys high. Simulated horizontal wind loading was applied to the end of the wall until failure occurred. Tests were carried out on walls both with and without a return at the loaded end and the results compared. Walls with returns

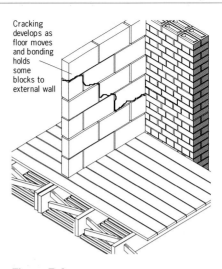

Figure 7.4
A rigid masonry partition built off a live timber floor often results in cracking

Cracking of phospho-gypsum blocks in partitions

The partitions were all supported on a timber first floor. The cracking was somewhat unusual in that it had taken diagonal forms rather than the more normal separation at the partition-to-ceiling junctions. No significant cracking other than in these partitions was seen, and it was concluded that it was associated with drying shrinkage movement of the first floor influenced by the way in which the partition was formed – that is to say, because an adhesive was used to form the joints between blocks and between the partitions and external walls.

exhibited greater racking strengths than plain walls. Brickwork walls were generally stronger than the blockwork walls and exhibited lower strains [219].

Dimensional stability, deflections etc

Ideally, to maintain stability, partitions should be restrained at both top and sides, but they may be found restrained only at the top or the sides. Frequently the wall or ceiling plaster provides the only restraint and even this may be removed during rehabilitation work (Figure 7.5). Storey height door linings were often used to stabilise masonry partitions and should be removed only with caution.

For coefficients of linear thermal and moisture expansion etc, see Chapter 1.2.

Rising damp and condensation

Partition walls should be at less risk of rising damp than external walls because of the normally drier environment around the foundations. However, where partitions are bonded to an external wall a discontinuity in dampproofing can exist if the two DPCs are at different levels and not linked. This condition is more likely if the partition and external wall were constructed at different times. As with other walls, the cause of rising damp is more likely to be bridging than a material failure or omission of the DPC.

Bridging of DPCs is commonly caused by plastering, solid floors and abutting external walls. Previously installed remedial DPC systems may be ineffective, poorly installed, or damaged.

Rising damp in partitions is more easily diagnosed than in external walls because condensation is unlikely to be a factor. Obvious signs of rising damp include salting and tidemarks. Moisture levels can be checked on walls at low level, but results obtained with an electric resistance-type meter should be interpreted with caution. Plaster samples may be taken for analysis if doubts about the source of moisture exist.

Fire

If conversion of an existing building is contemplated, compliance with building regulations will be required. Loadbearing partitions will need to have fire resistance for a period of between a half and two hours, depending on the size of the building and location of partitions. Walls protecting means of escape only will require at least a half hour protection. Where a partition supports a structural element with a designated fire resistance, the partition is required to meet at least the same criterion.

An unplastered half brick loadbearing partition should give one hour fire resistance. Sanded plasters give only marginal improvements, but lightweight plasters offer considerable improvement – for example increasing the resistance of half brick walls to six hours. However, they have to remain in place during a fire, and they may be too thin or lack adequate restraint, so that stability of the structure in fire might be jeopardised.

Figure 7.5
A disrupted brick partition during rehabilitation work

Services may pass through fire-resisting partitions without fire stopping and there may be gaps at perimeters. Cavities in construction above a partition (eg a suspended ceiling), below a partition (eg a floor void) or to sides of a partition (eg a wall cavity) may, in certain circumstances, provide a bypass route for fire in the absence of cavity barriers (Figure 7.6).

Noise and sound insulation

Sound insulation is often important in building types other than housing; for example in offices. Partitions which do not meet the underside of the structural floor, for example terminating at suspended ceilings, can be outflanked by sound travelling over the top of the partition within the suspended ceiling void. Remedial work to build up the partition head to the underside of the structural floor is sometimes feasible. Masonry partitions may develop perimeter gaps, possibly resulting from structural movement or lack of bonding or tying.

Even where adequate partitions exist, significant flanking sound may be transmitted through the floor void and gaps between plain edge floor boards. Buildings heated with warm air systems may have sound paths through return air grilles fitted in

doors, ducts that communicate between rooms and poorly isolated circulation fans. Services passing through partitions may be inadequately sealed. Glazing above doors or other elements can also be weaknesses. Lightweight and poorly fitting flush doors will reduce sound insulation performance of an otherwise adequate wall.

The sound insulation qualities of existing partitions in housing might easily be dismissed as unimportant. No mandatory criteria exist and, generally, limited advice will be found in guidance documents. BS 8233[214] does, though, include a requirement for partitions enclosing WCs (38 dB). In the early 1960s the Parker Morris report[220] highlighted the need for maintaining privacy and activity isolation within dwellings, but little progress has been made since then. BS 8233 discusses the control of internally generated noise in design and construction.

The airborne sound insulation of a plastered brick partition has been measured under laboratory conditions at around 39 dB for a half brick wall to the full height of the room, though sound insulation of around 35 dB may be more normal from single leaf partitions; where insulation values above this are required, it may be better to use a double leaf construction. The weakest link is usually the door, and often the value for the partition as a whole depends on the effectiveness of the door.

Durability

Masonry partition walls commonly suffer from cracking, almost irrespective of material. In the case of non-loadbearing walls, this could be from building off a timber suspended floor which has yet to dry down to its in-service moisture content. If the wall is not bonded into the external wall, the crack may easily occur at the wallhead-to-ceiling joint. If it is bonded into the external wall, the crack may occur at any point in the height, depending where the bonding-in occurs. Cracking could also be from shrinkage of the material, especially light weight

Cracking in dense concrete block partitions

The BRE Advisory Service was asked to investigate the cracking to the blockwork in several industrial units. The partition walls subdividing the main structure were constructed in 215 mm concrete blockwork. The steelwork in the line of the partition walls was enclosed by 140 mm blockwork which should have been constructed leaving a 10–12 mm clearance between the steelwork and the blockwork. However, the partition walls had been taken up to roof level and the blockwork built tight onto the structural steelwork at the junction of the stanchions and rafter members.

Cracking was mainly confined to where the blockwork had been built around the junction of the stanchions and roof trusses, but there was some cracking at lower levels. Cracks in general followed a vertical line through the perpendicular mortar joints and across the block stretcher face. Above the fourteenth course the cracks stopped and reappeared in the blockwork that was built around the roof truss.

Where the blockwork butted the steel trusses the gap had been filled with a hard mortar mix. Any movement in the steel stanchions, roof trusses and bracing members due to thermal movements or wind loading would have been transferred to the blockwork with resultant cracking. The gap should have been filled with a flexible filler to accommodate the movements in the steel frame.

There was little evidence of any structural instability within the blockwork examined during the site investigation. The majority of the cracks would fall into the very slight or slight category.

Cracking due to drying shrinkage could be filled and the walls redecorated, even though the cracks would be likely to reappear due to thermal movements. The cracked blockwork at high level around the roof trusses and bracing could be left if the appearance was acceptable. If not, the cracked blockwork would need to be removed and rebuilt and the gap between the steelwork and blockwork filled with a flexible filler.

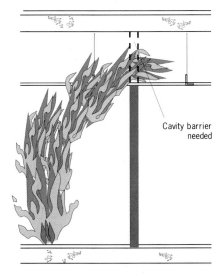

Figure 7.6
Partitions which do not reach the underside of the structural floor can provide routes for fire spread

Cavity barrier needed

Figure 7.7
Here the short return is inadequately bonded to carry the two lintel bearings, resulting in likely poor load carrying capacity for the massive floor trimmer over

concrete which has been allowed to become saturated on site.

In the case of loadbearing walls, one of the most likely causes of cracking is poor bonding of masonry. On many occasions, BRE investigators on site have seen very poor bricklaying practices (Figure 7.7). Also there is the risk of cracking through shrinkage, as in the non loadbearing case described above.

Work on site

Inspection

It is important to establish if a partition has any structural role, using drawings if available, or by site checking of joist spans and other adjacent structural elements. Crack patterns should be examined for indications of the cause of movement. It is often useful on site to record height and thickness of masonry partitions and timber sizes and spans of stud partition for later 'desk checking'. Plumbness and any evidence of bowing should be noted.

A partition should be examined to determine whether it is protecting an escape route. Fire resistance of the partition wall construction should be assessed and, where cavities exist, an optical probe examination might be considered desirable.

Likely sound reduction values should be estimated and compared with criteria. Holes, gaps, glazing and imperfections should be noted, together with voids that could conduct sound through or around a partition (Figure 7.8).

The problems to look for are:
◊ cracking (shrinkage of materials or of supporting suspended floors)

Figure 7.8
A hot air heating duct through a partition. There is no lintel and the infilling block has already dropped

◊ instability (no returns, pinning to ceiling, or support from full height framing)
◊ no lintels (walls carried on door frames)
◊ removal of existing supports to roofs or floors during rehab
◊ unintended loadings from roofs or floors
◊ inadequate fire resistance
◊ asymmetrical loadings (eg from water storage tanks)

Chapter 7.2

Framed, movable and relocatable partitions

This chapter deals with all kinds of framed partitions including both those which are static (such as ordinary timber studs with plasterboard facings) or movable (such as sliding and folding partitions) or relocatable partitions, sometimes called operable walls (such as those proprietary systems now widely used in office accommodation).

Before the nineteenth century, timber was frequently used to reinforce masonry nogged partition walls (Figure 7.9), especially structural partitions which commonly were trussed. Some of these trussed partitions will also be found to provide support for floors or roofs. During the nineteenth century, the increased use of metalwork enabled improvements to the design of trussed partitions. The twentieth century has seen enormous increases in the variety of solutions which are possible, particularly with moveable partitions some of which have excellent fire and acoustic performance.

Such flimsy constructions 100 mm thick have even been found as separating walls by BRE investigators on site (Figure 7.10).

Trussed partitions may also be encountered which span from loadbearing wall to loadbearing wall where the full load of the partition is carried on these walls in preference to being carried on the floors. Such framed partitions may also carry intermediate floors (Figure 7.11 on page 228) or form bearings for struts to purlin or butterfly roofs (see Figure 2.5 in *Roofs and roofing*[24]).

The simplest form of framed partition is that which consists of nogged timber studs carried on floor plates, and covered each side with lath and plaster or plasterboard which may or may not be skimmed.

Metal stud framed partitions, some of which may be demountable or relocatable but also faced with plasterboard, have been a later development. Steel stud systems may have the advantage of rapid erection compared with, say, timber stud partitions cut on site. The

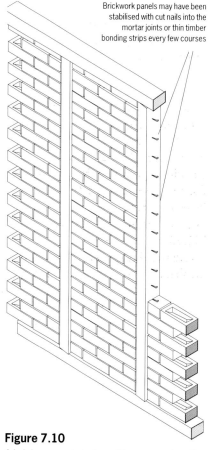

Brickwork panels may have been stabilised with cut nails into the mortar joints or thin timber bonding strips every few courses

Figure 7.10
A brick nogged stud partition, occasionally used as a separating wall

Characteristic details

Basic structure
Framed partitions may be loadbearing or non-loadbearing, just as with masonry partitions. In Victorian and Edwardian times partitions were commonly of timber stud lathed and plastered. Some of these may be found to be loadbearing, typically constructed in 100×50 mm or 75×50 mm timber studs, with the spaces bricknogged with brick either laid flat or on edge.

Figure 7.9
A stud partition with a plinth of brick nogging on one side and lath and plaster on the other

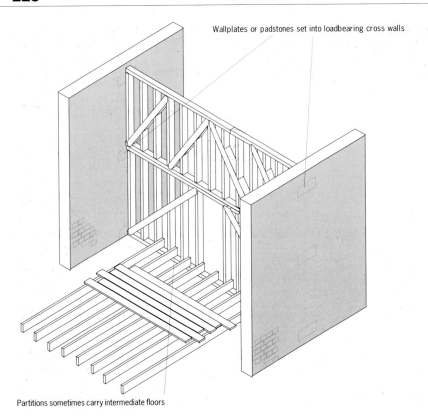

Wallplates or padstones set into loadbearing cross walls

Partitions sometimes carry intermediate floors

Figure 7.11
Trussed framed partitions sometimes carry intermediate floors

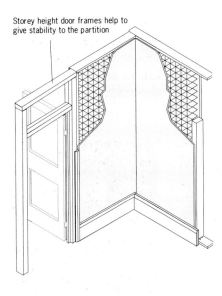

Storey height door frames help to give stability to the partition

Figure 7.12
Paper cored prefabricated plasterboard partitions

framing for these systems, although formed from galvanized steel pressed metal sections, is in principle assembled the same way as traditional timber stud construction using horizontal head and sole plates, vertical 'studs' between these and horizontal cross noggings at intermediate positions between the studs. Self-drilling self-tapping screws are used to connect the members together.

Non-loadbearing proprietary paper cored prefabricated plasterboard panels with taped and filled joints became available after the 1939–45 war (Figure 7.12). They have been widely used in housing in the years since then.

The number of different designs of proprietary partitions is legion, and it is not possible to cover all of them in this book. However, mention may be made of very lightweight hollow cored melamine units introduced after the 1939–45 war, and steel and mineral fibre reinforced cement composite panels. These latter panels consist of 0.5 mm thick galvanized steel sheets mechanically bonded under pressure to both sides

of a non-combustible fibre reinforced cement core so that the layers become an integral whole. The material was available in two thicknesses: 6 mm and 9.5 mm.

Yet other types might consist of prefabricated chipboard sheets faced with vinyl on the outer surface and a balancing layer, often paper mounted, on the inside.

Moveable partitions
Moveable partitions are of many types. The simplest consist of lightweight accordian or concertina-type fibreboard, or steel cored vinyl-covered narrow slats hinged together by the coverings. More substantial systems consist of hinged leaves, similar to doors, connected in pairs or multiples and carried on overhead tracks and, for the heavier versions, also floor tracks. Fully glazed types are also available which are held in upper and lower clamping rails, but no stiles, where the safety glass alone acts as the leaf (Figure 7.13). Most of the types described here may be hand operated or mechanically driven and may or may not have pass doors and vision panels.

External corners
These are the most vulnerable areas of lightweight framed or cored partitions, and should be protected, for example by cover moulds or other forms of armouring.

Accommodation of services
Partitions, even though nominally non-loadbearing, are frequently required to carry asymmetrical hanging loads (eg bookshelves and sanitary fittings). Failure to allow adequately for live loads associated with these has been noted on BRE site investigations.

In the past, support for fixtures and fittings, particularly cantilevered sanitary fittings fixed on stud partitions, often relied on wood noggings between studs; this produced mixed results. In prefabricated paper-cored panels, support is provided by means of noggings driven into the cores.

If service pipes or wires are to be installed in stud partitions these must be installed before the plasterboard facing is fixed on both sides. Where the partitions are of the relocatable type it is of course better to avoid siting service runs within or on their surfaces. Plasterboard is held in place using fixing strip and self-drilling self-tapping screws, the screw heads being concealed by clip fitting trim and skirting which is a push fit over the fixing strip.

Main performance requirements and defects

Strength and stability
Existing stud partitions may already have been weakened by removal or cutting of diagonal bracing members, or they may not have had them in the first place (Figure 7.14). Removal of brick infill can reduce lateral strength or racking resistance.

All kinds of partitions are subjected to impact loads, and can be subjected to tests similar to those used for door leaves referred to in Chapters 5 and 8.

Glazing
Glazing in all new moveable partitions in domestic accommodation, and in all moveable partitions in other building types, must be safety glazing to BS 6262 up to a height of 1.5 m above finished floor level. Annealed glass, for example in vision panels and pass doors, is permitted provided the smaller dimension of the pane is less than 250 mm. Fire doors, of course have additional requirements.

Dimensional stability, deflections etc
Non-loadbearing partitions built off suspended timber floors depend for their stability on the stability of the floor. When this moves, the partition will move with it (Figure 7.15).

For coefficients of linear thermal and moisture expansion etc, see Chapter 1.2.

Rising damp and condensation
DPC failure under a timber partition could result in rot in the sole plate and abutting studs, and to settlement of the partition.

Figure 7.13
Room height fully glazed operable wall

Figure 7.14
When stripped of its lath and plaster, this partition had no horizontal noggings or braces. When built, it was obviously considered that the laths and plaster provided sufficient stability on their own

Figure 7.15
In this pair of photographs of an Edwardian partition, the floor has sunk, due in part to foundation movements and in part to deterioration of the joist bearings over sleeper walls. The first picture shows that the door lining, sitting on the floor boards, has sunk by 15 mm in relation to the partition; the partition, though, has bridged the gap caused by the sinking of the floor. However, the second picture shows a diagonal crack and cracking beneath the skirting, indicating that the partition has failed to bridge the shrinkage elsewhere; even a new moulding has not tackled the fundamental problem. Stabilising the movements in the floor upon which the partition is founded is the only remedial solution

Figure 7.16
A stud partition after a fire in an institutional building

Noise and sound insulation
Timber stud partitions may be found to be constructed from undersized members lined with low mass boarding (eg medium board or softboard) which gives poor sound insulation.

The airborne sound insulation of lightweight hollow plastics partitions, both complete and also in association with a suspended ceiling, have been measured under laboratory conditions. The results gave around 23 dB for an unfilled plastics partition. Better performance, up to around 30 dB, can be obtained from paper cored prefabricated partitions. Even better performance can be obtained from ordinary stud partitions, up to around 37 dB and perhaps more, and up to or just over 60 dB (depending on facing thicknesses) if the studs are staggered – that is to say the facing boards are supported by every other

Fire
Fire can have a devastating effect on lightweight framed partitions. (Figure 7.16).

Timber stud partitions may have linings such as hardboard, plywood or medium board which do not satisfy fire resistance or surface spread of flame criteria; for example Class 0 where used adjacent to means of escape routes. Where plasterboard, or similar, is present it may have inadequate thickness, insufficient plaster finish or no noggings behind board joints. Strange as it may seem, it is possible to design fire resisting partitions which are composed entirely of combustible materials.

Loadbearing stud framed partitions with studs – a minimum of 44 mm wide, spaced at not more than 600 mm, covered with 12.5 mm plasterboard with all joints taped and filled – should give a half hour's fire resistance. Similar performance can be obtained from partitions with metal studs. Doors and frames

in partitions enclosing fire escape routes will normally require a minimum of a half hour fire resistance.

Fire resistance is most easily increased by adding plasterboard or an additional thickness of wet plaster to the existing construction. An extra layer of 12.5 mm plasterboard to each side of the partition already described will increase the fire resistance to a full hour. Current opinion is that existing lath and plaster will make only a minimal contribution to fire resistance.

Relocatable and movable partitions will need to conform to fire resistance and surface spread of flame requirements, just as for any other partition in the same situation. Concertina-type partitions have little fire resistance, but other types are available which give up to two hours, depending on construction and glazing. See also Chapter 6.3 for partitions adjoining a protected shaft.

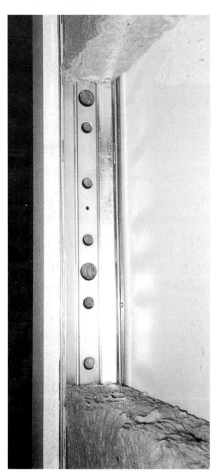

Figure 7.17
This relocatable chipboard faced partition is seen with one panel and part of the sound insulation filling removed for examination

stud and there is no connection between the leaves other than at sill and head. For even higher performance, an absorbent mat may be hung in the cavity (Figure 7.17).

Partitions are sometimes found to be nailed to a floating deck, with obvious consequences for cracking at the joints as the flooring deflects. This practice probably will not detract from the sound insulation of the floor. See the relevant discussion in Chapter 4.2 of *Floors and flooring*[30].

So far as sliding/folding partitions are concerned, sound insulation is frequently little better than doors of similar construction, and will be worse if the gaps are not sealed. Values as low as 10 dB have been measured. Sealing all edges with compressible gaskets and making the units heavier might improve performance to perhaps around 25 dB, but to obtain higher values with single construction is hardly practicable. Two simple partitions spaced apart by a 200 mm gap might give around 40 dB, while sophisticated double construction

with absorbent cavity fill and having a weight of, for example, 50–60 kg/m^2 of wall area, would give a worthwhile improvement.

Where the system is top hung only, and the bottom is sealed by retractable seals, the condition of the floor surface will affect performance to a considerable extent; particularly where, for example, there are recessed or eroded joints in the flooring allowing air gaps under the partition in the closed position.

Durability

Framed partitions, particularly in older buildings, sometimes get wet from leaking roofs. Timbers which are damp, but show no signs of decay, need to be dried out to 20% moisture content or below to avoid the risk of decay. Application of wood preservatives may be appropriate if drying time cannot be reduced to less than about six weeks[78].

When kept dry, of course, timber stud partitions last for many centuries, even though they may need a little restoration occasionally (Figure 7.18).

Apart from impact damage, satisfactory service from movable partitions depends to a considerable degree on wear and tear in the suspension systems.

Some wood based boards used for facing relocatable partitions may be treated with fire retardants which are hygroscopic in nature, leading to increases in moisture content and a risk of bowing if the moisture take-up is not uniform. Moisture contents of 20–22% have been measured by BRE on site. Whether this bowing is contained depends on the integrity of the fixings.

Work on site

Site conditioning of boards for new or replacement work may be necessary, depending on circumstances. Additionally, care should be taken to store them perfectly flat before use. BRE investigators on site have seen boards quite badly bowed because poor support conditions have allowed sagging of the stack. Stacks can also overload existing floors.

BS 8200-8[221] covers work on site for plasterboard partitions.

Figure 7.18
Timber stud partition dating, probably, from the sixteenth century. The plaster, which at one time covered the studs, has been renewed, and the studs treated with a dark stain. There is evidence of considerable former movement in the frame

Chapter 8 **Internal doors**

This chapter deals with all kinds of internal doors in all building types; from cupboard doors which see only occasional use, doors for able-bodied pedestrian use and those suitable for wheelchair users, doors used by hospital staff for trolleys, hospital beds and equipment, to doors providing access to lifts. The range of performance requirements is wide and sometimes quite onerous, from impact resistance to fire resistance – and all must operate with safety at all times (Figure 8.1).

Figure 8.1
A pair of double swing doors to a lobby, glazed with wired plate glass

Chapter 8.1   Doors for normal traffic

This chapter deals with all kinds of hinged and pivoting internal doors except fire doors, which are dealt with in the next chapter, Chapter 8.2.

Figure 8.2
A simple matched door in a sixteenth or seventeenth century artisan's house in Suffolk. The leading edge of the door has been repaired. The door is handed and hung unconventionally, and the strap hinges are on the other side

Characteristic details

Basic structure
In the simplest domestic construction, the strap hinged ledged and matched door in softwood reigned supreme until the advent of the flush door after the 1939–45 war. These doors were not always braced (Figure 8.2). In the more expensive domestic buildings, and in other non-utilitarian building types, the butt hinged framed and panelled door, in softwood or in hardwood according to availability of funds, was almost universal. Softwood was invariably painted.

Flush doors take many different forms, with cores which may range from rolled paper to solid timber, and veneers from thin cellulose fibre boards to plywoods faced with exotic species. All need to be provided with blocks of sufficient size to accommodate hinges and furniture, but many were not lipped and edged. The standards introduced after the Tatchell Report[222] gave us the well known leaf widths of 2 feet 3 inches, 2 feet 6 inches and 2 feet 9 inches, and a height of 6 feet 8 inches. They were used in vast numbers in the housing drives of the 1960s and 1970s.

Glazing of internal doors ranges from the provision of one or more vision panels in hinged leaves – to assist in the prevention of accidents – to half or fully glazed doors where light transmission is required. Fully glazed internal doors may also be frameless, and supported by pivots at head and foot rather than by hinges.

Door leaves protected against X-rays are available. The protection is given normally by thin lead sheet incorporated into the leaf, with X-ray resistant glass being specified for the vision panels.

For new construction, hinged doorsets are increasingly being specified complete with frame. This is becoming of more importance in relation to sound and fire performance (Figure 8.3).

Doors for special purposes, such as those sliding doors providing protection and access to lifts, are almost certain to be of prefinished steel, or, in older installations, may be of the expanding gate type. Some timber doors may be finished with melamine sheets.

Figure 8.3
A steel door and frame installed in the test rig, and about to be tested for strength and stability

Frames and linings

Internal doors depend for their support on the strength and fixity of the frame or lining, for the tightness of their fit on the position of the stop, whether it is integral with the frame or fitted loose in a lining, and for the effective operation of the latch. Frames traditionally were fixed to the partition with cut nails driven into wooden slips placed in the mortar joint; these fixings can deteriorate over the years. Linings may be one piece in thin partitions, but in thicker ones may be split, sometimes containing panelling in the reveals.

Openings and joints

Door leaves may need to be fitted with grilles, or to have gaps left under the leaves to allow ventilation between rooms, in order to feed air to combustion appliances. These grilles and gaps destroy most of the sound and fire resistance of doors.

Main performance requirements and defects

Choice of materials

As described above, most existing internal doors, before the invention of the flush door, were in wood, and framed and panelled in either European redwood or whitewood. The more expensive doors were made in imported tropical or European hardwoods (Figure 8.4). Mahogany was widely used in doors from about the middle of the eighteenth century.

Door frames and linings have almost invariably been in timber. Where the doorway is pierced through thick masonry, and the doors are heavy, the linings may be very complex, requiring careful restoration. Some rather utilitarian steel frames were used from the 1960s, often without lintels, which do not call for particular comment.

Strength and stability

In general, framed softwood leaves of up to 900 mm width should possess adequate strength for normal domestic situations,

provided the hinges are satisfactory and their number relates to the weight of the leaf. Leaves larger than 900 mm are more prone to distortion, and, for this reason, where clear openings of more than 1 m are needed (eg for the passage of trolleys), it is more usual to provide two leaves which may be of equal or unequal widths. Flush leaves are perhaps less prone to lozenging, but generally follow the above rule.

Door leaves may be tested for strength from both hard body (eg 50 mm steel ball) and soft body (eg 30 kg dry sand) impacts (Figure 8.5), and from slamming.

Main doors to flats

The main entrance door to individual flats requires special consideration from the point of view of strength and stability, although it does not have to withstand the weather. As it is often the only door leading to the outside, it has to provide:

● a barrier to intruders
● a means of escape for the occupants in an emergency
● access for the emergency services
● on occasions a barrier to fire and smoke

Figure 8.4
An internal six panelled door used by Nash at Ffynone, in the early nineteenth century. Two of the panels have split, probably as a result of changes in the heating regime. A near invisible repair would be possible by removing the beads and glueing the splits. Surprisingly for a heavy door, there is only a single pair of hinges (Photograph by permission of Earl Lloyd George of Dwyfor)

Figure 8.5
A test on a WC cubicle door. In this case it is the lock which is receiving particular attention

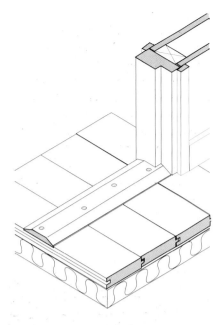

Figure 8.6
A timber threshold to an internal door

In blocks of flats where there is no access control over visitors, the main entrance door to the flat may be the only physical security measure available to the residents. As such, it has to be capable of resisting prolonged attack by intruders using tools, but also has to satisfy the requirements for rapid entry from the Fire Services.

The London Fire Brigade and Civil Defence Authority require a door to be capable of being breached in less than three minutes using hand held cutting tools. Police records indicate that doors in the more vulnerable estates, in this situation, have been subject to uninterrupted attack by intruders using sledgehammers and proprietary jacks for periods of up to 30 minutes.

As a minimum, the doors and frame should be of substantial unglazed wood or steel construction, to meet the requirements of BS 8220-1[69]. They must be fitted with hardware similar to that specified for external doors (see Chapter 5.1).

Using an escape mortice lock is recommended, since it needs to be locked with a key on leaving; this reduces the danger of accidentally locking a child in the flat. The lock can be opened quickly by a simple action from the inside without the use of a key. BRE and DETR believe this will meet the requirements of the England and Wales Building Regulations Approved Document B[52].

Glazing
Glazing in all new doors in domestic accommodation, and in all doors in other building types, must be safety glazing to BS 6262 up to a height of 1.5 m above finished floor level. Annealed glass is permitted provided the smaller dimension of the pane is less than 250 mm.

Dimensional stability, deflections etc
For coefficients of linear thermal and moisture expansion etc, see Chapter 1.2.

Safety
Thresholds
With the increase of factory made building components, such as complete door sets, thresholds to internal doors could become more common. These may potentially be dangerous and should be tapered off unless a fitted carpet is used each side of them (Figure 8.6). A threshold associated with a change of level is not so dangerous because the change in height is visible.

Glazed doors
Glazing in doors is covered by Approved Document N of the Building Regulations[186] which requires that glass in critical locations should either:
● break safely
● be robust or in small panes, or
● be permanently protected

If a door is glazed to the floor, people may try to walk through it, and in these circumstances clear glazing should not be placed lower than 900 mm above finished floor level. Obscured glass or a guard rail at 900–1000 mm will help to make the door obvious (although this is not strictly a requirement under Clause N2 of the Building Regulations with respect to dwellings).

Glass in doors obviously has to be strong enough to withstand everyday use and slamming. Toughened or laminated glass may need to be used in glazed doors in some cases. Glazing at low level is always a danger for a child or adult. Glazed doors should never be put at the bottom of stairs.

Dampness and condensation
Although dampness in the outdoor sense may not apply, there may sometimes be a risk of condensation from unheated areas, spillages of liquids leading to conditions of high humidity and some environments having relatively high humidities which have caused doors to warp. In hospitals, for example, the normal hygrothermal conditions ranging from 25–65% RH over a temperature range of 10–25 °C can in some areas become 25–100% RH over 10–30 °C. If these conditions occur for any length of time on either side of an unsuitable timber door, warping can occur.

Figure 8.7
Even this well fitting heavy door between a Regency dining room and kitchen would have required quiet servants. (The door is shown slightly ajar)

Thermal properties

Thermal properties are rarely called for in internal doors, though conditions may need to be controlled where the door separates rooms intended to have vastly different temperatures.

Thermal conductivity (W/mK):
- hardwoods 0.15
- softwoods 0.13
- aluminium 160
- steel 50

Fire

Those doors which specifically need to possess fire resistance are dealt with in Chapter 8.2.

See also Chapter 6.3 for doors in partitions adjoining a protected shaft.

Noise and sound insulation

The banging of internal doors is one of the noises easily transmitted to adjoining dwellings. The best remedy is to fit a suitable door closer. However, laboratory studies have shown that simple buffers round the door frame can also give an appreciable reduction in noise from moderate slams, but are less effective against hard slams[223].

So far as noise transmission through internal doors to adjacent spaces is concerned, the average lightweight hollow cored internal door with wide clearances provides very little sound insulation – perhaps as low as 10 dB, depending on construction and the size of the gaps between leaf and frame or lining. Close fitting heavier doors may give a little more (Figure 8.7).

Improvements may be effected by the fitting of draughtstripping to reduce the size of air gaps, but this will tend to make the door difficult to open. Where there is a paramount need to improve sound insulation between rooms, and a doorway is essential, consideration might be given to the construction of a lobby, with doors at each end. Lining the lobby with sound absorbent material is worthwhile. The larger the lobby the better. Hanging doors on each face of a thin partition is rarely worthwhile in sound reduction terms let alone convenience in operation, though the builders of some of the country's stately homes often used two pairs of heavy doors within the thickness of massive masonry walls with reasonably good results (Figure 8.8).

Durability

Since there are no external weather conditions affecting internal doors, the most likely major influence on their durability is resistance to impact damage in heavily trafficked routes. In hospitals, for example, it is not unknown for doors to be opened by trolleys and beds being pushed against them. The former Property Services Agency were said to have had problems with internal doors in barrack accommodation where the army boot was the preferred method of opening them. Warping and twisting of leaves in areas of different hygrothermal conditions have already been referred to. On the other hand, where conditions of use are not severe, door leaves can last indefinitely.

Steel sheet doors can dent relatively easily, and the dents are not removeable. In similar vein, although melamine finishes are highly resistant to wear and tear, they crack under severe impacts; they are not repairable and must be replaced. Painted timber, on the other hand, can be repaired.

Maintenance

The fashion for stripping painted framed doors by dipping them in caustic solution has ruined many a sound example. The timber may shrink unevenly and crack, the rails move away from the stiles and almost certainly the wedges loosen, leading to warping, not to mention the patches which become visible over cut-out knots (Figure 8.9). Old doors which were intended to be painted should be kept painted.

Figure 8.9
This softwood door has been stripped by immersion. Movement can be seen in the displacement of the tenon and the opening of the stile-to-rail joint

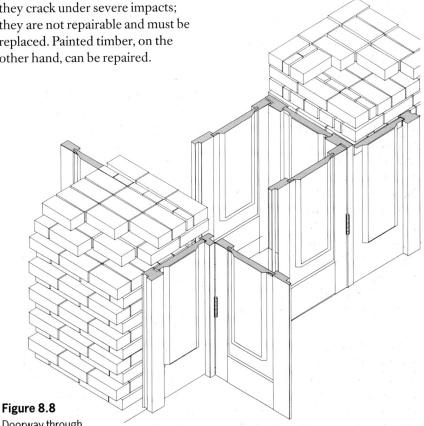

Figure 8.8
Doorway through
a massive masonry cross-wall

Work on site

Storage and handling of materials

New wood door leaves are normally supplied wrapped, and the wrapping should not be removed until the door is about to be fitted.

Workmanship

Site fitting of leaves to existing frames or linings and refitting furniture is a skilled operation, particularly in heritage buildings (Figure 8.10).

The problems to look for are:
◊ warping of leaves between areas of differing hygrothermal conditions
◊ inadequate sound insulation
◊ thresholds not tapered off
◊ non-safety glazing at low level
◊ impact damage
◊ splitting of panels in framed leaves
◊ lack of closers for controlling slamming
◊ linings splitting or twisting

Figure 8.10
A late nineteenth century mahogany door inserted into a circular-on-plan wall

Chapter 8.2   **Fire doors**

The main purpose of fire doors is to protect openings in fire walls built either to separate parts of a building or to enclose stairway walls, and which could, if not protected, be easily breached in fire (Figure 8.11).

Fire doors can greatly limit fire damage by keeping smoke and flames from spreading. However, they cannot function if they are left open, unless they are held by magnetic or other types of catches which release automatically in the event of a fire occurring. Accidental damage so that they will not close properly is also a possibility which must be guarded against.

The precise role of any fire door will depend on whether it is intended to protect escape routes from the effects of fire, or to protect the contents or the structure of a building by limiting fire spread. Some doors may therefore need to fulfil only the first function, others the second. In many cases both requirements will apply[224].

Characteristic details

Basic structure

The industry has successfully developed door assemblies using widely different forms of design, and it is therefore not appropriate to attempt to give a simple listing of the these various kinds of construction, although some indication of typical construction may be useful.

The door and its frame may be of any construction which passes a fire resistance test. However, in order to be satisfactory, the door should be reasonably straight and true and lie flush against the stop when closed; in earlier years, the gap between the door edge and the frame would not have exceeded 3 mm, and the door frame would have had a rebate or stop not less than 25 mm deep, but those criteria do not now necessarily apply.

The door furniture is of importance. There would usually have been at least one pair of metal hinges, all parts of which would have been non-combustible and which had a melting point not less than 800 °C. The lock or latch typically would have had a brass or steel tongue which engaged into a brass or steel latch plate or keep for a minimum distance of 10 mm when the door was closed. Rim locks would have been mounted on the non-risk side of the door or bolted through the thickness of the door.

Any plain glazing would typically have been of at least 6 mm wire reinforced glass not exceeding 1.2 m² in area and fitted with solid wood beading not less than 13 mm in cross-section.

Figure 8.11
The head of the doorway on the left has been breached, and the double doorway ahead shows evidence of smoke infiltration. The fire is out and the investigation is beginning

In order to achieve the full 30 minutes protection for fire resistance, if not also for integrity, an intumescent strip or strips will probably need to be provided in the edge of the door or frame, not in the door stop, to seal the gap in the event of fire. Indeed, timber doors cannot achieve more than 20 minutes unless an intumescent strip is fitted all round. An additional hinge will also typically be fitted, making one and a half pairs.

If there is a glazed panel, beading will probably be either of non-combustible material which will not melt or disintegrate at temperatures below 900 °C, or of timber with a metal capping, or of timber treated with intumescent paint.

In addition, there should be a self-closing device, although doors fitted with these devices may be held open by fusible links or by other electro-mechanical or electro-magnetic devices activated by smoke.

Rules for the construction and installation of fire break doors and shutters are also given by the Fire Officers Committee[225]. The doors and shutters covered by these rules are intended primarily for use in firebreak walls. Four rules are specified for various types of firebreak doors and shutters. Rules 1 to 3 are general construction specifications for doors and shutters to the Committee's own designs while rule 4 is essentially a performance specification for doors and shutters of proprietary design.

The use of two doors in tandem to close the same opening will sometimes be encountered. Under the Approved Document, it is permitted to fit two doors, each of lesser fire resistance than a single door, provided that together they comply with the requirement.

There are also special provisions for doors enclosing lift shafts.

Openings and joints

As already noted, the fit of the door in its frame is crucial to performance – the closer the fit, the more likely the door is to achieve its purpose.

Main performance requirements and defects

Choice of materials for structure

Many fire doors in existing buildings will be of solid wood, or perhaps in some cases of wood derivatives, or of iron or steel sheet. A few wooden doors may be found which in the past have been upgraded with asbestos insulation board; these doors will require special treatment for removal and disposal. Fire doors currently supplied may be of any kind of construction which passes the test.

Strength and stability

Fire doors should satisfy the impact requirements of non-fire doors. See also the same section of Chapter 8.1.

Dimensional stability, deflections etc

Where doors and frames are made of timber or timber derivatives, it is important that they are not subjected to dampness or condensation which will affect the fit of the door. Fire doors cannot usually be planed off to ease fit, for example, when edged with intumescent strips. Those which warp will also tend be ineffective in fire. However, it is as well to check with the manufacturer whether slight easing is permitted without prejudicing performance.

For coefficients of linear thermal and moisture expansion etc, see Chapter 1.2.

Thermal properties

Thermal insulation may be as appropriate for fire doors as for normal doors, depending on circumstances. See also the same section of Chapter 8.1

Fire

As already seen, the effectiveness of a fire door depends on the performance of the whole assembly of door, frame and door furniture. In the fire test, the assembly is judged on the ability of the assembly to:
● remain in position in the opening
● prevent the passage of flame through cracks or gaps
● restrict excessive transmission of heat

Figure 8.12
This pair of fire doors has survived quite well in a fire

Specifiers may have been familiar with the old classification of fire doors, as 30/20, 30/30, or 60/60. These figures denoted the numbers of minutes under test for stability and integrity. Those doors which were constructed to pass BS 459-3[226] were termed fire-check doors. Currently, only integrity is relevant, and doors are now for the most part either simply half hour or one hour.

Intumescent strips fitted into the edges of the door swell on heating (140–300 °C) and close the gap between door and frame.

Where fire doors are supplied not already fitted into frames, it will be necessary to check with the manufacturer on the appropriateness of the existing frame to give the required performance. In some cases an identification colour coded plug will be fitted into the door leaf edge to indicate the fire resistance of the leaf, and whether or not further protection is needed from the frame.

In the case of old timber fire doors, performance may not be up to current standards. Where timber is used on the exposed face of a door, the consequent movement of moisture within the section causes deformation of the door panel and a tendency to form a concave shape towards the heat source. Failure of such a door assembly will occur first at any weak point which may be exploited by flames. Causes of possible weakness are related therefore to the fit of the door within the frame, the size of the door stop or rebate, the quality of the door furniture, and the extent and fixing of any glazing. On the other hand, some old doors stand up remarkably well to fire (Figure 8.12).

A fire door is sometimes required in a separating or compartment wall to provide a means of escape, and in these circumstances it will need to have the same fire resistance as the wall into which it is fitted.

Noise and sound insulation

Sound insulation may be as appropriate for fire doors as for normal doors, depending on circumstances. See also the same section of Chapter 8.1.

Durability

Fire doors should be sufficiently robust to retain their integrity under constant use, so that they will be fully effective when and if they are called upon to perform the vital function for which they have been supplied. See also the same section in Chapter 8.1.

Maintenance

The proper mechanical operation of fire doors is similar to that for ordinary doors, though self-closing devices are even more important. Cases have been seen where building owners have removed self-closing devices in order to leave fire doors in the open position.

Work on site

Workmanship

It may be possible to upgrade an existing door to function as a fire door.

Panelled doors having stiles and rails not less than 37 mm in thickness may be suitable for upgrading; flush doors need to have a substantial sub-frame, giving a similar overall thickness. Unless the precise specification of a flush door is known, it is difficult to be sure about its suitability for upgrading. Achieving an acceptable standard does depend on the condition and specification of the door in the first place. The final decision on suitability must rest with the fire or building authority.

For protection to be of value, the importance of fixing cannot be over emphasised. Additional surfacing material must be fixed to an existing door so that, under fire conditions where thermal movement is likely to take place between the door and protective material, nails or screws are not stressed and therefore are not pulled out.

Existing planted stops may need to be replaced, or additional material screwed or pinned and glued.

Existing fire doors, supplied as such, should not be modified without the express approval of the manufacturers since modifications may invalidate test results. In particular, fire doors should not be fitted subsequently with vision panels.

Inspection
In addition to those listed in Chapter 8.1, the problems to look for are:
◊ old fire doors not acceptable under current standards
◊ distortion of leaves in frames
◊ hold-open devices malfunctioning
◊ self-closing devices made inoperative
◊ asbestos board on old doors

Chapter 9   **Applied external finishes**

Many external walls are self-finished, or at any rate finished as part of the process of construction of the wall – pointing of brickwork and stonework are obvious examples. Indeed, over two thirds of all masonry walls in dwellings in England are finished in this way; that is to say, neither rendered nor tile hung. It is not recorded how many of these walls were painted[2]. This chapter, however, deals with finishes which are applied after the construction of the wall, such as rendering, whether painted (Figure 9.1) or not, adhered tiling, and tile or slate hanging.

Where materials guarantees are required by the specifier, careful reading of the conditions could reveal precise requirements for the application of materials used in external finishes. Mastics, adhesives, paints, and concrete repair systems are often very sensitive to temperature, moisture, dust and dirt. It has been suggested, only half in jest, that conditions are so precise for some materials that they can be used on only a few days each year.

Figure 9.1
A newly painted terrace house in Cardiganshire

Chapter 9.1

Tiling

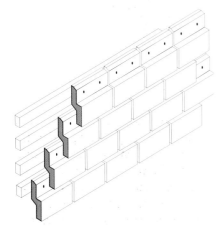

Figure 9.2
Mathematical tiling

This chapter deals with tile and slate hanging on vertical surfaces, as well as adhered non-overlapped ceramic tiling. (For tiling and slating on inclined surfaces, see *Roofs and roofing*[(24)]).

When taxes on bricks were levied in England between 1784 and 1850, this encouraged the use of rebated tiles which simulated bricks in appearance[(227)]. These tiles went under several names, the most common being mathematical tiles (Figure 9.2). When the tax on bricks was repealed in the nineteenth century, their popularity waned and cheaper non-rebated roofing tiles, which were primarily intended to be used on sloping surfaces, largely replaced them. Mathematical tiles, though, are occasionally found on twentieth century buildings.

Of the stock of dwellings in England, 131,000, or around 1 in 140 of the total, have tile or slate hanging as the main finish to the external walls[(2)].

Characteristic details

Basic structure
Tiling
Thin ceramic tiling covering large areas of the façades of buildings has largely been a phenomenon of the twentieth century, and the finish has proved extraordinarily popular with designers, though with mixed results (Figure 9.3). In Victorian times, where a precision facing in ceramics was required, terracotta or faience was specified (see Chapter 2.1). Since this material was very much thicker than the thin tiles which became popular in the 1970s and moreover tended to be bonded more thoroughly to the backing wall, there were far fewer problems than have been encountered with detachment of thin tiling adhered with cement mortars.

Tile and slate hanging
Tile or slate hanging is a weather resistant finish for low-rise buildings, but not at ground floor or access levels because of poor impact resistance. Neither would it be advisable to specify tile hanging over three storeys high, because of possible dislodgement by wind conditions, unless special arrangements were to be made for fixing and clipping. Mathematical tiles are more robust, and they may be available with integral shaped thermal insulation boards, but evidence would be needed of the security and longevity of attachment arrangements.

The techniques most commonly found will be similar to those used in roofing, with tiles and slates nailed to horizontal battens over a vapour permeable membrane, except that the laps may be much reduced in size.

Abutments, external and re-entrant corners
Tiling
Special corner tiles, finished on the edge as well as the face, undoubtedly make for a more satisfactory corner than trying to use a standard tile.

Tile and slate hanging
External corners in tiles are best formed by special corner units. This of course is not possible in slating, and virtually the only possibility is to

Figure 9.3
Adhered ceramic tiling on the then newly built BRE Low Energy Building

provide lapped undercloaks which are wrapped round the corner underneath the slates. Radiusing the corner is possible if very narrow slates are used, but this places an increased premium on the integrity of the underlay and battens.

It is frequently the abutment of tile and slate hanging against a brick or other return which is most vulnerable to rain penetration, with rain being driven sideways across the face of the relatively impervious materials. The end joint is commonly flashed or provided with soakers, but a better solution is to create a secret gutter, similar to those used in sloping roofs, in which the flashing is concealed underneath the overlapping units. There is little prospect of driving rain being blown over the vertical batten where the joint is protected by the re-entrant corner.

Main performance requirements and defects

Choice of materials for structure
Tiling
Ceramic tiles butt jointed and laid and pointed in cement or proprietary mortars form the most usual construction. The mortars are often compounded with rubber and plastics additives to reduce shear modulus and improve resistance to differential movement.

Tile and slate hanging
Roofing tiles of fired clay or of concrete, and natural and artificial slates nailed to timber battens are the most usual form of construction, though proprietary metal battens may sometimes be encountered. The principle of operation is the same as with roofing, that one tile or slate covers the joint in the next course below, though a single lap rather than the double lap common in roofing is more usual. Very occasionally, wood shingles may be found on vertical or near vertical surfaces.

For a description of the properties of the various materials, see *Roofs and roofing*.

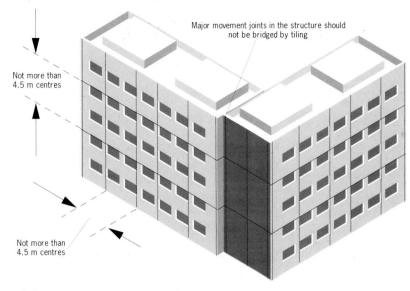

Figure 9.4
Movement joint requirements for external tiling

Strength and stability
Impact damage, particularly at low level, is crucial to the long term performance of all kinds of thin tiles and slates; however they may be fixed. For this reason, as noted already, they ought not to be used on the ground storey of buildings.

Dimensional stability, deflections etc
Since tile and slate hanging is comprised of overlapping small units, the thermal expansion of the material is taken up in the joints.

For coefficients of linear thermal and moisture expansion of tiles and other background materials, see Chapter 1.2.

Tiled walls
Loss of adhesion of tiles due to differential movement of the background and the tile has been the unfortunate consequence of inadequate specification on too many otherwise fine buildings. As with other fired clay products, ceramic tiles expand after manufacture and may continue to expand, though at a reducing rate, for some years afterwards. Crazing of any glaze can indicate this kind of expansion. Moisture-induced irreversible initial expansion, coupled with temperature and moisture-induced changes in the wall to which the tiles are applied,

can impose compressive loads in the plane of the tiling, leading to cracks and displacement. Movement joints not less than 6 mm wide are recommended to be incorporated at not more than 4.5 m centres, both horizontally and vertically, and at vertical external and re-entrant corners in large areas (Figure 9.4); major movement joints in the wall should not be bridged by the tiling. Naturally, it should be possible to match the colour of the ordinary joint to make the appearance of the movement joints acceptable.

Site practices that are most likely to lead to cracking or detachment of tiling are:
● failure to incorporate soft joints at adequate intervals
● the incorporation of soft joints of insufficient width
● the installation of tiling across movement joints in the wall
● failure to allow adequate time for substrate shrinkage before tiling: at least 14 days should elapse before tiling on render

Recommendations for other aspects of wall tiling can be found in BS 5385-2[228].

Tile and slate hanging
Dimensional stability is relatively unimportant for hung tiles and slates, since the laps take care of any such movements.

Figure 9.5
Tile hanging in progress on a housing site. Unfortunately the 'breather' membrane is not a breather membrane at all, but is a polyethylene sheet. This forms a vapour control layer on the cold side of the thermal insulation, and there is therefore a risk of condensation inside the polyethylene. There should be toe boards on the scaffolding

Weathertightness, dampness and condensation
Tiling
As with any comparatively impervious finish which is prone to cracking, the weathertightness of the wall will depend on the avoidance of cracks which allow rain water run-off to penetrate behind the finish. Once behind, it may not easily find a way out, and frost action can then occur.

Tile and slate hanging
Provided the tiling or slating is constructed without obvious faults, the wall should be weathertight for all exposures of driving rain, although wind damage may lift tiles or slates, as with roofing. A suitable breather membrane (Figure 9.5) to BS 4016[229] is important. See also *Roofs and roofing.*

Fire
Ceramic and concrete tiles, and natural and artificial slates, are all suitable for use adjacent to boundaries.

Durability
Tiling
Provided the tiles are suitable for use externally, durability depends for the most part on the provision of suitable movement joints (see the case study opposite) to overcome any differential movements between tile and substrate, and on the integrity of the adhesive[230] and grouting.

Tile and slate hanging
Tile or slate hanging may be found with corroded nailing and there may be rot in any supporting battens. (For the durability of the slates and tiles themselves, see *Roofs and roofing.*) The fact that tiles are laid almost vertically does not make them immune to biological growths. While in many cases growths may be thought desirable, removal by application of surface biocides will often enhance the durability of the material.

Man-made slates are available in a very wide range of materials – for example fibre reinforced cements and fillers such as natural slate bound with resins – and it is therefore not possible to give universally applicable guidance on the durability to be expected. Although shorter lifespans can be expected than are

normally achieved from the best natural slates, premature deterioration can happen with all kinds of slating materials.

Stability of colour may be a problem, although it is not thought to be widespread; distortion and cracking are more common but not in all products. Some slates may lose their surface coating but this is relatively rare in BRE's experience.

Work on site

Storage and handling of materials
See the relevant sections in *Roofs and roofing.*

Workmanship
Tiling
Adequacy of the bedding and grouting is of prime importance (Figure 9.7), together with providing sufficient movement joints.

Tile and slate hanging
As with roofs, in replacement work it could be advantageous to use nails of aluminium, austenitic stainless steel, copper or silicon bronze (BS 1202-2[231] and 1202-3[232]). Nails of steel (whether galvanised or not), aluminium or copper alloys should not be used where battens have been treated with CCA (copper-chromium-arsenic) preservatives.

External tiling failure on a health service building

The BRE Advisory Service was invited to establish the reasons for an external tiling failure. The building was concrete framed with blockwork or brickwork infill panels. The external wall tiling had been fixed to a render backing with a thin bed adhesive. Cracking, loss of adhesion and deterioration of the tiling had occurred some 15 years after the building had been completed.

The external tiling had cracked and the render coat had bulged away from the background (Figure 9.6) at the south-west corner. Away from the corner, tiles mainly were intact and still adhered to the background. The first movement joint was 7.5 m from the corner and then at 6.75 m spacing (the column positions).

Figure 9.6
The render coat bulging away from the background (right), loss of adhesion (centre) and cracking (left)

Tiles on a parapet had also become detached. The top of the parapet wall was constructed with a tiled concrete coping bedded on to common brickwork with no DPC, therefore any rain penetrating through the coping tiles could allow repeated wetting and drying of the brickwork and render backing. The internal face of the parapet wall was rendered. The horizontal tiles to the coping had cracked at positions corresponding to the joints in the concrete coping, and there was some frost damage.

The render backing to the tiles taken from the area where the tiling had become detached was analysed for sulfate content. The amount of sulfate, determined as the percentage of sulfur trioxide, was 3.5%. In a mix of 1:1:6 cement:lime:sand; approximately 0.3% sulfate would normally be expected if ordinary Portland cement was used in the mix. The sample was also analysed using differential thermal analysis, which showed there was a significant amount of ettringite present. The common bricks used as the base for the render backing to the tiling contained sufficient soluble salts for the reaction to occur.

The major cause of the failed tiling was sulfate attack in the brickwork mortar and render backing to the tiling. There was considerable evidence that rain had penetrated through the tiling and entered the render and brickwork background; this was shown by the stalactites on the soffit to the walls at ground level and salt deposits and staining of the concrete columns. The external wall tiling had not formed an impervious cladding to the building and rain ingress through the tile joints had occurred.

Some of the cracking in the tiling could be attributed to insufficient movement joints, particularly at junctions with dissimilar materials (ie between the brickwork and concrete columns).

It was doubtful if it would be possible to repair or replace the existing tiling. It was possible the building could be overclad with, for example, profiled sheeting, glass reinforced plastics panels, or weather boarding. In designing the overcladding it would be important that care be taken with the detailing at junctions to openings, parapet walls etc to ensure the water would be excluded from the existing walling. It might be possible to fix the overcladding to the existing tiling and thus avoid the noise and disruption in removing it.

Figure 9.7
Replacement of detached tiles in progress. Spacers are in place; a vertical movement joint can be seen above the little finger

Chapter 9.2   **Rendering**

This chapter deals with daub, renders, stucco, pargetting and other dashed finishes including roughcast and harling.

Renders have been used for many years on various types of external walls, either for reasons of weathertightness or of appearance, or both. Decorative renders and pargetting began to be widely used in the late sixteenth century, but, before that time, plain but whitewashed lime:sand mixes were normal. With appropriate choice of mixes and finishes, external renderings should provide good service for many years, but their longevity is very dependent on the quality of the background as well as the quality of materials and workmanship (Figure 9.8).

Around 1 in 5 of all dwellings in England have rendered finishes as the predominant wall finish[2]. The practice of rendering walls has been even more common in Scotland, where nearly three quarters of all dwellings are rendered[4].

Stucco has been a favoured external finish to unfaced masonry for several centuries. The Romans certainly used it, and Vitruvius writing a few years before the Birth of Christ referred to the technique of slaking quick lime to be used in stucco. After the fall of Rome the use of stucco declined until it was revived in Italy in the fifteenth century. In the UK in 1538, Nonsuch Palace was decorated in stucco by master plasterers from Italy. The recipes in common use varied widely, though were basically of lime as the cementing agent, with additions of materials mostly of organic origin such as dung and dairy products[13]. The Adam brothers were said to have used a 1:1:4 mixture of lime, bone ash and clean sand. Stucco was normally finished with an oil based paint.

The ideal rendering prevents the penetration of water, is free from cracks, well keyed to the wall surface and has a pleasing and durable finish. Generally, renders perform well, but it is important to ensure that they are carefully specified, properly mixed and skilfully applied, and that wall surfaces are adequately prepared.

Observations have shown that pebble dash, roughcast and harling are least likely to change in appearance over long periods, though there may be local problems. By using coloured cements in the final coat and 'dashes' made from crushed spar, pebbles and even shells, pebble dash can provide pleasing variety as well as improving the shedding of water driven onto the wall. Pebble dash does not become as dirt stained as roughcast, although the pebbles may become loosened by weathering. Both pebble dash and roughcast maintain their appearance longer than smooth finishes, which tend to suffer from surface crazing. In rural areas algae and lichens appear to thrive better on wet dash than dry-dash surfaces; perhaps providing a better medium for spores to germinate. Smooth finishes today usually give a suitable base for the external paints now available, but satisfactory behaviour of these paint finishes is closely related to the durability of the rendering itself.

Figure 9.8

Osborne House – a veritable tour de force of 'Roman cement' rendering

Figure 9.9
Adhesion failure on a rather rough looking brick substrate

Relief or incised decorative modelling in renders, sometimes in contrasting colours, may sometimes be encountered, particularly on half-timbered work. This technique, now rare, is called pargetting or parge-work. Specialist advice will be necessary on its conservation.

Characteristic details

Basic structure
As already noted in Chapter 2.6, daub used over wattles in early times essentially consisted of a mixture of clay and chalk, and possibly lime, reinforced with chopped straw or reed. These mixtures by themselves were not durable in rain or frost, but when covered with a coating of lime plaster became reasonably durable if the annual limewashing was attended to and the building had generous overhanging eaves. Any shrinkage gaps opening between timbers and the daub would tend to be closed by the dry clay swelling when wetted.

Since medieval times, changes in local practices with respect to mixes have developed only slowly. Until the invention of Portland cement, mixes were extensively lime based,

with some pozzolanas added to give a render which developed considerable strength and durability over time.

Practice on the continent of Europe has over the years arguably been well in advance of UK practice with specification and application of external renders. The relevant features of good practice are well known – the protection by a flashing of even the smallest projection, throwing the render onto the wall surface, ensuring a rough textured finished surface, using cement:lime:sand (not cement:sand) mixes, careful selection of well graded gritty sands, and using gauge boxes rather than the inaccurate and inconsistent batching 'by the shovel' still commonly practised today! Failure of renders continues to be a fairly frequent subject of requests for help from the BRE Advisory Service (Figure 9.9).

In Scotland, building materials, designs and traditional building practices differ somewhat from those used in other parts of the United Kingdom. The traditional Scottish rendering mix, used successfully on brickwork for many years, was a 1:3 cement:sand mix finished with a roughcast or harling in which the final coat, containing a proportion of fairly coarse aggregate, was thrown on as a wet mix and left as a rough textured finish[233].

Insulating renders
Insulating renders can combine a measure of insulation with a more or less acceptable rendered finish. These renders, incorporating lightweight aggregates such as polystyrene or expanded minerals (perlite) in a cement-based mix in thicknesses of 25–75 mm and without reinforcement, have had some application in low-rise housing in the UK since 1978 and in high-rise since 1983. They have had mixed results.

Abutments
Eaves and verges
Notwithstanding the effects of high wind loadings, the larger the overhang the better, particularly in conditions of severe exposure. Large

overhangs protect the top joint and avoid rainwater percolating behind the render (Figure 9.10). Conversely, great care needs to be taken with both specification and workmanship when using clipped eaves and verges where the top joint is not protected.

Window sills and heads
Hood moulds and window surrounds may be found constructed in brick and rendered over to simulate stone, and this practice dates at least from the mid–sixteenth century.

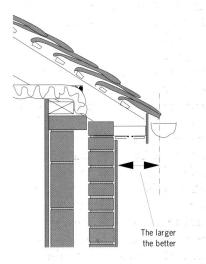
The larger the better

Figure 9.10
Eaves should offer protection to the head of the rendering

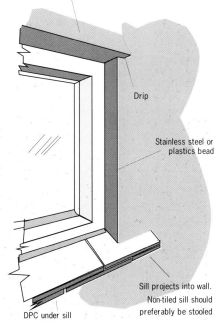
Heads can with advantage be bellcast.
Drip
Stainless steel or plastics bead
Sill projects into wall. Non-tiled sill should preferably be stooled
DPC under sill

Figure 9.11
Rendering around a window opening

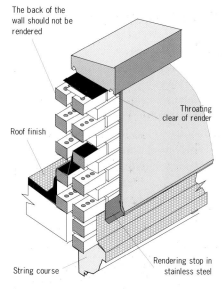

The back of the wall should not be rendered

Throating clear of render

Roof finish

String course

Rendering stop in stainless steel

Figure 9.12
Rendering parapets

Figure 9.13
Rendering below the DPC has failed and fallen away, perhaps understandably in view of the condition of the substrate!

Hoods, bellcasts and sills should overhang the remainder of the rendering by at least 50 mm with the throating clear of the surface. Sills should project beyond reveals (Figure 9.11 on page 249). Rendered sills are not recommended, especially those under large windows, as these tend to be continuously damp saturated by run-off from the window above.

Parapets and screen walls
Parapet walls to be rendered should be of cavity construction. Rendering should not be applied to the backs of walls which are rendered on the front (Figure 9.12).

Base of walls
Walls below DPC level will be wetter than those above, so specifications for rendering below the DPC must be chosen with care (Figure 9.13). Renderings below DPCs should be physically separated from those above to avoid salts carried by rising damp crystallising behind the rendering (Figure 9.14 on page 252).

External corners
External corners can be reinforced with expanded metal beading in stainless steel, or even plastics; if, though, differential movement has not been allowed for, it may just transfer the inevitable crack from the arris to the edge of the lathing.

Main performance requirements and defects

Choice of materials for structure

The types of background, their strength and condition, are crucial to successful rendering as much as to appropriate specification, including render mix (see the feature panel below), numbers of coats, and the use of additives and special surface treatments.

If the background is deficient in some respect, it may be possible to introduce a surface treatment, or overlay with an armature of some kind which allows a wider choice of solutions. For a background of varying materials, a mix suitable for the weakest material should be specified or the background covered with expanded metal lathing.

Masonry backgrounds to be rendered directly should be stronger than the rendering. Each subsequent coat should be not stronger than that preceding. Weak or eroded backgrounds may still be rendered, though they will need surface preparation (see the feature panel opposite).

Cements and limes
● Cements should comply with BS 12[234], BS 146[235] or BS 4027[236]
● Limes should comply with BS 890[237]

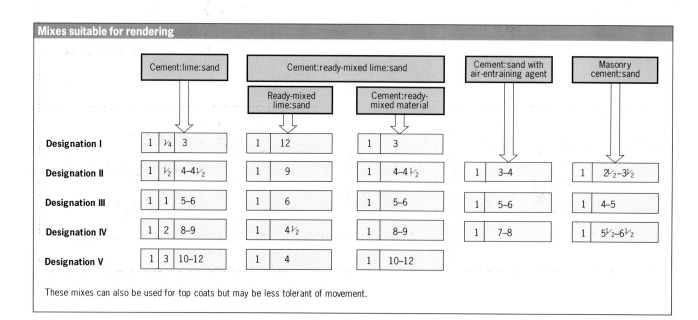

Mixes suitable for rendering

	Cement:lime:sand			Cement:ready-mixed lime:sand					Cement:sand with air-entraining agent		Masonry cement:sand	
				Ready-mixed lime:sand		Cement:ready-mixed material						
Designation I	1	¼	3	1	12	1	3						
Designation II	1	½	4–4½	1	9	1	4–4½		1	3–4	1	2½–3½
Designation III	1	1	5–6	1	6	1	5–6		1	5–6	1	4–5
Designation IV	1	2	8–9	1	4½	1	8–9		1	7–8	1	5½–6½
Designation V	1	3	10–12	1	4	1	10–12						

These mixes can also be used for top coats but may be less tolerant of movement.

Matching the rendering system to the background

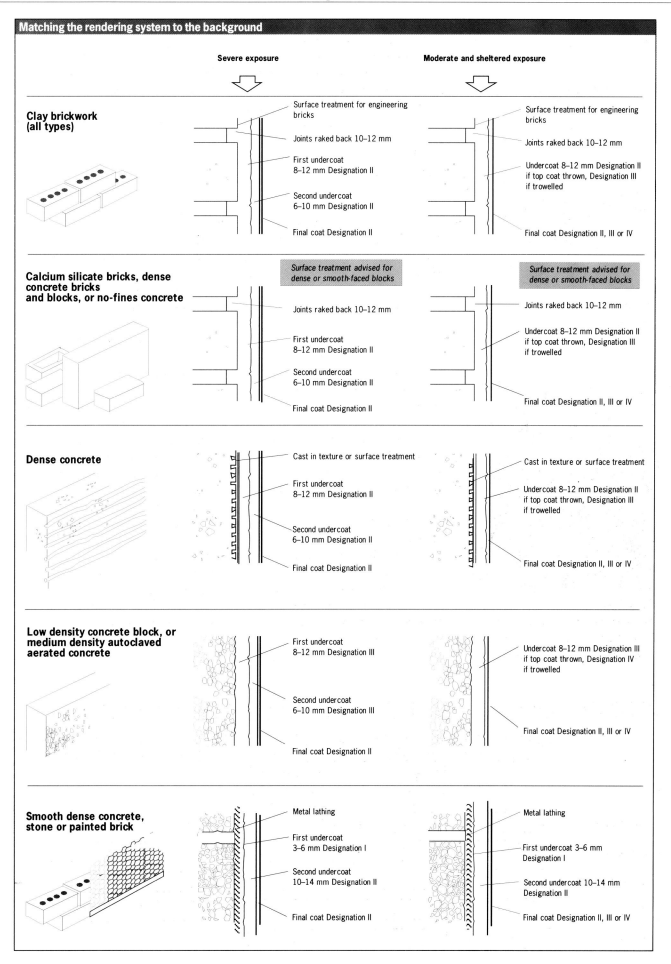

	Severe exposure	Moderate and sheltered exposure

Clay brickwork (all types)

Severe exposure:
- Surface treatment for engineering bricks
- Joints raked back 10–12 mm
- First undercoat 8–12 mm Designation II
- Second undercoat 6–10 mm Designation II
- Final coat Designation II

Moderate and sheltered exposure:
- Surface treatment for engineering bricks
- Joints raked back 10–12 mm
- Undercoat 8–12 mm Designation II if top coat thrown, Designation III if trowelled
- Final coat Designation II, III or IV

Calcium silicate bricks, dense concrete bricks and blocks, or no-fines concrete

Severe exposure:
- *Surface treatment advised for dense or smooth-faced blocks*
- Joints raked back 10–12 mm
- First undercoat 8–12 mm Designation II
- Second undercoat 6–10 mm Designation II
- Final coat Designation II

Moderate and sheltered exposure:
- *Surface treatment advised for dense or smooth-faced blocks*
- Joints raked back 10–12 mm
- Undercoat 8–12 mm Designation II if top coat thrown, Designation III if trowelled
- Final coat Designation II, III or IV

Dense concrete

Severe exposure:
- Cast in texture or surface treatment
- First undercoat 8–12 mm Designation II
- Second undercoat 6–10 mm Designation II
- Final coat Designation II

Moderate and sheltered exposure:
- Cast in texture or surface treatment
- Undercoat 8–12 mm Designation II if top coat thrown, Designation III if trowelled
- Final coat Designation II, III or IV

Low density concrete block, or medium density autoclaved aerated concrete

Severe exposure:
- First undercoat 8–12 mm Designation III
- Second undercoat 6–10 mm Designation III
- Final coat Designation II

Moderate and sheltered exposure:
- Undercoat 8–12 mm Designation III if top coat thrown, Designation IV if trowelled
- Final coat Designation II, III or IV

Smooth dense concrete, stone or painted brick

Severe exposure:
- Metal lathing
- First undercoat 3–6 mm Designation I
- Second undercoat 10–14 mm Designation II
- Final coat Designation II

Moderate and sheltered exposure:
- Metal lathing
- First undercoat 3–6 mm Designation I
- Second undercoat 10–14 mm Designation II
- Final coat Designation II, III or IV

Sands, aggregates[(238)] and pigments

- Sands should be well-graded; that is to say, having a good mix of coarse and fine particles, and ideally should comply with Clause 5 of BS 1199[(239)]
- Although sands not complying with the above can give good results, great care should be taken when contemplating the use of fine sands
- Finer sands should be used for final coats with tooled finishes
- Coarser sands should be used for final coats with scraped finishes

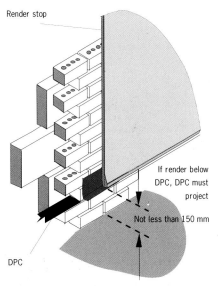

Render stop

If render below DPC, DPC must project

Not less than 150 mm

DPC

Figure 9.14
Criteria for rendering at the base of a wall

Figure 9.15
Movement joints have been provided on this newly rendered façade, and this should avoid cracking

- Coarse aggregates for dashed coats should comply with BS 882[(64)] or BS 63-1[(240)], or, perhaps more appropriately, BS 63-2[(241)]
- Pigments should comply with BS 1014[(242)]

Water

- Water of potable quality is suitable

Reinforcement

- Natural fibres used in specialist applications should be dry, clean and grease and oil free
- Mineral and polymer fibres should be specified only in accordance with manufacturers' recommendations

Admixtures

- Plasticisers should comply with BS 4887-1[(243)]
- Plasticisers should not be used with renderings made with masonry cement
- Waterproofing agents should be used in accordance with manufacturers' instructions
- Bond strength, resistance to rain penetration and durability are improved by polymer dispersions such as SBR (styrene-butadiene rubber) and acrylics, which should be used in accordance with manufacturers' instructions

Strength and stability

Impact resistance of the insulating renders is the main concern, and this should be checked before specification. Those renders incorporating polymers minimise the risks of cracking.

Dimensional stability, deflections etc

Loss of adhesion of renders due to differential movement of the background is frequent.

For coefficients of linear thermal and moisture expansion of rendering and other background materials see Chapter 1.2.

Map-pattern cracking is characteristic of shrinkage in the render itself. Cracks are often fine enough to be both obscured and sealed by the application of a masonry paint coating, but clearly-defined cracks – often the result of applying a strong finish coat over a weak backing coat – may well penetrate the full thickness of the render, with the risk of rain penetration and, if on clay brickwork, of sulfate attack.

Predominantly linear vertical cracks in a render are likely to indicate a change of substrate material at those locations. Since differential size changes in the substrate are likely to continue, these cracks should be raked out and filled with a suitable sealant.

Cracks due to sulfate attack and to corrosion of wall ties are readily distinguishable from those due to moisture-induced size changes. If movement joints are provided at centres appropriate to the movements expected of both render and substrate, cracking should be avoided (Figure 9.15).

Weathertightness, dampness and condensation

Although renderings are often used to enable weather resistance of a basic wall material to be increased, the more exposed the wall to be rendered, the more restricted becomes the choice of specifications available. These are the factors.

- Rendering is one method of increasing the exposure rating to wind driven rain of a wall by one or two categories, that is to say a wall which would otherwise be suitable only for sheltered exposures[20] can be upgraded to severe or very severe
- Exposure to marine or polluted environments may lead to attack on the cement content of renders and an increased rate of corrosion of metal lathing
- Rendered walls exposed to driving rain as well as pollution may streak differentially. Flint or calcareous dry dashes have the best self-cleaning properties, though some flints contain iron which can lead to staining

Renders undoubtedly reduce rain penetration into walls. The following points observed during experimental work by BRE will have a bearing on their effectiveness.

- Renderings with mix proportions of 1:1:6 and 1:½:4½ were very effective in reducing the passage of water into brick backgrounds, and the addition of a dry-dash finish further improved the performance. Cracking, or loss of dash by erosion, might significantly alter this protective property

- Rendering did not reduce the passage of water into aerated concrete backgrounds to the same extent as observed with clay brick backgrounds. Rendering does, however, help to prevent rain penetration through the joints
- Although evaporation rates were generally much lower than rates of absorption, there was no significant build-up of water within the materials
- Rainwater absorbed intermittently did not penetrate deeply before it was lost by evaporation

Thermal properties

Insulating renders are available to increase the thermal insulation value of solid walls, and at the same time provide increased resistance to deterioration (Figures 9.16 and 9.17, and Figure 9.18 on page 254). Precautions to be taken include the avoidance of water vapour transfer from inside the building condensing on the external insulation system by the careful selection of permeabilities or the provision of vapour control layers as appropriate[40].

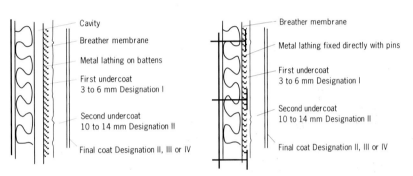

Figure 9.17
Two methods of applying thermal insulation

Figure 9.16
Applying external thermal insulation to solid walled houses

Stucco failure on heritage dwellings

During one particular winter, large areas of stucco became detached from a number of properties. From the information available, the weather had been of a higher than normal rainfall accompanied by freezing conditions. At this time some of the stucco became detached and fell to the ground. In the majority of cases the stucco had failed at the more exposed parts of the buildings (ie chimneys, parapets, mouldings, sills and porticos).

The BRE Advisory Service was asked to investigate. It was apparent also that the properties had suffered structural movements which had resulted in the cracking of walls, particularly around window openings, parapet abutments and porticos.

Subsequent investigations revealed that the remainder of the stucco was badly cracked and hollow. Cracks in the stucco had allowed the ingress of water into the brickwork and behind the render. The freezing of this water had produced the damage to the stucco, as well as deterioration to the brickwork.

The original designs of the chimneys, parapets, sills, mouldings and porticos did not provide adequate protection from the weather. The parapet walls should have been

protected by a coping that had a good overhang with a drip to throw the water clear of the top of the stucco. The brickwork and the stucco below the coping should have been protected by a damp proof course. The parapets had no protection, in fact the stucco was taken up over the top of the parapet and had cracked in a manner that was typical of shrinkage movements, sulfate attack and structural movements.

To avoid a recurrence of the frost damage to the stucco would require careful inspection. The defective areas should be hacked off and re-rendered while at the same time reconsidering the design of the exposed detailing. After completion of the repairs, inspections should be repeated at regular intervals.

It might be possible to effect temporary repairs to loose, hollow and cracked areas by injecting polymer resins or cementitious grouts, and protecting the upper faces of the cornices, mouldings and string courses with a proprietary flashing material; but the satisfactory performance, even of such an extensive repair programme, could not be guaranteed.

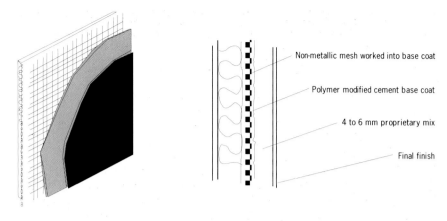

Figure 9.18
A third method of applying thermal insulation

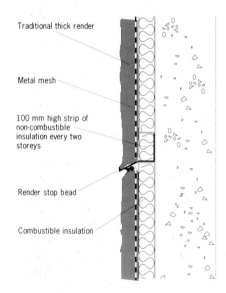

Figure 9.19
A rendered metal lathing system – if thermoplastic insulants are used, cavity fire barriers are needed every two storeys

Fire

Plastics materials used for fixing external rendered thermal insulation systems to walls are likely to melt or burn in a fire and this can lead to deformation of the cladding and extension of damage. Loaded fixings fail at much lower temperatures than unloaded fixings.

Experience of real fires is limited, but laboratory tests on multi-storey rigs have been carried out to assess the fire performance of systems incorporating combustible insulants fixed directly to a masonry wall and to compare these with that of timber cladding. From the tests it is possible to make the following general points.

- Metal reinforced cementitious render, 25 mm thick, provides effective protection to combustible insulation. The reinforcement should be independently supported
- Glass fibre reinforced thin renders perform reasonably well over non-combustible insulation, but offer little protection to combustible insulants
- Polymeric insulants protected only by unreinforced thin resin coatings should not be used where there is any risk of direct flame attack

There will be some risk in fire of parts of the cladding becoming detached and falling to the ground, particularly where plastics fixings play a significant role. Metal fixings are less likely to lose their integrity in fire. A proportion of metal fixings should therefore be included.

Insulating render systems of the types given above do not appear to cause an exposure risk to an adjoining building, and there is little risk of fire in overcladding entering a building through window openings.

Where there is no cavity, but combustible materials are used for insulation purposes, it is advisable to interrupt the sheets with a cavity barrier, particularly where the insulation is thermoplastic (eg polystyrene) (Figure 9.19). Cavity barriers are not required for systems where there is no cavity as such and non-combustible materials are used for insulation.

Recommendations to reduce fire spread are as follows:
- rendered metal lathing, thermoplastic insulant: sufficient metal fasteners to stabilise the cladding, and fire barriers every two storeys
- rendered metal lathing, thermosetting insulant: sufficient metal fixings to stabilise the cladding
- glass fibre fabric-reinforced thin renders, thermoplastic insulant: fire barriers, which also support the cladding, are required at every storey

Noise and sound insulation

Rendered walls depend for the most part on the mass of the background, and reference should be made to appropriate earlier chapters.

Durability

It has already been noted that few surfaces are truly self-cleaning under the action of rainwater, since streams of run-off will preferentially follow certain routes rather than others. Some rendered surfaces will tend to dirty unevenly, and may be prone to algal growth.

Field observations of renderings have revealed that some failures occur very early in the life of a new building, while others only become serious a year or more after the building is occupied. The early failures have mostly been shown to be due to unsatisfactory site practices and to the drying shrinkage characteristics of the rendering mixes, which could produce initial cracking. Some types of failure have been much more difficult to remedy and and have been due to a combination of faults.

Unsatisfactory storage of materials on site, inadequate protection of newly constructed walling against inclement weather, and defective workmanship all produce poor adhesion between the rendering and its background. The rough texture of brickwork provides a good key for the rendering undercoat, but newly completed brickwork which has been allowed to become rain-soaked has a film of water at the brick and rendering interface which impairs the key, particularly on walls where large areas depend on suction for their adhesion. On subsequent drying shrinkage, the rendering pulls away from the background, becomes hollow and may spall. There is evidence that on such wet backgrounds the use of a waterproofer in the undercoat **increases** the likelihood of water remaining at the interface. Well raked-out joints in the brickwork could provide some mechanical support. Difficulties experienced in rendering on background materials of very high or very low suction could have been eliminated if the European practice of providing a spatterdash coat had been more common.

Rendering may be cracked (Figure 9.20), unbonded, bulging, very soiled or with a deteriorated aggregate finish. Corrosion may have weakened any metal lathing, indeed on some Dorlonco houses, the lathing had virtually disappeared. Corner reinforcement beads may also have corroded.

The traditional construction of walls, rendered externally and well protected by roofs of moderate pitch, give few complaints, and examination of some quite old (40-year) cement-rich renderings have shown them to be relatively crack-free and resistant to moisture penetration. The rendering fills the capillaries in the mortar joints and the strong mixes were durable. Where poor quality clay bricks have been used for parapets and free-standing walls, these tend to show horizontal cracks typical of sulfate attack from soluble salts present in the bricks, but to initiate this, persistent dampness is essential and this should not occur in well designed and heated buildings.

Renderings may provide the only acceptable solution to problems of poor appearance. Contaminated backgrounds eg with soluble salts or oil, or combustion products should in general not be rendered direct, but should be covered with eml separated from the surface by a cavity on a breather membrane.

Figure 9.20
Cracked render, most probably due to shrinkage of a cement-rich mix

Features giving protection against weather

Renderings which are protected by features of the wall's surface, such as string courses, hood moulds and bell-mouths over window and other openings, will tend to be more durable than those on elevations without such surface modelling.

The main problem with insulating renders has been delamination either of the protective coat or within the lightweight render itself. Some systems have been used successfully for several years, whereas other systems have been withdrawn from the market following serious failures. They are not normally recommended for use directly onto walls subject to large amounts of movement unless great care is taken in their specification and completion; there is no completely satisfactory way of avoiding cracking over joints and the consequent risk of detachment. Only by supporting the render on a reinforced mesh and bridging the joints with a slip layer supported from the body of the panels, can these renders find application; for example on LPS dwellings.

The lightweight insulating renders do not perform well under impact tests, and should not be used in areas accessible to vandals.

Cracking of a spar dash render following cavity fill
Semi-detached bungalows had been built in an area of high exposure some 30 years prior to inspection by the BRE Advisory Service. The external walls were of cavity construction with spar dash rendered clay brick outer leaves. The cavities had been filled with a rock fibre insulation 10 years prior to the inspection. The pitched roofs were of traditional timber construction with concrete interlocking roof tiles. The gable end walls had the rendering taken to the undersides of the verge tiles.

The bungalows all showed cracks. As the cracked render was carefully removed, it was apparent the cracks continued in the mortar joints. The cracks in the mortar joints were either in the mortar itself or between the mortar and the bricks. The movement causing the cracking in the render had originated in the brickwork outer leaves. Bricks were removed to expose the cavity fill and further areas of render were cut away. It was not possible to find wall ties in the bed joints to either side of the exposed mortar beds.

Samples of the brickwork mortar were removed and analysed for sulfate content. The SO_3 content in the mortar was between 0.22% and 0.35%. These results showed that there was little evidence of excessive amounts of sulfate in the brickwork mortar.

In the opinion of the Advisory Officer the cracking in the brickwork and rendering was only slight. There was a strong possibility that the filling of the cavity with the insulation material could increase the thermal movements associated with the outer leaves. These movements could be concentrated along lines of weakness, that is at window openings or at DPC level. Although a definite diagnosis for the movements had not been possible, the external walls to the bungalows remained structurally sound. There was no evidence seen during the site inspection to suggest the walls were unstable and required strengthening or rebuilding.

The horizontal cracks should have been repaired in the following way: the rendering cut back on both sides of the cracks for at least 150 mm, and light expanded metal over building paper or polyethylene sheet fixed to the brickwork and embedded in the undercoats of the new rendering. This method might not entirely prevent further cracking but it would reduce its severity. The cracks at DPC level could be cut out to form permanent movement joints and filled with a suitable flexible mastic.

Cracking of insulating render

Some non-traditional houses had been thermally upgraded by the application of a thermally insulating render which had cracked. The BRE Advisory Service was invited to examine the render and advise on likely effects of the cracking on weathertightness and durability.

On examination it was observed that the majority of the cracks were less than 0.1 mm wide, with the maximum width 0.3 mm. A sample of the render was removed from the position of a crack, using a 100 mm diameter coring bit. The render separated at the wire of the steel cage (Figure 9.21). The crack, which continued through both coats of the render, had occurred at the thinnest part of the render. The remainder of the first coat was removed from behind the cage wires to expose the polystyrene insulation. It was apparent that the cages joined at this point.

Figure 9.21
Core taken through insulating render which had cracked. The crack is just visible above the core

The render was strong and hard, and it was concluded that the cracking had resulted from moisture movements with a contribution from thermal movements.

In the opinion of the Advisory Officer the render was durable, and the fine cracking had no significant effect on its performance.

Maintenance and repair

Detached areas of render will clearly need to be replaced, and defective render should be removed over the whole of an affected area.

Cracking is another matter. Inconspicuous cracking which is not permitting rain penetration is usually best left alone. With more noticeable cracks the decision to repair will depend largely on aesthetic

considerations and no two cases will call for the same treatment. However carefully it is undertaken, cutting and filling cracks up to 1 mm wide will produce a result which is more conspicuous than the original crack. Cracks in the 1–2 mm range could be injected using an epoxy resin pigmented to match the background, but the manufacturer's advice should be sought on a suitable material and procedure. For wider cracks, filling and re-rendering in accordance with BS 5262[244] may be suitable, but sometimes the only satisfactory solution may be to cut back the render and refinish. In very bad cases it may even be necessary to cut out and replace blocks on either side of the crack before refinishing. To ensure that colour variations will be masked, any area of new roughcast may have to extend sufficiently to line up with a natural break in the elevation, such as a rainwater down pipe or a return in the wall. Consideration should be given to forming a permanent movement joint to coincide with a vertical crack, particularly when cracks recur at the same location in a number of houses. Fine cracks left unfilled will function as movement joints unless they become blocked with debris.

Work on site

Restrictions due to weather conditions
- Rendering should not proceed when the air temperature is below 2 °C or when it is expected to fall below that figure in the next few hours
- Sands etc should not be frost-bound
- Rendering should not proceed when rainfall exceeds the lightest of showers – rendering applied to saturated walls will become detached
- Newly rendered walls should be protected from frost
- In sunny weather, rendering should be carried out in the shade wherever possible

- Protection of the rendering from sun and wind , or spraying with water, while curing, may be necessary in hot, dry weather

Workmanship
- Plain finishes should be wood floated; steel floated finishes should be avoided
- Scraped finishes should be made after the final coat has hardened for several hours

Preparation of backgrounds
Clay brickwork
- If new brick walls are built with recessed joints to provide a key for the render, normally no other modification will be necessary. Old walls, especially with strong mortar joints, will probably need spatterdash, stipple or lathing to provide a key
- If soluble salts are present (eg in type MN and FN bricks[91]), sulfate resisting cements should be used
- Brick types OL and ON should be covered with metal lathing

Blockwork and in situ dense concrete
- Most blocks with open textures will not need modification
- Provision of movement joints needs to be checked. If not provided, they need to be cut, and render stops provided
- Dense, smooth blocks will need spatterdash or stipple
- If dense concrete is cast against textured shuttering, no further modification should be necessary. An alternative is to use a retarder on the shuttering, and the surface then washed
- Hard concrete may need bush hammering, or alternatively a spatterdash or stipple coat
- No-fines concrete needs no further modification

Stone
- Preparatory treatment is needed if the stone is of low suction, has an eroded surface or a poor key

Sheet or board insulation
- All external insulation boards should have a reinforcement or lathing applied over the insulation

Timber frame

● It will be necessary to adapt the rendering method to suit the original form of construction (Figure 9.22)

Metal frame

● As for timber frame. It may be necessary to provide secondary framing members between the stanchions of the metal frame, and they should be of stainless steel (preferred) or galvanised steel. Lathing should be fixed with stainless steel wire ties of 1.2 mm diameter at 100 mm centres (Figure 9.23).

Control of suction

● Open textured surfaces may have such high suction in hot weather that the final water:cement ratio will be affected and the coat weakened
● The wall surface should be wetted but not soaked in dry weather to reduce suction
● Walls to be rendered must not have been subject to prolonged rain in the previous 48 hours

Mechanical keys

● All raked joints should be recessed 10–12 mm

Metal lathing

● Expanded metal lathing fixed in accordance with BS 8000-10[245] can be used over most surfaces
● For exposed conditions, austenitic stainless steel to BS 1449[246] Grade 304 will be suitable

Spatter or stipple coats

● Spatterdash is thrown on the wall surface whereas stipple is stiff brush applied, well scrubbed in before the stipple is raised
● The mix proportions for both are 1 part cement to 1–2 parts coarse sand, mixed to a thick cream
● A bonding agent is necessary for stipple, but optional for spatterdash
● Curing is by dampening over two days

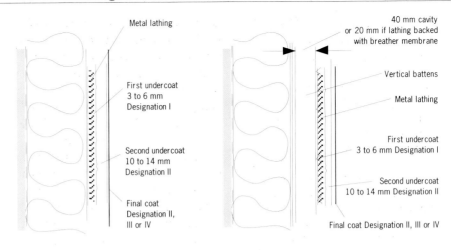

Rendering over cross-braced studs

Rendering over rigid sheathing

Figure 9.22
Rendering on timber framed building

Bonding agents

● To promote adhesion, SBR latexes, or acrylic polymers should be applied, mixed with cement or cement and sand, and well scrubbed into the surface
● Render should be always applied directly over the wet bonding agent

Where subsequent coats are to be hand applied, it will be necessary to provide a key. This should be done by combing or scratching the set surface of the rendering with a comb set with teeth about 20 mm apart. The scratch lines should be wavy in shape, and should penetrate half way through the coat.

Daywork and movement joints

● since daywork joints will normally show in most kinds of unpainted renders, consideration should be given to their siting (eg behind or alongside architectural features). Movement joints should be provided to coincide with any movement joints provided in the main structure
● In the case of rendering on metal lathing, the frequency of movement joints should be in no case more than bays of 5 m width

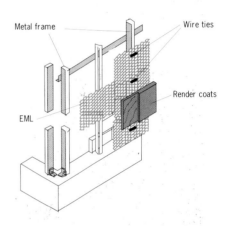

Figure 9.23
Rendering on metal framed building

Batching

● When rendering mixes are batched with dry sand (as opposed to damp sand), the mix tends to end up weaker than mixes batched with damp sand. Cement content should therefore be increased about 20% to compensate
● Where renderings are to be used on metal lath – and there is therefore no suction in the background, and coats are of the same nominal mix – the undercoat tends to be weaker than subsequent coats because of a higher water:cement ratio. More cement should be added to the undercoat mix (as above) or a slightly weaker second undercoat used

Repair of cracks

● Where cracks could be caused by size changes in the substrate, site practices should ensure that movement joints are carried through the thickness of the render with no hard material bridging a joint at any point. Otherwise, at changes of substrate material, site work should ensure that the bond between render and substrate is interrupted so that the render is fully isolated from the substrate at those locations (Figure 9.24). Non-corrodible reinforcement should be securely fixed to the substrate on both sides of the isolated area.

Inspection
The problems to look for are:
◊ detachment – incorrect mixes or poor surface preparation
◊ cracking
◊ peeling of finishes
◊ friable surfaces
◊ incorrect mixes
◊ incorrect lath fixing
◊ poor surface preparation – scratch coats inadequate
◊ lack of proper curing
◊ impact damage on thin renders over thermal insulation
◊ sulfate attack
◊ inadequacy of existing sill projections when adding insulating renders

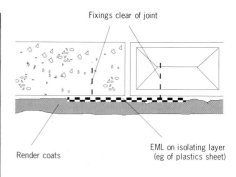

Fixings clear of joint

Render coats

EML on isolating layer (eg of plastics sheet)

Figure 9.24

The bond between render and substrate should be interrupted at changes of material

Inspection
The problems to look for in ceramic tiled walls are:
◊ tiles cracked or broken due to substrate movements or vandalism
◊ tiles loose due to adhesive failure
◊ sulfate attack in substrates
◊ tiles deteriorating, being not suitable for exposure of site
◊ no movement joints where required
The problems to look for in slated and tile hung walls are:
◊ slates or tiles broken, missing or delaminating
◊ tile nibs broken
◊ slates or tiles incorrectly fixed
◊ nails which have corroded
◊ flashings missing or damaged
◊ flashings too thin for durability
◊ snow or rain penetration

Scottish Office. *The Building Standards (Scotland) Regulations 1981–1987. Regulation: J3, Resistance to the transmission of heat on means to conserve energy – walls, floors and roofs.* Edinburgh, The Stationery Office, 1987

Department of the Environment for Northern Ireland. *The Building Regulations (Northern Ireland) 1990 Regulation: F3(1), Maximum U-value of walls, floors and roofs.* Belfast, The Stationery Office, 1990

British Standards Institution. Thermal insulating materials. Specification for bonded man-made mineral fibre slabs. *British Standard* BS 3958-5:1986. London, BSI, 1986

British Standards Institution. Specification for boards manufactured from expandable polystyrene beads. *British Standard* BS 3837-1:1986. London, BSI, 1986

British Standards Institution. Rigid polyurethane (PUR) and polyisocyanurate (PIR) foam for building applications – laminated board for general purposes. *British Standard* BS 4841-1:1975. London, BSI, 1975

British Standards Institution. Specification for rigid phenolic foam (PF) for thermal insulation in the form of slabs and profiled sections. *British Standard* BS 3927:1986. London, BSI, 1986

British Standards Institution. Specification for rigid expanded polyvinyl chloride for thermal insulation purposes and building application. *British Standard* BS 3869:1965. London, BSI, 1965

British Standards Institution. British Standard method of test for inorganic thermal insulating materials. *British Standard* BS 2972:1989. London, BSI, 1989

British Standards Institution. Methods for determining thermal insulating properties. *British Standard* BS 874-1 to 3:1986–90. London, BSI, 1986–90

British Standards Institution. Specification for urea-formaldehyde (UF) foam systems suitable for thermal insulation of cavity walls with masonry or concrete inner and outer leaves. *British Standard* BS 5617:1985. London, BSI, 1985

British Standards Institution. Code of practice for thermal insulation of cavity walls (with masonry or concrete inner and outer leaves) by filling with urea-formaldehyde (UF) foam systems. *British Standard* BS 5618: 1965. London, BSI, 1965

British Standards Institution. Thermal insulation of cavity walls by filling with blown man-made mineral fibre. *British Standard* BS 6232:1982. London, BSI, 1982

British Standards Institution. Thermal insulation of cavity walls using man-made mineral fibre batts (slabs). *British Standard* BS 6676:1986. London, BSI, 1986

British Board of Agrément. Index of current publications for list of certified cavity-wall insulation systems. Garston, British Board of Agrément

BRE. *Thermal insulation: avoiding risks.* BRE Report. Garston, Construction Research Communications Ltd, 1994

BRE. External walls: reducing the risk from interstitial condensation. *BRE Defect Action Sheet (Design)* DAS 6. Garston, Construction Research Communications Ltd, 1982

Southern J R. Summer condensation on vapour checks: tests with battened, internally insulated, solid walls. *BRE Information Paper* IP 12/88. Garston, Construction Research Communications Ltd, 1988

BRE. Minimising thermal bridging when upgrading existing housing. *Building Research Energy Conservation Support Unit Good Practice Guide* 183. Garston, BRECSU, 1996

Chapter 1.5
[43] **Chartered Institution of Building Services Engineers.** *CIBSE Guide,* Volume A. London, CIBSE, 1982

Further reading
Perera M D A E S and Tull R. BREFAN – a diagnostic tool to assess the envelope air leakages of large buildings. *Procs CIB W67 Symposium on Energy, Moisture and Climate in Buildings*

Chapter 1.6
[44] **Chartered Institution of Building Services Engineers.** *Window design manual.* London, CIBSE, 1987

[45] **BRE.** Estimating daylight in buildings, Part 2. *BRE Digest* 310. Garston, Construction Research Communications Ltd, 1986

[46] **Littlefair P J.** Average daylight factor: a simple basis for daylight design. *BRE Information Paper* IP 15/88. Garston, Construction Research Communications Ltd, 1988

[47] **British Standards Institution.** Specification for external cladding colours for building purposes. *British Standard* BS 4904: 1978. London, BSI, 1978

[48] **Littlefair P J.** Solar dazzle reflected from sloping glazed facades. *BRE Information Paper* IP 3/87. Garston, Construction Research Communications Ltd, 1987

Further reading
Littlefair P J. *Designing wih innovative daylighting.* BRE Report. Garston, Construction Research Communications Ltd, 1996

Chapter 1.7
[49] **BRE.** *Environmental design manual.* Garston, Construction Research Communications Ltd, 1988

[50] **BRE.** Sunlight availability protractor. Garston, Construction Research Communications Ltd, 1989

[51] **Anderson B R, Clark A J, Baldwin R and Milbank N O.** The BRE Domestic Energy Model. *BRE Information Paper* IP 16/85. Garston, Construction Research Communications Ltd, 1985

Further reading
Chartered Institution of Building Services Engineers. *Window design manual.* London, CIBSE, 1987

BRE. Estimating daylight in buildings, Part 2. *BRE Digest* 310. Garston, Construction Research Communications Ltd, 1986

Littlefair P J. Innovative daylighting systems. *BRE Information Paper* IP 22/89. Garston, Construction Research Communications Ltd, 1989

Chapter 1.8
[52] **Department of the Environment and The Welsh Office.** *The Building Regulations 1991 Approved Document B: Fire safety* (1992 edition). London, The Stationery Office, 1991

[53] **Scottish Office.** *The Building Standards (Scotland) Regulations 1981–87.* Edinburgh, The Stationery Office, 1987

[54] **Department of the Environment for Northern Ireland.** *The Building Regulations (Northern Ireland) 1994. Statutory Rules of Northern Ireland 1994 No 243.* Belfast, The Stationery Office, 1994

[55] **Hamilton S B.** A short history of structural fire protection of buildings, particularly in England. *National Building Studies Special Paper* 27. London, The Stationery Office, 1958

[56] **Loss Prevention Council.** *Code of practice for the construction of buildings. Insurers' rules for the fire protection of industrial and commercial buildings.* London, LPC, 1992

[57] **Fire Precautions Act 1971.** London, The Stationery Office, 1971

Further reading
Morris W A, Read R E H, Cooke G M E. *Guidelines for the construction of fire-resisting structural elements.* BRE Report. Garston, Construction Research Communications Ltd, 1988

British Standards Institution. Fire tests on building materials and structures. Test methods and criteria for the fire resistance of elements of building construction. *British Standard* BS 476-8:1972. London, BSI, 1972. (Now withdrawn but still referred to)

British Standards Institution. Fire tests on building materials and structures. Method for determination of the fire resistance of elements of construction (general principles). *British Standard* BS 476-20:1987. London, BSI, 1987

British Standards Institution. Fire tests on building materials and structures. Methods for determination of the fire resistance of loadbearing elements of construction. *British Standard* BS 476-21:1987. London, BSI, 1987

British Standards Institution. Fire tests on building materials and structures. Method for determination of the fire resistance of non-loadbearing elements of construction. *British Standard* BS 476-22:1987. London, BSI, 1987

British Standards Institution. Design of masonry structures. Eurocode 6, Parts 1.1 and 1.2. DD ENV 1996-1-1:1996 and DD ENV 1996-1-2:1997. London, BSI, 1996 and 1997

Rogowski R, Ramaprasad R and Southern J. *Fire performance of external thermal insulation for walls of multi-storey buildings.* BRE Report. Garston, Construction Research Communications Ltd, 1988

British Standards Institution. Code of practice for the protection of structures against lightning. *British Standard* BS 6651: 1985. London, BSI, 1985

BRE. Fire terminology. *BRE Digest* 225. Garston, Construction Research Communications Ltd, 1981

BRE. Toxic effects of fires. *BRE Digest* 300. Garston, Construction Research Communications Ltd, 1985

BRE. Human behaviour in fire. *BRE Digest* 388. Garston, Construction Research Communications Ltd, 1993

BRE. Protecting buildings against lightning. *BRE Digest* 428. Garston, Construction Research Communications Ltd, 1998

Chapter 1.9

[58] **Fothergill L C and Hargreaves N.** The effect of sound insulation between dwellings when windows are close to a separating wall. *Applied Acoustics* (1992), **35** (3) 253–61

[59] **BRE.** Sound insulation: basic principles. *BRE Digest* 337. Garston, Construction Research Communications Ltd, 1988

[60] **BRE.** The acoustics of rooms for speech *BRE Digest* 192. Garston, Construction Research Communications Ltd, 1976

Further reading

BRE. Improving the sound insulation of separating walls and floors. *BRE Digest* 293. Garston, Construction Research Communications Ltd, 1985

BRE. *Sound control for homes.* BRE Report. Garston, Construction Research Communications Ltd, 1993

BRE. Traffic noise and overheating in offices. *BRE Digest* 162. Garston, Construction Research Communications Ltd, 1979

BRE. Sound insulation of separating walls and floors. Part 1: walls. *BRE Digest* 333. Garston, Construction Research Communications Ltd, 1988

Chapter 1.10

[61] **Davis K and Tomasin K.** *Construction safety handbook.* London, Thomas Telford Services, 1990

[62] **Health and Safety Executive.** *Designing for health and safety in construction.* Sheffield, HSE

[63] **British Standards Institution.** Glazing for buildings. Code of practice for safety. Human impact. *British Standard* BS 6262-4: 1994. London, BSI, 1994

[64] **British Standards Institution.** Specification for aggregates from natural sources for concrete. *British Standard* BS 882:1992. London, BSI, 1992

[65] **British Standards Institution.** Specification for lightweight aggregates for masonry units and structural concrete. *British Standard* BS 3797:1990. London, BSI, 1990

[66] **British Standards Institution.** Specification for air-cooled bast furnace slag for use in construction. *British Standard* BS 1047:1983. London, BSI, 1983

[67] **Advisory Committee on Asbestos.** *Final Report.* London, The Stationery Office, 1979

[68] **Curwell S R and March C G.** Hazardous building materials – a guide to the selection of alternatives. London, Spon, 1986

[69] **British Standards Institution.** Guide for security of buildings against crime. *British Standard* BS 8220-1 to 3. London, BSI, 1986–90

[70] **British Standards Institution.** Bullet-resistant glazing. Specification for glazing for interior use. *British Standard* BS 5051-1: 1988. London, BSI, 1988(1994)

[71] **British Standards Institution.** Specification for anti-bandit glazing (glazing resistant to manual attack). *British Standard* BS 5544:1978. London, BSI, 1978(1994)

[72] **British Standards Institution.** Specification for impact performance requirements for flat safety glass and safety plastics for use in buildings. *British Standard* BS 6206:1981. London, BSI, 1981

[73] **Bromley A K R.** Electromagnetic screening of offices for protection against electrical interference and security against electronic eavesdropping. *Procs of The CIBSE 1991 National Conference, University of Kent, Canterbury, 7–9 April 1991*

Further reading

BRE. Lists of excluded materials: a change in practice. *BRE Digest* 425. Garston, Construction Research Communications Ltd, 1997

Health and Safety Executive. *Alternatives to asbestos products: a review.* London, The Stationery Office, 1986

Construction Industry Research and Information Association. *A guide to the safe use of chemicals in construction.* London, CIRIA, 1981

British Standards Institution. Code of practice for glazing for buildings. *British Standard* BS 6262:1982. London, BSI, 1982

Department of Trade and Industry, Consumer Safety Unit. *Home and leisure accidents.* London, DTI, 1990

Cox S J and O'Sullivan E F. *Building regulation and safety.* BRE Report. Garston, Construction Research Communications Ltd, 1994

Chapter 1.11

[74] **British Standards Institution.** Guide to durability of buildings and building elements, products and components. *British Standard* BS 7543:1992. London, BSI, 1992

[75] **BRE.** Bird, bee and plant damage to buildings. *BRE Digest* 418. Garston, Construction Research Communications Ltd, 1996

[76] **BRE.** Reducing the risk of pest infestation in buildings. *BRE Digest* 415. Garston, Construction Research Communications Ltd, 1996

[77] **Bravery A F, Berry R W, Carey J K and Cooper D E.** *Recognising wood rot and insect damage in buildings.* BRE Report. Garston, Construction Research Communications Ltd, 1992

[78] **Berry R W.** *Remedial treatment of wood rot and insect attack in buildings.* BRE Report. Garston, Construction Research Communications, 1994

[79] **The Institution of Structural Engineers.** *Appraisal of existing structures* (2nd edition). London, SETO Ltd, 1996

[80] **British Standards Institution.** Windows and rooflights. Durability and maintenance. *British Standard* CP153-2: 1970. London, BSI, 1970

[81] **BRE.** Cleaning external surfaces of buildings. *BRE Digest* 280. Garston, Construction Research Communications Ltd, 1983

Further reading

Ministry of Agriculture, Fisheries and Food. United Kingdom atmospheric corrosivity values. London, MAFF. (Available from MAFF, Cartographic Branch, Lion House, Willowburn Estate, Alnwick, Northumberland NE66 2PF)

BRE. Zinc coated steel. *BRE Digest* 305. Garston, Construction Research Communications, 1986

British Standards Institution. Commentary on corrosion at bimetallic contacts and its alleviation. *British Standard* PD 6484:1979. London, BSI, 1979

Lea R G. Cockroach infestation of dwellings in the UK. *BRE Information Paper* IP 1/95. Garston, Construction Research Communications Ltd, 1995

Chapter 1.12

[82] **Garvin S L.** *Domestic automatic doors and windows for use by elderly and disabled people.* BRE Report. Garston, Construction Research Communications Ltd, 1997

Further reading

British Standards Institution. Windows and rooflights. Durability and maintenance. *British Standard* CP153-2:1970. London, BSI, 1970

British Standards Institution. Guide to specifying performance requirements for hinged or pivoted doors (including test methods). *British Standard* DD 171:1987. London, BSI, 1987

British Standards Institution. Performance of windows. Specification for operation and strength characteristics. *British Standard* BS 6375-2:1987. London, BSI, 1987

Covington S A. Ergonomic requirements for windows and doors. *BRE Information Paper* IP 2/82. Garston, Construction Research Communications Ltd, 1982

Chapter 2.1

[83] **Bonnell D G and Butterworth B.** *Clay building bricks of the United Kingdom.* London, The Stationery Office, 1950

[84] **Stratton M.** *The nature of terracotta and faience, in architectural ceramics: their history, manufacture and conservation.* London, James and James, 1996

[85] **British Standards Institution.** Specification for masonry cement. *British Standard* BS 5224:1995. London, BSI, 1995

[86] **British Standards Institution.** Masonry cement. Test methods. *British Standard* BS EN 413-2:1995. London, BSI, 1995

[87] **BRE.** Connecting walls and floors: Parts 1 and 2. *BRE Good Building Guide* GBG 29. Garston, Construction Research Communications Ltd, 1997

[88] **BRE.** *Results of fire resistance tests on elements of building construction.* Volumes 1 and 2. BRE Report. Garston, BRE, 1975 and 1976

[89] **Hopkinson J S.** Fire spread in buildings. *BRE Information Paper* IP 21/84. Garston, Construction Research Communications Ltd, 1984

[90] **McIntyre W A.** Investigations into the durability of architectural terra-cotta and faience, *Department of Scientific and Industrial Research Special Report* No 12. London, The Stationery Office, 1929

[91] **British Standards Institution.** Specification for clay bricks. *British Standard* BS 3921:1985. London, BSI, 1985

[92] **Crammond N J and Dunster A M.** *Avoiding deterioration of cement-based building materials. Lessons from case studies: 1.* BRE Report. Garston, Construction Research Communications Ltd, 1997

[93] **BRE.** Temporary support: assessing loads above openings in external walls. *BRE Good Building Guide* GBG 10. Garston, Construction Research Communications Ltd, 1992

[94] **BRE.** Providing temporary support during work on openings in external walls. *BRE Good Building Guide* GBG15. Garston, Construction Research Communications Ltd, 1992

[95] **BRE.** Safety of large masonry walls. *BRE Digest* 281. Garston, Construction Research Communications Ltd, 1984

Further reading

Harrison W H. Sticking together. *Brick Bulletin* (The Brick Development Association, Windsor), Winter 1996

Handiside C C and Haseltine B A. *Bricks and brickwork.* Windsor, The Brick Development Association, 1982

Newman A J. *Rain penetration through masonry walls: diagnosis and remedial measures.* BRE Report. Garston, Construction Research Communications Ltd, 1988

Arora S K. Design of masonry walls subjected to concentrated vertical loads. *BRE Information Paper* IP 10/92. Garston, Construction Research Communications Ltd, 1992

BRE. Strength of brickwork and blockwork walls: design for vertical load. *BRE Digest* 246. Garston, Construction Research Communications Ltd, 1981

BRE. Masonry and concrete structures: measuring in-situ stress and elasticity using flat jacks. *BRE Digest* 409. Garston, Construction Research Communications Ltd, 1995

BRE. Brickwork: prevention of sulfate attack. *BRE Defect Action Sheet (Design)* DAS 113. Garston, Construction Research Communications Ltd, 1987

BRE. External masonry walls: assessing whether cracks indicate progressive movement. *BRE Defect Action Sheet (Design)* DAS 102. Garston, Construction Research Communications Ltd, 1987

BRE. External walls: joints with windows and doors – application of sealants. *BRE Defect Action Sheet (Site)* DAS 69. Garston, Construction Research Communications Ltd, 1985

BRE. External walls: joints with windows and doors – detailing for sealants. *BRE Defect Action Sheet (Design)* DAS 68. Garston, Construction Research Communications Ltd, 1985

BRE. Suspended timber floors: joist hangers in masonry walls – installation. *BRE Defect Action Sheet (Site)* DAS 58. Garston, Construction Research Communications Ltd, 1984

BRE. Suspended timber floors: joist hangers in masonry walls – specification. *BRE Defect Action Sheet (Design)* DAS 57. Garston, Construction Research Communications Ltd, 1984

BRE. External masonry walls: vertical joints for thermal and moisture movements. *BRE Defect Action Sheet (Design)* DAS 18. Garston, Construction Research Communications Ltd, 1983

BRE. Substructure: DPCs and DPMs – specification. *BRE Defect Action Sheet (Design)* DAS 35. Garston, Construction Research Communications Ltd, 1983

European Union of Agrément (UEAtc). The assessment of damp proof course systems for existing buildings. *Method of Assessment and Test* No 39. Garston, British Board of Agrément, 1988

Beech J C. The selection and performance of sealants. *BRE Information Paper* IP 25/81. Garston, Construction Research Communications Ltd, 1981

BRE. Building mortar. *BRE Digest* 362. Garston, Construction Research Communications Ltd, 1991

BRE. Diagnosing the causes of dampness. *BRE Good Repair Guide* GRG 5. Garston, Construction Research Communications Ltd, 1997

BRE. Treating rising damp in houses. *BRE Good Repair Guide* GRG 6. Garston, Construction Research Communications Ltd, 1997

BRE. Treating condensation in houses. *BRE Good Repair Guide* GRG 7. Garston, Construction Research Communications Ltd, 1997

BRE. Treating rain penetration in houses. *BRE Good Repair Guide* GRG 8. Garston, Construction Research Communications Ltd, 1997

Harrison W H. Conditions for sulfate attack on brickwork. *Chemistry and Industry,* 19 September 1981

BRE. Cleaning external surfaces of buildings. *BRE Digest* 280. Garston, Construction Research Communications Ltd, 1983

Chapter 2.2

[96] British Standards Institution. Specification for cast stone. *British Standard* BS 1217:1986. London, BSI, 1986

[97] Templeton W, Edgell G J and de Vekey R C. The robustness of the domestic house. Part 4: accidental damage. *Procs of the British Masonry Society, April 1988*, **2**, 127–30

[98] Templeton W, Edgell G J and de Vekey R C. The robustness of the domestic house. Part 3: positive wind pressure test on gable wall. *Procs of the British Masonry Society, April 1988*, **2**, 121–26

[99] Bonshor R B and Bonshor L L. *Cracking in buildings.* BRE Report. Garston, Construction Research Communications Ltd, 1995

[100] BRE. Testing bond strength of masonry. *BRE Digest* 360 Garston, Construction Research Communications Ltd, 1991

[101] BRE. External masonry walls: vertical joints for thermal and moisture movements. *BRE Defect Action Sheet (Design)* DAS 18. Garston, Construction Research Communications Ltd, 1983

[102] British Standards Institution. Specification for urea-formaldehyde (UF) foam systems suitable for thermal insulation of cavity walls with masonry or concrete inner or outer leaves. *British Standard* BS 5617: 1985. London, BSI, 1985

[103] British Standards Institution. Code of practice for thermal insulation of cavity walls (with masonry or concrete inner and outer leaves) by filling with urea-formaldehyde (UF) foam systems. *British Standard* BS 5618:1985. London, BSI, 1985

[104] British Standards Institution. Thermal insulation of cavity walls by filling with blown man-made mineral fibre. *British Standard* BS 6232:1982. London, BSI, 1982

[105] Rothwell G W. Effects of external coatings on moisture contents of autoclaved aerated concrete walls. Autoclaved aerated concrete, moisture and properties. *Procs of RILEM Symposium, Lausanne, March 1982*, 101–16

[106] BRE. Fire risk from combustible cavity insulation. *BRE Digest* 294 Garston, Construction Research Communications Ltd, 1985

[107] Bromley A and Pettifer K. *Sulfide related degradation of concrete in south west England* (the Mundic problem). BRE Report. Garston, Construction Research Communications Ltd, 1997

[108] de Vekey R C, Tarr K and Worthy M. Workmanship and the performance of wall ties: effect of depth of embedment. *Masonry International* (1988), **2** (2) 43–6

[109] de Vekey R C. Corrosion of steel wall ties: recognition and inspection. *BRE Information Paper* IP 13/90. Garston, Construction Research Communications Ltd, 1990

[110] BRE. Installing wall ties in existing construction. *BRE Digest* 329. Garston, Construction Research Communications Ltd, 1988

[111] BRE. Replacing wall ties. *BRE Digest* 401. Garston, Construction Research Communications Ltd, 1995

[112] Newman A J. Workmanship, rain penetration and cavity wall insulation. *Procs of British Masonry Soc Masonry (3): Workmanship in Masonry Construction, 1989*, 58–61

Further reading

de Vekey R C. Cavity walls – still a good solution. *Procs of British Masonry Soc Masonry (5) 1993*, p35

Simms L G. Frog-up and frog-down brickwork compared. *The Builder* (1956), **191** (5917) 329–31

BRE. External masonry cavity walls: wall tie replacement. *BRE Defect Action Sheet (Design)* DAS 21. Garston, Construction Research Communications Ltd, 1983

Energy Efficiency Office. Cavity wall insulation in existing dwellings: polystyrene bead insulation. *Building Research Energy Conservation Support Unit (BRECSU) Good Practice Case Study* 64. London, EEO, 1993

Energy Efficiency Office. Cavity wall insulation in existing housing. *Building Research Energy Conservation Support Unit (BRECSU) Good Practice Guide* 26. London, EEO, 1993

de Vekey R C. Ties for cavity walls and masonry cladding. *Structural Survey* (1990), **8** (4) 384–92

BRE. External masonry cavity walls: wall ties – installation. *BRE Defect Action Sheet (Site)* DAS 20. Garston, Construction Research Communications Ltd, 1983

British Board of Agrément, BRE, National Federation of Building Trades Employers, National House-Building Council (England and Wales). Cavity insulation of masonry walls – dampness risks and how to minimise them. Garston, BBA, 1983

BRE. Cavity trays in external cavity walls: preventing water penetration. *BRE Defect Action Sheet (Design)* DAS 12. Garston, Construction Research Communications Ltd, 1982

European Union of Agrément (UEAtc). The assessment of precast insulating concrete blocks for general use in building. *Method of Assessment and Test* No 12. Garston, British Board of Agrément, 1977

Pountney M T, Maxwell R and Butler A J. Rain penetration of cavity walls: report of a survey of properties in England and Wales. *BRE Information Paper* IP 2/88. Garston, Construction Research Communications Ltd, 1988

Cockram A H and Arnold P J. Urea-formaldehyde foam cavity wall insulation: reducing formaldehyde vapour in dwellings. *BRE Information Paper* IP 7/84. Garston, Construction Research Communications Ltd, 1984

Newman A J *Rain penetration through masonry walls: diagnosis and remedial measures.* BRE Report. Garston, Construction Research Communications Ltd, 1988

BRE. Strength of brickwork and blockwork walls: design for vertical load. *BRE Digest* 246. Garston, Construction Research Communications Ltd, 1981

BRE. Safety of large masonry walls. *BRE Digest* 281. Garston, Construction Research Communications Ltd, 1984

BRE. Cavity insulation. *BRE Digest* 236. Garston, Construction Research Communications Ltd, 1984

BRE. Choosing between cavity, internal and external wall insulation *BRE Good Building Guide* GBG 5. Garston, Construction Research Communications Ltd, 1990

BRE. Repairing brick and block masonry. *BRE Digest* 359. Garston, Construction Research Communications Ltd, 1991

Harrison W H. Avoiding latent mortar defects in masonry. *BRE Information Paper* IP 10/93. Garston, Construction Research Communications Ltd, 1993

BRE. Calcium silicate (sandlime, flintlime) brickwork. *BRE Digest* 157. Garston, Construction Research Communications Ltd, 1981

BRE. Perforated clay bricks. *BRE Digest* 273. Garston, Construction Research Communications Ltd, 1983

BRE. Building mortar. *BRE Digest* 362. Garston, Construction Research Communications Ltd, 1991

Southern J R. Bremortest: a rapid method of testing fresh mortars for cement content. *BRE Information Paper* IP 8/89. Garston, Construction Research Communications Ltd, 1989

Fletcher K E. The conformance of masonry mortars and their constituent sands with British Standards. *BRE Information Paper* IP 10/85. Garston, Construction Research Communications Ltd, 1985

Stupart A W and Skandarmoorthy J S. Mortars for blockwork: improved thermal performance. *BRE Information Paper* IP 2/98. Garston, Construction Research Communications Ltd, 1998

British Ceramic Research Association.
Model specifications for clay and calcium
silicate structural brickwork. *Special
Publication* SP56. Stoke-on-Trent, BCRA,
1990
de Vekey R C and Reed W J. The structural
performance of cavity wall ties. *Procs of
British Masonry Soc M(I) 8th BCS, 1986*, pl22
de Vekey R C. Ties for cavity walls: new
developments. *BRE Information Paper*
IP 16/88. Garston, Construction Research
Communications Ltd, 1988
BRE. Cleaning external surfaces of buildings.
BRE Digest 280. Garston, Construction
Research Communications Ltd, 1983

Chapter 2.3
[113] **Schaffer R J.** The weathering of
natural building stones. London, The
Stationery Office, 1931. (Also available as a
reprint. Garston, Construction Research
Communications Ltd, 1991)
[114] **BRE.** Selecting natural building stones.
BRE Digest 420. Garston, Construction
Research Communications Ltd, 1997
[115] **BRE.** Decay and conservation of stone
masonry. *BRE Digest* 177. Garston,
Construction Research Communications Ltd,
1984

Further reading
Hart D. *The building magnesian limestones of
the British Isles.* BRE Report. Garston,
Construction Research Communications Ltd,
1988
Leary E. *The building limestones of the
British Isles.* BRE Report. London, The
Stationery Office, 1989
Leary E. *The building sandstones of the
British Isles.* BRE Report. Garston,
Construction Research Communications Ltd,
1986
Hunter C A and Berry R W. Control of
biological growths on stone. *BRE Information
Paper* IP 11/95. Garston, Construction
Research Communications Ltd, 1995
BRE. Cleaning external surfaces of buildings.
BRE Digest 280. Garston, Construction
Research Communications Ltd, 1983
Ashurst J and Dimes F G. *Conservation of
building and decorative stone*, Volumes 1 and
2. London, Butterworth-Heinemann, 1990

Chapter 2.4
[116] **Diamant R M E.** Reema Conclad
system. *Architect and Building News* (1967)
232 (7) 1960–69
[117] **Cornish J P, Henderson G, Uglow C
E, Stephen R K, Southern J R and
Sanders C H.** *Improving the habitability of
large panel system dwellings.* BRE Report.
Garston, Construction Research
Communications Ltd, 1989

[118] **Currie R J.** *Carbonation depths in
structural quality concrete: an assessment of
evidence from investigations of structures
and from other sources.* BRE Report.
Garston, Construction Research
Communications Ltd, 1986
[119] **BRE.** Carbonation of concrete and its
effects on durability. *BRE Digest* 405.
Garston, Construction Research
Communications Ltd, 1995
[120] **BRE.** The durability of steel in
concrete: Part 1 – mechanism of protection
and corrosion. *BRE Digest* 263. Garston,
Construction Research Communications Ltd,
1982
[121] **BRE.** The durability of steel in
concrete: Part 2 – diagnosis and assessment
of corrosion-cracked concrete. *BRE Digest*
264. Garston, Construction Research
Communications Ltd, 1982
[122] **BRE.** The durability of steel in
concrete: Part 3 – the repair of reinforced
concrete. *BRE Digest* 265. Garston,
Construction Research Communications Ltd,
1982

Further reading
Reeves B R. *Large panel system
dwellings:preliminary information on
ownership and condition.* BRE Report.
Garston, Construction Research
Communications Ltd, 1986
Edwards M J. Weatherproof joints in large
panel systems: 2 Remedial measures. *BRE
Information Paper* IP 9/86. Garston,
Construction Research Communications Ltd,
1986
Edwards M J. Weatherproof joints in large
panel systems: 3 Investigation and diagnosis
of failures. *BRE Information Paper* IP 10/86.
Garston, Construction Research
Communications Ltd, 1986
Currie R J and Reeves B R. *Guidance on
inspection and appraisal of the quality of
construction and materials in large panel
system dwellings.* BRE Report. Garston,
Construction Research Communications Ltd,
1987
Roberts M H. Carbonation of concrete made
with dense natural aggregates. *BRE
Information Paper* IP 6/81. Garston,
Construction Research Communications Ltd,
1986
Armer G S T. Overcladding. *Symposium:
Assessment and repair of large panel
concrete structures, University of
Strathclyde, Glasgow, 29 May 1985.* (Papers
published by The Institution of Structural
Engineers, Scottish Branch)
Beech J C. The selection and performance of
sealants. *BRE Information Paper* IP 25/81.
Garston, Construction Research
Communications Ltd, 1981

Davies H and Rothwell G W. The
effectiveness of surface coatings in reducing
carbonation of reinforced concrete. *BRE
Information Paper* IP 7/89. Garston,
Construction Research Communications Ltd,
1989
BRE. Concrete: cracking and corrosion of
reinforcement. *BRE Digest* 389. Garston,
Construction Research Communications Ltd,
1993

Chapter 2.5
[123] **Reeves B R and Martin G R.** *The
structural condition of Wimpey no-fines low-
rise dwellings.* BRE Report. Garston,
Construction Research Communications Ltd,
1989
[124] **Williams A W and Ward G C.** *The
renovation of no-fines housing.* BRE Report.
Garston, Construction Research
Communications Ltd, 1991
[125] **BRE.** *Universal houses.* BRE Report.
Garston, Construction Research
Communications Ltd, 1989
[126] **Currie R J.** *The structural condition of
Easiform cavity-walled dwellings.* BRE Report.
Garston, Construction Research
Communications Ltd, 1988
[127] **British Standards Institution.** Code
of practice for the structural use of normal
reinforced concrete in buildings. *British
Standard* BS CP 114:1948. London, BSI,
1948
[128] Building Technical File No 4, January
1984

Further reading
Glick D and Reeves B R. *The structural
condition of cast in situ concrete low-rise
dwellings.* BRE Report. Garston, Construction
Research Communications Ltd, 1996
Glick D and Reeves B R. *The structural
condition of cast in situ concrete high-rise
dwellings.* BRE Report. Garston, Construction
Research Communications Ltd, 1996

Chapter 2.6
[129] **Building Research Board.** Building in
cob and pisé de terre. *National Building
Studies Special Report* No 5, London, The
Stationery Office, 1922

Further reading
Smith R G. Building with soil-cement bricks.
Building Research and Practice, March/April
1974
Trotman P M. Dampness in cob walls. *Procs
of Out of Earth Conference, Dartington,
Devon, May 1995*

Chapter 3.1
[130] **Diamant R M E.** The Trusteel system
of building. *Architect and Building News*
(1964), **225** (24) 1045–8

[131] Yates T J S and Chakrabarti B. External cladding using thin stone. *BRE Information Paper* IP 6/97. Garston, Construction Research Communications Ltd, 1997

[132] British Standards Institution. Code of practice for design and installation of natural stone cladding and lining. *British Standard* BS 8298:1994. London, BSI, 1994

[133] BRE. Reinforced concrete framed flats: repair of disrupted brick cladding. *BRE Defect Action Sheet (Design)* DAS 2. Garston, Construction Research Communications Ltd, 1985

[134] BRE. *Establishing the structural response of metal members subjected to heat from one side.* BRE Report. Garston, Construction Research Communications Ltd, 1983

Further reading
BRE. Wall cladding: Designing to minimise defects due to inaccuracies and movements. *BRE Digest* 223. Garston, Construction Research Communications Ltd, 1979

de Vekey R C. Ties for masonry cladding. *BRE Information Paper* IP 17/88. Garston, Construction Research Communications Ltd, 1988

Harrison H W. *Steel-framed and steel-clad houses: inspection and assessment.* BRE Report. Garston, Construction Research Communications Ltd, 1987

Centre for Window and Cladding Technology. *Guide to the selection and testing of stone panels for external use.* Bath, Centre for Window and Cladding Technology, 1997

Lewis M D. Modern stone cladding. *American Society for Testing and Materials Manual* No MNL 21. Philadelphia, ASTM, 1995

Yates T J S, Matthews S L and Chakrabarti B. External cladding: how to determine the thickness of natural stone panels. *BRE Information Paper* IP 7/98. Garston, Construction Research Communications Ltd, 1998

Chapter 3.2
[135] Diamant R M E. The Firmcrete and Quickbuild systems of industrialized building. *Architect and Building News* (1965), **227** (20) 956–60

[136] Finch P. *Livett-Cartwright steel framed houses.* BRE Report. Garston, Construction Research Communications Ltd, 1987

[137] British Standards Institution. Code of practice for design of joints and jointing in building construction. *British Standard* BS 6093:1993. London, BSI, 1993

[138] Harrison H W. *Steel-framed and steel-clad houses: inspection and assessment.* BRE Report. Garston, Construction Research Communications Ltd, 1987

Further reading
Mayo A P. *A cladding bibliography of selected publications 1970–1990, and current national and international standards concerning cladding.* BRE Report. Garston, Construction Research Communications Ltd, 1991

BRE. Wall cladding: Designing to minimise defects due to inaccuracies and movements. *BRE Digest* 223. Garston, Construction Research Communications Ltd, 1979

Chapter 3.3
[139] Covington S A, McIntyre I S and Stevens A J. *Timber frame housing 1920–1975: inspection and assessment.* BRE Report. Garston, Construction Research Communications Ltd, 1995

[140] BRE. Specifying structural timber. *BRE Digest* 416. Garston, Construction Research Communications Ltd, 1996

[141] de Vekey R C. Timber-framed houses: interaction between frame and cladding brickwork subject to lateral loads. *Masonry International* (1987), **1** (1), 29–35

[142] BRE. The structural use of wood based panels. *BRE Digest* 423. Garston, Construction Research Communications Ltd, 1997

[143] British Standards Institution. Structural use of timber. Code of practice for the preservative treatment of structural timber. *British Standard* BS 5268-5:1989. London, BSI, 1989

[144] Timber Research and Development Association. Surveys of timber frame houses. *Wood Information Sheet* 0/10, High Wycombe, TRADA, 1992

[145] Timber and Brick Information Council. *Timber and brick homes handbook.* Rickmansworth, TBIC, 1992 (reprint)

[146] BRE. Supplementary guidance for assessment of timber-framed houses. Part 1 Examination. *BRE Good Building Guide* GBG 11. Garston, Construction Research Communications Ltd, 1993

[147] BRE. Supplementary guidance for assessment of timber-framed houses: Part 2. Interpretation. *BRE Good Building Guide* GBG 12. Garston, Construction Research Communications Ltd, 1993

Further reading
McIntyre I S and Stevens A J. *Timber frame housing systems built in the UK 1920–1965.* BRE Report. Garston, Construction Research Communications Ltd, 1995

McIntyre I S and Stevens A J. *Timber frame housing systems built in the UK 1966–1975.* BRE Report. Garston, Construction Research Communications Ltd, 1995

Covington S A, Bravery A F and Wynands R H. *Moisture conditions in the walls of timber-framed housing.* BRE Report. Garston, Construction Research Communications Ltd, 1992

European Union of Agrément (UEAtc). The assessment of structural sheathing for wall panels for timber framed dwellings. *Method of Assessment and Test* No 26. Garston, British Board of Agrément, 1988

Thorogood R P. Vapour diffusion through timber-framed walls. *BRE Information Paper* IP 1/81. Garston, Construction Research Communications Ltd, 1981

Mayo A P. Assessing the performance of timber frame wall panels subject to racking loads. *BRE Information Paper* IP 12/84. Garston, Construction Research Communications Ltd, 1984

McIntyre I S. Moisture relations in timber-framed walls. *BRE Information Paper* IP 21/82. Garston, Construction Research Communications Ltd, 1982

BRE. Dry rot: its recognition and control. *BRE Digest* 299. Garston, Construction Research Communications Ltd, 1985

BRE. External and separating walls: cavity barriers against fire – location. *BRE Defect Action Sheet (Design)* DAS 29. Garston, Construction Research Communications Ltd, 1983

BRE. External and separating walls: cavity barriers against fire – installation. *BRE Defect Action Sheet (Site)* DAS 30. Garston, Construction Research Communications Ltd, 1983

Chapter 3.4
[148] Harrison H W, Hunt J H and Thomson J. *Overcladding external walls of large panel system dwellings.* BRE Report. Garston, Construction Research Communications Ltd, 1986

[149] Health and Safety Executive. Work with asbestos cement. *HSE Guidance Note* EH 36. London, HSE, 1990

[150] Redfearn D. A test rig for proof-testing building components against wind loads. *BRE Information Paper* IP 19/84. Garston, Construction Research Communications Ltd, 1984

[151] Thorogood R and Saunders C. Metal skinned sandwich panels for external walls. *BRE Current Paper* CP 6/79. Garston, Construction Research Communications Ltd, 1979

[152] West J M and Majumdar A J. Durability of non-asbestos fibre-reinforced cement. *BRE Information Paper* IP 1/91. Garston, Construction Research Communications Ltd, 1991

[153] **Bonshor R B and Eldridge L L.** *Graphical aids for tolerances and fits. Handbook for manufacturers, designers and builders.* BRE Report. London, The Stationery Office, 1974

Further reading
BRE. Wall cladding: Designing to minimise defects due to inaccuracies and movements. *BRE Digest* 223. Garston, Construction Research Communications Ltd, 1979
BRE. Wall cladding defects and their diagnosis. *BRE Digest* 217. Garston, Construction Research Communications Ltd, 1978
Morris W A, Colwell S, Smit D, Andrews A and Connolly R. Test method to assess the fire performance of external cladding systems. *FRS Fire Note* 3. Garston, Construction Research Communications Ltd, 1998
European Union of Agrément (UEAtc). Directive for rigid PVC products used externally in building. *Method of Assessment and Test* No 8. Garston, British Board of Agrément, 1973
European Union of Agrément (UEAtc). UEAtc guide for the assessment of prefabricated units for external wall insulation (insulating cladding panels). *Method of Assessment and Test* No 45, Garston, British Board of Agrément, 1990
European Union of Agrément (UEAtc). Pre-coated metal sheet roofing and cladding. *Method of Assessment and Test* No 34. Garston, British Board of Agrément, 1986
European Union of Agrément (UEAtc). Technical Report for the assessment of installations using sandwich panels with a CFC-free polyurethane core. *Method of Assessment and Test* No 59. Garston, British Board of Agrément, 1996
European Union of Agrément (UEAtc). UEAtc directives for impact testing of opaque vertical building components. *Method of Assessment and Test* No 43. Garston, British Board of Agrément, 1987
Mayo A P. *A cladding bibliography of selected publications 1970–1990, and current national and international standards concerning cladding.* BRE Report. Garston, Construction Research Communications Ltd, 1991
Ward T I. Metal cladding: assessing thermal performance. *BRE Information Paper* IP 5/98. Garston, Construction Research Communications Ltd, 1998

Chapter 3.5
[154] **Moore J F A.** *The cracking of glass-reinforced cement in cladding panels.* BRE Report. Garston, Construction Research Communications Ltd, 1984

Further reading
European Union of Agrément (UEAtc). Directive for the assessment of products in glass-reinforced polyester for use in building. *Method of Assessment and Test* No 9. Garston, British Board of Agrément, 1973
BRE. Reinforced plastics cladding panels. *BRE Digest* 161. Garston, Construction Research Communications Ltd, 1974
BRE. *The use of glass reinforced cement in cladding panels.* BRE Report. Garston, Construction Research Communications Ltd, 1984
BRE. GRC. *BRE Digest* 331. Garston, Construction Research Communications Ltd, 1988
Moore J F A. The use of glass-reinforced cement in cladding panels. *BRE Information Paper* IP 5/84. Garston, Construction Research Communications Ltd, 1984
Majumdar A J. Non-Portland cement GRC. *BRE Information Paper* IP 7/82. Garston, Construction Research Communications Ltd, 1982
Majumdar A J and West J M. Polymer modified GRC. *BRE Information Paper* IP 10/87. Garston, Construction Research Communications Ltd, 1987
Majumdar A J and Singh B. Properties of GRC containing PFA. *BRE Information Paper* IP 11/86. Garston, Construction Research Communications Ltd, 1986

Chapter 4.1
[155] **BRE.** Timber for joinery. *BRE Digest* 407. Garston, Construction Research Communications Ltd, 1995
[156] **British Standards Institution.** Wood windows. Specification for factory assembled windows of various types. *British Standard* BS 644-1:1989. London, BSI, 1989
[157] **British Standards Institution.** Steel windows generally for domestic and similar buildings. *British Standard* BS 990:1945. London, BSI, 1945
[158] **British Standards Institution.** Specification for powder organic coatings for application and stoving to hot dip galvanised hot rolled steel sections and preformed steel sheet for windows and associated external architectural purposes, and for the finish on galvanised steel sections and preformed sheet coated with powder organic coatings. *British Standard* BS 6497:1984. London, BSI, 1984
[159] **British Standards Institution.** Specification for steel windows, sills, window boards and doors. *British Standard* BS 6510: 1984. London, BSI, 1984
[160] **British Standards Institution.** Specification for aluminium alloy windows. *British Standard* BS 4873:1986. London, BSI, 1986

[161] **British Standards Institution.** Specification for powder organic coatings for application and stoving to aluminium alloy extrusions sheet and preformed steel sheet for external architectural purposes, and for the finish on aluminium alloy sheet and preformed sections coated with powder organic coatings. *British Standard* BS 6496: 1984. London, BSI, 1984
[162] **British Standards Institution.** Specification for white PVC-U extruded hollow profiles with heat welded corner joints for plastics windows: materials type A. *British Standard* BS 7413:1991. London, BSI, 1991
[163] **British Standards Institution.** Specification for white PVC-U extruded hollow profiles with heat welded corner joints for plastics windows: materials type B. *British Standard* BS 7414:1991. London, BSI, 1991
[164] **British Standards Institution.** Specification for surface covered PVC-U extruded hollow profiles with heat welded corner joints for plastics windows. *British Standard* BS 7722:1994. London, BSI, 1994
[165] **British Standards Institution.** Specification for plastics windows made from PVC-U extruded hollow profiles. *British Standard* BS 7412:1991. London, BSI, 1991
[166] **BRE.** PVC-U windows. *BRE Digest* 404. Garston, Construction Research Communications Ltd, 1995
[167] **Garvin S L and Blois-Brooke T R E.** *Double-glazing units: a BRE guide to improved durability.* BRE Report. Garston, Construction Research Communications Ltd, 1996
[168] **Garvin S L.** *Domestic automatic doors and windows for use by elderly and disabled people.* BRE Report. Garston, Construction Research Communications, 1997
[169] **British Standards Institution.** Glazing for buildings. *British Standard* BS 6262:1982. London, BSI, 1982
[170] **British Standards Institution.** Windows and roof lights: cleaning and safety. *British Standard* BS CP 153-1:1969. London, BSI, 1969
[171] **British Standards Institution.** Windows, doors and rooflights. Code of practice for safety in use and during cleaning of windows and doors (including guidance on cleaning materials and methods). *British Standard* BS 8213-1:1991. London, BSI, 1991
[172] **British Standards Institution.** Protective barriers in and about buildings. *British Standard* BS 6180:1982. London, BSI, 1982
[173] **British Standards Institution.** Specification for enhanced security performance of casement and tilt/turn windows for domestic applications. *British Standard* BS 7950:1997. London, BSI, 1997

[174] British Standards Institution. Performance of windows. Part 2 Specification for operation and strength characteristics. *British Standard* BS 6375-2:1987. London, BSI, 1987

[175] British Standards Institution. Methods of testing windows: air permeability, watertightness and wind resistance. *British Standards* BS 5368-1, 2 and 3 (ENs 42, 86 and 77). London, BSI, 1976, 1980 and 1978

[176] Department of the Environment and The Welsh Office. *The Building Regulations 1991 Approved Document F: Ventilation* (1995 edition). London, The Stationery Office, 1994

[177] British Standards Institution. Code of practice for ventilation principles and designing for natural ventilation. *British Standard* BS 5925:1991. London, BSI, 1991

[178] BRE. Selecting windows by performance. *BRE Digest* 377. Garston, Construction Research Communications Ltd, 1992

[179] British Standards Institution. Specification for one part gun-grade silicone based sealants. *British Standard* BS 5889: 1989. London, BSI, 1989

[180] Rayment R, Fishwick P J, Rose P M and Seymour M J. Heat losses through windows. *BRE Information Paper* IP 12/93. Garston, Construction Research Communications Ltd, 1993

[181] BRE. Specifying preservative treatments: the new European approach *BRE Digest* 393. Garston, Construction Research Communications Ltd, 1994

[182] British Standards Institution. Workmanship on building sites: code of practice for glazing. *British Standard* BS 8000-7:1990. London, BSI, 1990

[183] British Standards Institution. Specification for hermetically sealed flat double-glazing units. *British Standard* BS 5713:1979. London, BSI, 1979

[184] International Standards Organisation. Quality assurance. *International Standard* ISO 9000:1987. Geneva, ISO, 1987

[185] The Glass and Glazing Federation. *The glazing manual.* London, GGF, 1990

Further reading

Beckett H E and Godfrey J H. *Windows.* London, Crosby Lockwood Staples, 1974

British Standards Institution. Specification for impact performance requirements for flat safety glass and safety plastics for use in buildings. *British Standard* BS 6206:1981. London, BSI, 1981

British Standards Institution. Code of practice for the design and installation of sloping and vertical patent glazing. *British Standard* BS 5516:1991. London, BSI, 1991

British Standards Institution. Guide for security of buildings against crime. Dwellings. *British Standard* BS 8220-1:1986. London, BSI, 1986

British Standards Institution. Guide for security of buildings against crime. Offices and shops. *British Standard* BS 8220-2:1987. London, BSI, 1987

British Standards Institution. Guide for security of buildings against crime. Warehouse and distribution units. *British Standard* BS 8220-3:1989. London, BSI, 1989

British Standards Institution. Code of practice: installation of security glazing. *British Standard* BS 5357:1976. London, BSI, 1976

British Standards Institution. Specification for thief resistant locks. *British Standard* BS 3621:1980. London, BSI, 1980

British Standards Institution. Code of practice for preservation of timber. *British Standard* BS 5589:1989. London, BSI, 1989

BRE. Wood windows: resisting rain penetration at perimeter joints. *BRE Defect Action Sheet (Design)* DAS 98. Garston, Construction Research Communications Ltd, 1987

British Woodworking Federation. *Double glazing timber windows on site.* London, BWF, 1994

BRE. Double glazing for heat and sound insulation. *BRE Digest* 379. Garston, Construction Research Communications Ltd, 1993

BRE. *Thermal insulation: avoiding risks.* BRE Report. Garston, Construction Research Communications Ltd, 1994

Tinsdeall N J. The sound insulation provided by windows. *BRE Information Paper* IP 6/94. Garston, Construction Research Communications Ltd, 1994

The Glass and Glazing Federation. *System design and glazing considerations for insulating glass units.* London, GGF, 1995

The Steel Window Association. *Specifier's guide to steel windows* (6th Edition). London, SWA, 1992

European Union of Agrément (UEAtc). Directive for the assessment of windows. *Method of Assessment and Test* No 1. Garston, British Board of Agrément, 1974

European Union of Agrément (UEAtc). UEAtc technical guide for the assessment of windows in PVC-U. *Method of Assessment and Test* No 17. Garston, British Board of Agrément, 1990

European Union of Agrément (UEAtc). Guidelines for the assessment of thermal break metal windows. *Method of Assessment and Test* No 47. Garston, British Board of Agrément, 1990

European Union of Agrément (UEAtc). UEAtc technical report for the assessment of windows in coloured PVC-U. *Method of Assessment and Test* No 57. Garston, British Board of Agrément, 1995

Carruthers J F S and Bedding D. The weatherstripping of windows and doors. *BRE Information Paper* IP 16/81. Garston, Construction Research Communications Ltd, 1981

BRE. Domestic draughtproofing: materials, costs and benefits. *BRE Digest* 319. Garston, Construction Research Communications Ltd, 1987

BRE. The role of windows in domestic burglary. *BRE Information Paper* IP 18/94. Garston, Construction Research Communications Ltd, 1994

BRE. Hardwoods for construction and joinery: current and future sources of supply. *BRE Digest* 417. Garston, Construction Research Communications Ltd, 1997

Garvin S L. *Domestic automatic doors and windows for use by elderly and disabled people.* BRE Report. Garston, Construction Research Communications Ltd, 1997

BRE. Repairing timber windows. *BRE Good Repair Guide* GRG 10. Garston, Construction Research Communications Ltd, 1997

Orsler R J. Preservative-treated timber for exterior joinery: applying the new European standards. *BRE Information Paper* IP 4/97. Garston, Construction Research Communications Ltd, 1997

Chapter 4.2

Many of the references and further reading items given under Chapter 4.1 also apply to this chapter

[186] Department of the Environment and The Welsh Office. *The Building Regulations 1991 Approved Document N: Glazing – materials and protection* (1992 edition). London, The Stationery Office, 1991

[187] Health and Safety Commission. Workplace (Health, Safety and Welfare) Regulations 1992. *Approved Code of Practice and Guidance* L24. London, The Stationery Office, 1992

[188] Beckett H E and Godfrey J A. *Windows.* London, Crosby Lockwood Staples, 1974

[189] British Standards Institution. Use of elements of structural fire protection with particular reference to the recommendations given in BS 5588 'Fire precautions in the design and construction of buildings'. Guide to the fire performance of glass. *British Standard* PD 6512-3:1987. London, BSI, 1987

Further reading

British Standards Institution. Steel windows for industrial buildings, *British Standard* BS 1787:1951. London, BSI, 1951
British Standards Institution. Specification for impact performance requirements for flat safety glass and safety plastics for use in buildings. *British Standard* BS 6206:1981. London, BSI, 1981

Chapter 4.3
[190] Cook A. The hole truth. *Building*, 18 April 1997
[191] The Glass and Glazing Federation. Non-vertical overhead glazing: guide to the selection of glass from the point of view of safety. *Data Sheet* No 7.1. London, GGF, 1994

Further reading

European Union of Agrément (UEAtc). Directives for the approval of insulating glazing. *Method of Assessment and Test* No 35. Garston, British Board of Agrément, 1985
European Union of Agrément (UEAtc). Technical guide for the approval of structural sealant glazing systems (bonded external glazing systems). *Method of Assessment and Test* No 52. Garston, British Board of Agrément, 1994

Chapter 5.1
[192] British Standards Institution. Internal and external wood doorsets, door leaves and frames. Dimensional requirements. *British Standard* BS 4787-1: 1980. London, BSI, 1980
[193] British Standards Institution. Specification for aluminium framed sliding glass doors *British Standard* BS 5286:1978. London, BSI, 1978
[194] British Standards Institution. Code of practice for the installation of replacement windows and doorsets in dwellings. *British Standard* BS 8213-4:1990. London, BSI, 1990
[195] European Union of Agrément (UEAtc). Internal and external doorsets. *Method of Assessment and Test* No 7. Garston, British Board of Agrément, 1970
[196] British Standards Institution. Guide to specifying performance requirements for hinged or pivoted doors (including test methods). *British Standard* BS DD 171:1987. London, BSI, 1987
[197] BRE. Inward opening external doors: resistance to rain penetration. *BRE Defect Action Sheet (Design)* DAS 67. Garston, Construction Research Communications Ltd, 1985
[198] British Standards Institution. Timber for and workmanship in joinery: specification for timber. *British Standard* BS 1186-1:1991. London, BSI, 1991

[199] BRE. Domestic draughtproofing: materials, costs and benefits. *BRE Digest* 319. Garston, Construction Research Communications Ltd, 1987

Further reading

European Union of Agrément (UEAtc). Directive for the assessment of doors. *Method of Assessment and Test* No 11. Garston, British Board of Agrément, 1969
Kingsbury P W. Ergonomic requirements for windows and doors. *BRE Information Paper* IP 2/82. Garston, Construction Research Communications Ltd, 1982
Garvin S L. *Domestic automatic doors and windows for use by elderly and disabled people.* BRE Report. Garston, Construction Research Communications Ltd, 1997

Chapter 5.2
[200] British Standards Institution. Code of practice for drainage of roofs and paved areas. *British Standard* BS 6367:1983. London, BSI, 1983

Chapter 5.3
General reading
European Union of Agrément (UEAtc). Directive for the Assessment of external roller shutters with horizontal slats. *Method of Assessment and Test* No 13. Garston, British Board of Agrément, 1969

Chapter 6.1
[201] Reddaway T F. *The rebuilding of London after the Great Fire.* London, Jonathan Cape, 1940
[202] BRE. Pitched roofs: separating wall/roof junction – preventing fire spread between dwellings. *BRE Defect Action Sheet (Design)* DAS 8. Garston, Construction Research Communications Ltd, 1982
[203] Cooke G M E. Fire engineering of tall fire separating walls. Part 1. *Fire Surveyor* (1987), **16** (3) 13–29
[204] Cooke G M E. Fire engineering of tall fire separating walls. Part 2. *Fire Surveyor* (1987), **16** (4) 19–29
[205] Parkin P, Purkis H J and Scholes W E. Sound insulation between dwellings, *Building Research Station National Building Studies Research Paper* 33. London, The Stationery Office, 1960
[206] Department of the Environment and the Welsh Office. *The Building Regulations 1991 Approved Document E: Resistance to the passage of sound* (1992 edition). London, The Stationery Office, 1991
[207] The Scottish Office. *The Building Standards (Scotland) Regulations Part H.* Edinburgh, The Stationery Office

[208] Department of the Environment for Northern Ireland. *The Building Regulations (Northern Ireland) 1990 Technical Booklet G: Sound.* Belfast, The Stationery Office, 1990
[209] Sewell E C. Sound insulation of stepped or staggered walls of plastered masonry. *Journal of Sound and Vibration* (1982), **84** (4) 463–80
[210] The Anglian house. *The Builder* (1964), **207** (6322) 121–2
[211] BRE. Sound insulation of separating walls and floors. Part 1: walls. *BRE Digest* 333. Garston, Construction Research Communications Ltd, 1988
[212] BRE. Improving the sound insulation of separating walls and floors. *BRE Digest* 293. Garston, Construction Research Communications Ltd, 1985
[213] BRE. Masonry separating walls: improving airborne sound insulation between existing dwellings. *BRE Defect Action Sheet* DAS 105. Garston, Construction Research Communications Ltd, 1987
[214] British Standards Institution. Code of practice for sound insulation and noise reduction for buildings. *British Standard* BS 8233:1987. London, BSI, 1987

Further reading

BRE. Rising damp in walls: diagnosis and treatment. *BRE Digest* 245. Garston, Construction Research Communications Ltd, 1981
BRE. External and separating walls: cavity barriers against fire – location. *BRE Defect Action Sheet (Design)* DAS 29. Garston, Construction Research Communications Ltd, 1983
BRE. External and separating walls: cavity barriers against fire – installation. *BRE Defect Action Sheet (Site)* DAS 30. Garston, Construction Research Communications Ltd, 1983
BRE. Masonry separating walls: airborne sound insulation in new-build housing. *BRE Defect Action Sheet (Design)* DAS 104. Garston, Construction Research Communications Ltd, 1987

Chapter 6.2
General reading
Timber Research and Development Association. Timber frame separating walls *TRADA Wood Information Sheet* 19. High Wycombe, TRADA, 1983

Chapter 6.3
[215] *Mandatory rules for means of escape in case of fire.* London, The Stationery Office, 1985

Further reading
Department of the Environment and The Welsh Office. *The Building Regulations 1991 Approved Document B: Fire safety* (1992 edition). London, The Stationery Office, 1991

Chapter 7.1
[216] **British Standards Institution.** Hollow clay building blocks. *British Standard* BS 1190:1951. London, BSI, 1951
[217] **British Standards Institution.** Guide to accuracy in building. *British Standard* BS 5606:1978. London, BSI, 1978
[218] **Kumar S and de Vekey R C.** Narrow lightweight block walls: performance under vertical load. *International Journal of Masonry Construction* (1981), **2** (3) 91–102
[219] **Cavanagh C J, Edgell G J and de Vekey R C.** The racking strength of lightly loaded partition walls. *8th International Symposium on Loadbearing Brickwork, London, 22–23 November 1983 (Technical Section 2, Structures)*
[220] *Homes for today and tomorrow* (Parker Morris). London, The Stationery Office, 1956

Further reading
BRE. Selecting wood based panel products. *BRE Digest* 323. Garston, Construction Research Communications Ltd, 1992

Chapter 7.2
[221] **British Standards Institution.** Workmanship on building sites: code of practice for plasterboard partitions and dry linings. *British Standard* BS 8000-8:1989. London, BSI, 1989

Further reading
Timber Research and Development Association. Internal walls. *Wood Information Sheet* 11. High Wycombe, TRADA, 1987

Chapter 8.1
[222] **Ministry of Works.** *The use of standards in buildings. Standards Committee First Progress Report.* London, The Stationery Office, 1944
[223] **Fothergill L C and Savage J E.** Reduction of noise nuisance caused by banging doors. *Applied Acoustics* (1987), **21** (1) 39–52

Further reading
European Union of Agrément (UEAtc). Internal and external doorsets. *Method of Assessment and Test* No 7. Garston, British Board of Agrément, 1970
European Union of Agrément (UEAtc). Directive for the assessment of doors. *Method of Assessment and Test* No 11. Garston, British Board of Agrément, 1969

Chapter 8.2
[224] **BRE.** Fire doors. *BRE Digest* 320. Garston, Construction Research Communications Ltd, 1988
[225] **Fire Officers Committee.** Rules for the construction and installation of fire break doors and shutters. London, FOC, 1985
[226] **British Standards Institution.** Fire-check flush doors and wood and metal frames (half hour and one hour types). *British Standard* BS 459-3:1951. London, BSI, 1951

Further reading
Webster C J and Ashton L A. Investigations on building fires. Part IV. Fire resistance of timber doors. *National Building Studies Technical Paper* 6. London, The Stationery Office, 1951

Chapter 9.1
[227] **Lloyd N A.** History of English brickwork. London, Architectural Press
[228] **British Standards Institution.** Wall and floor tiling: code of practice for the design and installation of external ceramic wall tiling and mosaics (including terra cotta and faience tiles). *British Standard* BS 5385-2:1991. London, BSI, 1991
[229] **British Standards Institution.** Specification for building papers (breather type). *British Standard* BS 4016:1972. London, BSI, 1972
[230] **British Standards Institution.** Specification for adhesives for use with ceramic tiles and mosaics. *British Standard* BS 5980:1980 (1991). London, BSI, 1980 (1991)
[231] **British Standards Institution.** Specification for nails: copper nails. *British Standard* BS 1202-2:1974. London, BSI, 1974
[232] **British Standards Institution.** Specification for nails: aluminium nails. *British Standard* BS 1202-3:1974. London, BSI, 1974

Chapter 9.2
[233] **Lawson E M.** External rendering – Scottish experience. *BRE Current Paper* CP 29/76. Garston, Construction Research Communications Ltd, 1976
[234] **British Standards Institution.** Specification for Portland cement. *British Standard* BS 12:1996. London, BSI, 1996
[235] **British Standards Institution.** Specification for Portland blastfurnace cements. *British Standard* BS 146:1996. London, BSI, 1996
[236] **British Standards Institution.** Specification for sulfate resisting Portland cement *British Standard* BS 4027:1991. London, BSI, 1991

[237] **British Standards Institution.** Specification for building limes. *British Standard* BS 890:1972. London, BSI, 1972
[238] **British Standards Institution.** Glossary of building and civil engineering terms. Concrete and plasters. Aggregates. *British Standard* BS 6100-6.3:1984 (1995). London, BSI, 1984 (1995)
[239] **British Standards Institution.** Specification for sands. *British Standard* BS 1199:1976. London, BSI, 1976
[240] **British Standards Institution.** Specification for single size aggregate for general purposes. *British Standard* BS 63-1: 1987. London, BSI, 1987
[241] **British Standards Institution.** Specification for single size aggregate for surface dressing. *British Standard* BS 63-2: 1987. London, BSI, 1987
[242] **British Standards Institution.** Specification for pigments for Portland cement and Portland cement products. *British Standard* BS 1014:1975 (1992). London, BSI, 1975 (1992)
[243] **British Standards Institution.** Mortar admixtures: specification for air entraining (plasticising) admixtures. *British Standard* BS 4887-1:1986. London, BSI, 1986
[244] **British Standards Institution.** Code of practice for external renderings. *British Standard* BS 5262:1991. London, BSI, 1991
[245] **British Standards Institution.** Workmanship on building sites. Code of practice for plastering and rendering. *British Standard* BS 8000-10:1989. London, BSI, 1989
[246] **British Standards Institution.** Specification for stainless and heat resisting steel plate, sheet and strip. *British Standard* BS 1449-2:1983. London, BSI, 1983

Further reading
BRE. Choosing external rendering. *BRE Good Building Guide* GBG 18. Garston, Construction Research Communications Ltd, 1994
BRE. External walls: rendering – application. *BRE Defect Action Sheet (Site)* DAS 38. Garston, Construction Research Communications Ltd, 1983
BRE. External walls: rendering – resisting rain penetration. *BRE Defect Action Sheet (Design)* DAS 37. Garston, Construction Research Communications Ltd, 1983

Murray I H. The adhesion of cementitious render to a brick background . *Procs of 8th Int Symp on Loadbearing Brickwork, 22–23 November 1983 (Technical Session 4, British Ceramic Society)*
British Cement Association. *External rendering appearance matters* 2. London, BCA, 1988
Ashurst J and Ashurst N. *Mortars, plasters and renders: practical building conservation.* Volume 3. London, English Heritage

Eldridge H J. *Common defects in buildings* London, The Stationery Office, 1976

European Union of Agrément (UEAtc). UEAtc directives for the assessment of external insulation systems for walls (expanded polystyrene insulation faced with a thin rendering). *Method of Assessment and Test* No 22. Garston, British Board of Agrément, 1988

European Union of Agrément (UEAtc). Directive for the assessment of plastic renderings. *Method of Assessment and Test* No 24. Garston, British Board of Agrément, 1983

European Union of Agrément (UEAtc). Technical guide for the assessment of external wall insulation systems faced with mineral render. *Method of Assessment and Test* No 51. Garston, British Board of Agrément, 1992

Stirling C M and Southern J R. Pull-out tests on fixing pins for external insulation. *Insulation Journal*, January 1987

BRE. Cementitious renders for external walls. *BRE Digest* 410. Garston, Construction Research Communications Ltd, 1995

BRE. Assessing external rendering for replacement or repair. *BRE Good Building Guide* GBG 23. Garston, Construction Research Communications Ltd, 1997

BRE. Repairing external rendering. *BRE Good Building Guide* GBG 24. Garston, Construction Research Communications Ltd, 1997

Chapter 9.3

[247] **British Standards Institution.** Plywood. *British Standard* BS 6566-1 to 8: 1985 (1991). London, BSI, 1985 (1991)

[248] **BRE.** Hardwoods for construction and joinery: current and future sources of supply. *BRE Digest* 417. Garston, Construction Research Communications Ltd, 1997

[249] **BRE.** Selecting wood based panel products. *BRE Digest* 323. Garston, Construction Research Communications Ltd, 1992

[250] **BRE.** Plywood. *BRE Digest* 394. Garston, Construction Research Communications Ltd, 1994

Further reading

BRE. Preservative treating veneer plywood against decay. *BRE Information Paper* IP 24/86. Garston, Construction Research Communications Ltd, 1986

BRE. Timbers: their natural durability and resistance to preservative treatment. *BRE Digest* 296. Garston, Construction Research Communications Ltd, 1985

Smith G A. The effect of weathering on the properties of fibre building board. *BRE Information Paper* IP 20/81. Garston, Construction Research Communications Ltd, 1981

Chapter 9.4

[251] **BRE.** Painting walls. Part 1: choice of paint. *BRE Digest* 197. Garston, Construction Research Communications Ltd, 1982

[252] **British Standards Institution.** Specification for water repellents for masonry surfaces. *British Standard* BS 6477:1992. London, BSI, 1992

[253] **British Standards Institution.** Specification for knotting. *British Standard* BS 1336:1988. London, BSI, 1988

[254] **British Standards Institution.** Specification for water-borne priming paints for woodwork. *British Standard* BS 5082: 1993. London, BSI, 1993

[255] **British Standards Institution.** Specification for solvent-borne priming paints for woodwork. *British Standard* BS 5358: 1993. London, BSI, 1993

[256] **British Standards Institution.** Specification for undercoat and finishing paints. *British Standard* BS 7664:1993. London, BSI, 1993

[257] **BRE.** Painting exterior wood. *BRE Digest* 422. Garston, Construction Research Communications Ltd, 1997

[258] **BRE.** Natural finishes for exterior wood. *BRE Digest* 387. Garston, Construction Research Communications Ltd, 1993

[259] **British Standards Institution.** Code of practice for painting of buildings. *British Standard* BS 6150:1991. London, BSI, 1991

[260] **British Standards Institution.** Specification for lead based priming paints. *British Standard* BS 2523:1983. London, BSI, 1983

[261] **British Standards Institution.** Specification for permissible limit of lead in low-lead paints and similar materials. *British Standard* BS 4310:1979. London, BSI, 1979

[262] **Boxall J.** Exterior wood stains. *BRE Information Paper* IP 5/91. Garston, Construction Research Communications Ltd, 1991

[263] **Whitely P.** Difficulties in painting Fletton bricks. *BRE Information Paper* IP 22/79. Garston, Construction Research Communications Ltd, 1979

[264] **Butlin R N, Russel C and McCaig I.** The removal of graffiti. *Procs of International Stone Cleaning Conference, Edinburgh, April 1992*

[265] **Wallace J and Whitehead C.** Graffiti removal and control. *Construction Industry Research and Information Association Special Publication* 71. London, CIRIA, 1989

Further reading

Miller E R. Safer paint: water-borne alternatives. *Architects Journal* (1992), **196** (5) 40–2

BRE. Painting walls. Part 2: failures and remedies. *BRE Digest* 198. Garston, Construction Research Communications Ltd, 1984

European Union of Agrément (UEAtc). Directive for the assessment of building sealants. *Method of Assessment and Test* No 14. Garston, British Board of Agrément, 1976

European Union of Agrément (UEAtc). The assessment of masonry coatings. *Method of Assessment and Test* No 33. Garston, British Board of Agrément, 1986

Whiteley P and Gardiner D. Painting plastics. *BRE Information Paper* IP 11/79. Garston, Construction Research Communications Ltd, 1979

BRE. Maintaining exterior wood finishes. *BRE Good Building Guide* GBG 22. Garston, Construction Research Communications Ltd, 1995

Miller E R and Boxall J. Water-borne coatings for exterior wood. *BRE Information Paper* IP 4/94. Garston, Construction Research Communications Ltd, 1994

Boxall J and Smith G A. Maintaining paintwork on exterior timber *BRE Information Paper* IP 16/87. Garston, Construction Research Communications Ltd, 1987

Miller E R, Boxall J and Carey J K. External joinery: end grain sealers and moisture control. *BRE Information Paper* IP 20/87. Garston, Construction Research Communications Ltd, 1987

Whitford M J. *Getting rid of graffiti.* London, Spon, 1992

Chapter 10.1

[266] **British Standards Institution.** Ceramic floor and wall tiles. *British Standard* BS 6431-1 to 23:1983–86. London, BSI, 1983–86

[267] **British Standards Institution.** Code of practice for the design and installation of ceramic wall tiling and mosaics in normal conditions. *British Standard* BS 5385-1: 1990. London, BSI, 1990

Further reading

BRE. Internal walls: ceramic wall tiles – loss of adhesion. *BRE Defect Action Sheet (Site)* DAS 137. Garston, Construction Research Communications Ltd, 1989

European Union of Agrément (UEAtc). Directive for the assessment of ceramic tile adhesives. *Method of Assessment and Test* No 21. Garston, British Board of Agrément, 1982

Chapter 10.2

[268] **British Standards Institution.** Code of practice for internal plastering. *British Standard* BS 5492:1990. London, BSI, 1990

Chapter 10.3

[269] **BRE.** Fire performance of walls and linings. *BRE Digest* 230. Garston, Construction Research Communications Ltd, 1979

[270] **Fire Protection Association.** Fibre reinforced calcium silicate insulating boards. *Information Sheets on Building Products* B11. London, FPA, 1979

[271] **Mapes P.** Historic wallpaper conservation. *The Building Conservation Directory 1997*, 155–7

Further reading

BRE. *Results of surface spread of flame tests on building products.* BRE Report. London, The Stationery Office, 1976

Fire Protection Association. *Fire tests on building products: fire propagation.* London, FPA, 1980

BRE. Selecting wood based panel products. *BRE Digest* 323. Garston, Construction Research Communications Ltd, 1992

Chapter 10.4

General reading

Miller E R. Safer paint: water-borne alternatives. *Architects Journal* (1992), **196** (5) 40–2

BRE. Painting walls. Part 1: choice of paint. *BRE Digest* 197. Garston, Construction Research Communications Ltd, 1982

BRE. Painting walls. Part 2: failures and remedies. *BRE Digest* 198. Garston, Construction Research Communications Ltd, 1984

Hunter C A. Controlling mould growth by using fungicidal paints. *BRE Information Paper* IP 12/95. Garston, Construction Research Communications Ltd, 1995

Index

Walls, windows and doors is systematically structured to enable the reader seeking particular information to identify quickly the parts of the book relevant to his or her search. The broad structure of chapter and sub-chapter titles will be seen in the Contents list on page iii.

Chapter 0, the introduction to *Walls, windows and doors*, describes the performance of these elements in buildings in the UK, sources of information and data, and a brief review of industry problems.

Chapter 1 and its sub-chapters are self-explanatory.

Chapters 2–8 describe the main types of external and internal walls, cladding, windows and doors.

Chapters 9 and 10 describe the main applied finishes for external and internal walls.

Section headings within the sub-chapters of Chapters 2–10 broadly link to the functions shown as sub-chapters in Chapter 1. Much of the structure of sub-chapters is standardised, with section headings drawn from a standard list (see below). Not all of these section headings are used in every sub-chapter, and some have been modified to fit particular circumstances. There may be additional headings not listed below.

Section headings in sub-chapters of Chapters 2–10
Characteristic details
- Basic structure
- Abutments
- Corners
- Openings and joints
- Accommodation of services

Main performance requirements and defects (not necessarily in the order shown)
- Choice of materials for structure
- Strength and stability
- Dimensional stability, deflections etc
- Weathertightness, dampness and condensation
- Ventilation
- Thermal properties
- Daylighting and light transmission
- Fire
- Noise and sound insulation
- Durability
- Maintenance

Work on site
- Storage and handling of materials
- Restrictions due to weather conditions
- Workmanship
- Inspection (in panel)

To avoid what would otherwise be considerable repetition of text (because a lot of the information given for one type of vertical element will apply to one or more other vertical elements), many sections refer the reader to other sub-chapters.

Using the Index
The Index excludes words and expressions that are already presented in the list of Contents or the list of section headings. Therefore the reader should undertake his or her search in the following order.
1 list of Contents (pages iii and iv)
2 list of standard section headings (previous column)
3 Index (starts next page)

Where words and expressions which appear in the Contents and/or list of section headings also appear in other contexts, they may be shown in the Index.

Page references to captions to illustrations are shown in bold.